DATING AND AGE DETERMINATION OF BIOLOGICAL MATERIALS

CROOM HELM APPLIED BIOLOGY SERIES
Edited by Peter J. Baron, BSc. Hons, PhD

NUTRITION AND DISEASE
Edited by R. J. Jarrett

THE NATURE OF ENZYMOLOGY
R. L. Foster

NUTRITION AND FOOD PROCESSING
H. G. Muller and G. Tobin

THE MASTERPIECE OF NATURE:
The Evolution and Genetics of Sexuality
G. Bell

TOXIC HAZARDS IN FOOD
D. M. Conning and A. B. G. Lansdown

CELL LOCOMOTION IN VITRO
C. A. Middleton and J. A. Sharp

DATING AND AGE DETERMINATION OF BIOLOGICAL MATERIALS

Edited by MICHAEL R. ZIMMERMAN and
J. LAWRENCE ANGEL

CROOM HELM
London • Sydney • Dover, New Hampshire

Croom Helm Ltd, Provident House, Burrell Row,
Beckenham, Kent BR3 1AT
Croom Helm Australia Pty Ltd, Suite 4, 6th Floor,
64–76 Kippax Street, Surry Hills, NSW 2010 Australia

British Library Cataloguing in Publication Data

Dating and age determination of biological
materials.——(Applied biology series)
1. Biomedical materials——Dating
I. Zimmerman, Michael R. II. Angel, J. Lawrence
III. Series
930.1′028′5 R857.M3

ISBN 0–7099–0470–3

Croom Helm, 51 Washington Street,
Dover, New Hampshire, 03820 USA

Library of Congress Cataloging-in-Publication Data

Dating and age determination of biological materials.

Includes index.
1. Archaeological dating. 2. Man—Age determination.
3. Mammals—Age determination. I. Zimmerman, Michael R.
II. Angel, J. Lawrence (John Lawrence)
CC78.D37 1986 930.1′028′5 86–4563
ISBN 0–7099–0470–3

Printed and bound in Great Britain
by Billing & Sons Limited, Worcester.

CONTENTS

PREFACE

In the course of archaeological investigations, human and nonhuman biological materials are often unearthed. Such remains, which may be or become the primary objects of study, can be either skeletal or mummified. Mummification, the artificial or natural preservation (by drying, freezing or tanning) of soft tissues, may be partial or complete. Another type of biological material often found in archaeological contexts is wood, which may have been structural or converted to charcoal by burning. Other organic materials may include clothing or mummy wrappings.

Such materials are amenable to a variety of dating techniques. Accurate dating of biological materials may be the only means of dating a site or can provide correlation with other, more conventionally archaeologic techniques, such as the evaluation of pottery, hieroglyphic texts or other artefacts. Before the discovery and application of the constant rate of decay of radiocarbon, dating was based on a variety of techniques, including lake mud varves and associated pollen, vegetation and palaeoclimatic links, sea level changes and debatable pluvial-glacial links, tree-ring sequences, geological and archaeological stratification, and historical records.

The shortcomings of these techniques considerably obscured the timing of the various events of human biological and cultural evolution and microevolution, particularly of the Pleistocene and Recent periods. With the gain in accuracy achieved by radiocarbon dating, pre-Bronze Age developments, previously compressed into one millenium, were now known to have covered seven millenia. There was now time for the origin of farming, the invention of transportation, and other features of the Neolithic and Urban Revolutions. The explanation of such biological phenomena as the postglacial spread of anopheline mosquitoes and the new falciparum malaria they carried, with the resulting evolution of the various child-protecting abnormal haemoglobins, also benefited from the expansion of the time frame.

In palaeopathologic studies of skeletal or mummified material, particularly when using museum collections, there may be little or no direct archaeological association. Dating is essential in providing an historical context for the evaluation of the significance of any disease processes detected.

The first section of the book covers direct and indirect dating techniques. Radiocarbon dating is the most definitive technique available, and is covered in the first chapter by Reiner R. R. Protsch. While the method is not as straightforward as it first seemed, it remains the most powerful tool for date determination.

Two newer techniques are covered in the next two chapters. Patricia A. Masters explains the intricacies of amino acid racemisation and illustrates its application to a variety of materials in Chapter 2. Motoji Ikeya reviews the physics of electron spin resonance, in Chapter 3, and demonstrates the extraordinary breadth of materials to which the technique is applicable, ranging from stalactites and fossilised bones to last month's potato chips.

Dendrochronology, as discussed by Jeffrey S. Dean in Chapter 4, can be applied directly to the dating of wood found in archaeological sites, and these dates can be extrapolated to animal and human remains found in the same site or stratum. His discussion leads to two ancillary techniques, presented by Michael R. Zimmerman. Both techniques vary in accuracy and are only occasionally useful. There were successive changes in mummification styles over the millenia of ancient Egyptian prehistory and history, but there was little consistency of practice, with much socioeconomic variation, and only gross distinctions can be made. Tattooing of the face and body, practised in only a few areas of the world, is a relative method. As discussed by Zimmerman and Jens Peder Hart Hansen, the technique depends on the dating of decorated artefacts with which the tattoos are compared.

Age determination is another important facet of the study of human remains. It then becomes possible to construct population profiles and develop palaeoepidemiological approaches to such topics as infant and maternal mortality, relative fertility and nutrition in different groups, trauma, and so on. On the individual level, many diseases occur in specific age ranges and, given the obligatory tentative nature of many palaeopathologic diagnoses, a proper age 'fit' is most encouraging. Age determination can be a crucial factor in the differential diagnosis for certain disorders.

In contrast to dating, age determination is an area in which a variety of techniques yield roughly comparable ranges, without any one method being recognised as most definitive. The multiple skeletal factors described by J. Lawrence Angel, Judy M. Suchey, and M. Yaşar Işcan in Chapter 5, must be taken as a constellation yielding a narrow age range, while the outline, in the same chapter, by Zimmerman of gross visceral changes results only in the broadest of age range classifications.

Much more specific techniques follow. Radiologic studies, especially of the rate of calcification of the bones of the wrist and hand, are discussed by Evelyn J. Bowers in Chapter 6, with the qualification that the bones must be intact and *in situ*, limiting the technique, in this context, to the study of mummies. Douglas Ubelaker reviews the determination of microscopic changes in the osteon density of bone with ageing in Chapter 7.

Age-specific changes of the deciduous and permanent teeth are particularly useful in the evaluation of immature individuals, and Raymond L. Costa's review in Chapter 8 includes the correlation of dental wear and other changes with age, applicable to both young and old. Finally, Chapter 9 returns to amino acid racemisation; Patricia A. Masters explains how the technique is applied to living individuals and, under certain specific conditions, to ancient remains.

A book of this nature cannot be completed without the contributions of many individuals, to whom the editors express the deepest gratitude. Of first importance are the authors, and we thank those who contributed the earliest chapters for their patience and those who came in later for the promptness with which they responded to our requests. The editors of Croom Helm are also thanked for their patience in dealing with this international effort. Special thanks go to M. Yaşar Işcan and Susan R. Loth for their assistance in the editing of Chapter 5, and to James Seaver and Harold Ott for their help with the illustrations for Chapter 6. Finally, we thank our families for their patience with this demanding and time-consuming project.

CONTRIBUTORS

J. Lawrence Angel, PhD
Department of Anthropology
National Museum of Natural History
Smithsonian Institution
Washington DC 20560
USA

Evelyn Bowers, PhD
Surgical Associates of the Children's Hospital of Philadelphia
One Children's Center
34th St and Civic Center Blvd
Philadelphia, PA 19104
USA

Raymond L. Costa, Jr, PhD
Department of Oral Anatomy, College of Dentistry
University of Illinois at the Medical Center, Chicago
801 South Paulina Street
PO Box 6998
Chicago, IL 60680
USA

Jeffrey S. Dean, PhD
Laboratory of Tree Ring Research
Building # 58
The University of Arizona
Tucson, AZ 85721
USA

Jens Peder Hart Hansen, MD
Department of Pathology
Gentofte Hospital
DK-2900 Hellerup Copenhagen
Denmark

Motoji Ikeya, PhD
Department of Electrical Engineering
Technical College,
Yamaguchi University
Tokiwadai, Ube 755
Japan

M. Yaşar Işcan, PhD
Department of Anthropology
Florida Atlantic University
Boca Raton, FL 33431
USA

Patricia M. Masters, PhD
A-012, University of California
Scripps Institution of Oceanography
La Jolla, CA 92093
USA

Reiner R. R. Protsch, PhD
Institut der Anthropologie und Humangenetik für Biologie
J. W. Goethe Universität Frankfurt
Siesmayerstrasse 70
6000 Frankfurt/Main
Federal Republic of Germany (FGR)

Judy Myers Suchey, PhD
Department of Anthropology
California State University, Fullerton
Fullerton, CA 92634
USA

Douglas Ubelaker, PhD
Department of Anthropology
National Museum of Natural History
Smithsonian Institution
Washington DC 20560
USA

Michael R. Zimmerman, MD, PhD
Department of Anthropology
University of Pennsylvania
Philadelphia, PA 19104
USA and
Department of Pathology
Coney Island Hospital
Brooklyn, NY 11235
USA

PART ONE

DATING TECHNIQUES

1 RADIOCARBON DATING OF BONES

Reiner R. R. Protsch

Bones and teeth are often the only material adequately preserved in archaeological sites. Direct dating of these can solve important anthropologic questions, in that it can pinpoint the precise period of time to which a fossil hominid belongs. Such a period may cover, as in the case of 'anatomically modern man', a timespan ranging from 30,000 years up to possibly 200,000 years BP. The Neanderthals, too, are known to date from the same time period. Designating a hominid as an early or very late representative of any one hominid group could be crucial in solving an evolutionary controversy.

Similarly, problems related to more recently dated materials from Mesolithic, Neolithic (Protsch and Berger, 1973), Bronze Age, Iron Age or even historic times can be answered. Indeed, precise dating is often the only means by which an archaeologic interpretation is made possible, especially when archaeological materials are either unassociated with the burials or totally absent.

In time periods older than 10,000 years BP, amino acid dating provides, under controlled conditions, a reliable tool to cross-check C^{14} dating. Amino acid dating is the only direct dating technique available for use on bones older than 70,000 years (see Chapter 2). However, amino acid dating cannot be used to solve dating questions below 10,000 years BP; radiocarbon dating presents the only physicochemical means of absolute time placement within this crucial period.

It is important to note that C^{14} dating should not be used simply to supply dates at random. It is often its application to specific anthropologic and archaeologic problems that provides the sole means of answering anthropologic questions. Unfortunately, absolute dating is occasionally viewed simply as additional information. If radiocarbon dates do not coincide with relative, purely archaeologic estimations, then absolute dates are often viewed with scepticism by archaeologists and anthropologists.

Under unfavourable environmental conditions, bones may provide the only material capable of being dated. Such was the case in several large-scale dating projects that sought to solve the question of the earliest appearance of 'anatomically modern man' in Europe and his presence in Africa over 30,000 years BP (Beaumont, De Villiers and Vogel, 1978; Glowatzki and Protsch, 1973; Henke and Protsch, 1978; Leakey, Hay, Thurber, *et al.*, 1972; Leakey, Protsch and Berger, 1968; Protsch, 1974a, b, c, 1975, 1976a, b; Protsch, 1978a, b, c, d, e; Protsch, 1981; Protsch and

De Villiers, 1974; Protsch and Glowatzki, 1974; Protsch and Oberholzer, 1975; Protsch, and Semmel, 1978; Vogel and Beaumont, 1972), early man in the Americas (Berger, *et al.*, 1971; and in references McNeish, Berger and Protsch, 1970; Protsch, 1978c; Bada and Protsch, 1980), and questions of the domestication of certain faunal species in the Near East and south eastern Europe (Berger and Protsch, 1973; Protsch and Berger, 1973).

Bones can be dated by using both the organic and inorganic portions, but it has been found that the organic portion most often supplies more accurate dates. With very few exceptions, dates on the collagen portion of bone are accurate in historical time periods, in studies including late Middle Age dates (around the fourteenth and fifteenth centuries AD), early Middle Age dates (fourth to seventh centuries AD), and dates from Roman, Bronze Age and dynastic Egyptian times. When, however, other organics such as wood and charcoal are found in association with osteological materials, comparative dates are run on both. One is often confronted with the argument that absolute radiocarbon dates going beyond historical times, i.e. older than 5,000 years BP, could be wrong, but why should contamination exist only in older materials? Here, relative geologic, palaeontologic, or archaeologic association correlating with absolute radiocarbon dates can be used to support accuracy of an absolute dating technique. If one compares dates on all organic materials commonly used for radiocarbon dating up to now, collagen radiocarbon dating proves to be as exact as dates derived from wood, charcoal, or other non-osteological organics.

There is a severe shortcoming in the radiocarbon dating of collagen, however. In about half of the samples processed, an adequate amount of osteological material is not available for dating. On the other hand, only very little osteological material is needed in many cases. This distinction depends upon the conditions of preservation. Samples from extremely dry and hot environments are usually well preserved and only a few grams of bone are needed for dating. The same is true for preservation under extremely cold conditions, such as permafrost. In a number of cases from Arctic locations a few grams of bone are adequate for a radiocarbon date. The La Brea Tar Pits in California are an example of a third situation providing excellent preservation of bone collagen. In all three cases, the amount of organic material in the bone, the collagen, is comparable to that in fresh bone, i.e. 23 per cent to 27 per cent.

On the other hand, the inorganic portion may often be leached out of samples *in situ*. For example, in the Egyptian samples, a total of 20g of bone contained as much as 16g of collagen, the inorganic portion being virtually absent; only one quarter of the sample was sufficient for radiocarbon dating at the Frankfurt radiocarbon facility. The organic components of other osteologic materials, found in and completely covered by calcareous deposits such as loess and travertine, are often exceptionally well preserved. Here again, relatively little material is needed for collagen radiocarbon dating.

For these reasons, it is necessary for the dating specialist to collect detailed geologic and geomorphologic information on the site before accepting osteologic material for dating. If possible, one should collect the material personally, especially if a large series of dates is planned.

In an environment that quickly dissolves either organic or inorganic portions of bones, such as jungle or other hot and humid areas, the bones are poorly preserved, if at all. An alkaline environment generally preserves only the inorganic portion of the bone, rapidly dissolving the collagen. In such a case, only a minimal portion of the organic matter is still preserved, and a date can be obtained only by processing large quantities of bones. A skull or postcranial bones preserved under such conditions might yield a date, but the total skull would have to be sacrificed just to obtain the date. As explained below, the dating specialist can estimate with fair accuracy the amount of osteologic material needed for radiocarbon collagen dating.

Just as there are contaminating factors in the dating of organic materials such as wood and charcoal (rootlets, etc), bone samples present the same and many additional problems. When a date does not comply with the 'relative' estimation of the archaeologist or anthropologist who submitted the sample for dating, the laboratory date is often cast into doubt. Careful re-examination of a large series of samples yielding such 'questionable' dates often reveals that errors occur during selection, excavation and handling before submission to the laboratory. False results may be caused by sloppy marking and collection procedures, including the handling of bone with fatty or oily hands, packing with other organic materials such as cotton or paper, or marking a stratigraphic position and origin incorrectly.

Whenever possible, samples should be collected by the dating specialist, who should also keep his own collection notes and a detailed record of the stratigraphic and geomorphologic peculiarities of the stratigraphic horizon. Bones should be packed in cellophane, aluminium foil or similar nonorganic materials. It is not meant to suggest that the archaeologist has insufficient knowledge of the dating procedures to be followed. These precautions simply serve to eliminate any errors that can occur before the material is even submitted to the radiocarbon specialist.

On occasion, a sufficient amount of soft tissue is preserved in mummified remains so that radiocarbon dating is possible. The tissue is lylophilised and the collagen extracted for dating — subject, of course, to the same principles and cautions noted for bone.

Having outlined the above principles, this chapter now concentrates solely on the dating of bone samples. It is divided into the following sections:

Principles of radiocarbon dating;
Principles and application of microanalysis;
Dating the inorganic portion of bone (apatite);
Dating bone by accelerators;
Dating the organic content of bone (collagen);
Kelsterbach — a case study in C^{14} collagen and amino acid dating.

The section dealing with collagen dating also includes the processing and dating of samples by conventional means, using a gas proportional CO_2 counting system.

Principles of Radiocarbon Dating

Radiocarbon measurements have been used for a wide variety of purposes, only a small fraction of which involve age determinations useful in palaeoanthropology and archaeology. Familiar materials used for dating are wood, charcoal, peat, leather, paper and other organic materials. Some inorganic materials like shell and mortar have been used, but the organic and inorganic portion of teeth and bones are most often used. Of these, one should preferentially date the organic portion, as will be explained later. Before going into detail relating to dating the organic portion of bones and teeth, it is necessary to comment upon some of the basic principles of radiocarbon dating.

The same theoretical principles involved in the dating of the materials listed above apply to the dating of the organic portions of teeth and bones. These principles were originally described by Libby (1952, 1955) and were covered thoroughly in the literature. A few corrections were added concerning the exact half-life computations and conversion of the absolute C^{14} age to a range of ages, for example, those of historic times. Only a brief review will follow here.

Three isotopes of carbon are contained in the total carbon reservoir of the earth (atmosphere, biosphere and oceans): C^{12} and C^{13}, the two stable isotopes, and the radioactive isotope, C^{14}. There is, however, a difference in the relationship of these three isotopes to one another, depending on whether they occur in the biosphere, the atmosphere, or the oceans, which mostly contain inorganic carbon. Regardless of these differences, the mean ratio among C^{12}, C^{13} and C^{14} is approximately 100:1:0.01.

Of the three isotopes, the stable isotopes C^{12} and C^{13} are not subject to physical changes, while C^{14} undergoes radioactive decay at a constant rate of about 1 per cent per 80 years. At the same time, the supply of C^{14} is continually replenished in the upper atmosphere by nuclear reactions associated with cosmic radiation. Highly energetic cosmic rays infuse into the upper atmosphere and interact with atmospheric nuclei to produce free secondary neutrons. These collide again with nitrogen atoms, specifically those of the isotope N^{14}, to form C^{14}:

$$N_7^{14} + n_0^1 = C_6^{14} + H_1^1$$

The C^{14} produced in this reaction has a half-life of 5,730 ± 30 years (Mann, Marlow and Hughes, 1961), by the most recent calculation. However, in the radiocarbon date list reports, the 'Libby half-life' value of 5,568 ± 30 years is still used. The chemical properties of C^{14} do not differ from those of the stable isotopes. The C^{14} atoms are oxidised by atmospheric oxygen and form radioactive carbon monoxide molecules. They are again oxidised, mostly by direct action with atmospheric oxygen or, to a lesser extent, through the action of bacteria in the biosphere (Pandow, MacKay and Wolfgang, 1960; MacKay, Pandow and Wolfgang, 1963). Thus, there is a continuous supply of radioactive carbon entering the total carbon reservoir of the earth.

The three individual reservoirs obtain radioactive carbon dioxide through its formation in the atmosphere, where it becomes a radioactive part of all atmospheric carbon dioxide. C^{14} then enters the oceans, mainly by interaction with surface water, which dissolves the C^{14} and eventually transports it to the deeper waters. It is then absorbed into the carbonate- and bicarbonate-containing portions of mollusc shells.

C^{14} also enters the organic portion of the biosphere, but by a somewhat different process. Carbon dioxide, including the C^{14} isotope, is incorporated into plants by photosynthesis. Herbivorous fauna accumulate the same amount of C^{14} when they eat plants. The same applies to carnivorous fauna, which prey upon the herbivores. Omnivorous animals, like man, ingest both sources and subsequently show the same activity of radiocarbon in all soft and skeletal components as do plants, herbivores and carnivores.

The mean life of a C^{14} atom is about 8,000 years. When it disintegrates, the products of decay are, once again, a nitrogen atom and an electron, in this case a beta particle. The equation is as follows:

$$C_6^{14} = N_7^{14} + \beta_{-1}$$

It is estimated that the production of C^{14} in the atmosphere is in a state of equilibrium with its radioactive decay in the total carbon reservoir of the earth. Libby (1955, p. 5) calculated the equilibrium and its stabilisation in the reservoir at about 81 metric tons. Just as the equilibrium is maintained in the total reservoir, it is also maintained in every living organism. A living organism exchanges carbon with the atmosphere so that its specific radiocarbon activity is identical to that of the atmosphere. At death, an organism — be it plant or animal — is no longer able to obtain additional

C^{14}: at that moment its C^{14} activity begins to decrease through radioactive decay. The measurement of the C^{14} activity of the organism at any one time allows the calculation of that moment in time at which the organism ceased to exchange carbon with the total carbon reservoir. The rate of decay is known and one is thus able to calculate the total period lapsed since the death of the organism. Averaging the present activity and comparing it to a sample of modern activity (oxalic acid derived from Hawaiian sugar cane) gives the equation:

$$t = -\lambda \log_e \frac{I}{I_0}$$

It is now possible to calculate the age of the sample, t being the time of death, I the actual activity of the test sample, I_0 a modern sample's activity, and λ the decay constant of C^{14}. For λ there are presently two half-life values: 5,568 ± 30 with a 8,033-year lifespan, and another of 5,730 ± 30 with a 8,267-year lifespan.

It was originally assumed that the radiocarbon activity in the carbon reservoir was constant over time and that the half-life of C^{14} was accurate. Although adjustments were found necessary, they did not have a noticeable effect on the relative accuracy of the radiocarbon dating method, as such. New corrections made it possible to use this technique with a high degree of accuracy when compared to other techniques presently available. A number of specialists supplied these corrections (de Vries and Barendsen, 1953; Suess, 1965; Damon, Long and Grey, 1966). As the technique becomes increasingly more refined, one can expect more corrections.

The following assumptions are crucial for accurate age calculations using radiocarbon dating. (Detailed explanations are mentioned elsewhere and are not the point of concentration of this particular chapter.) It is crucial to observe the assumption of: first, an accurate half-life calculation (Libby, 1955, p. 36) (convert the old Libby half-life of 5,568 ± 30 to the new half-life of 5,730 ± 30 by multiplying the former by 1.03); second, the knowledge and detection of all forces causing variations in the production of C^{14} and making a detection possible (de Vries, 1958; Willis, Tauber and Münnich, 1960; Ralph and Stuckenrath, 1960; Stuiver, 1961, 1965; Elsasser, Ney and Winkler, 1956; Kigoshi and Hasegawa, 1965); third, the carbon reservoir exchange rates (Suess, 1965); and, fourth, the interference of C^{14} distribution and activity caused by man. Small variations in atmospheric C^{14} concentration, which can deviate up to 2 per cent over a few thousand years, were adequately corrected by the research of Stuiver (1961, 1965), Suess (1965), and Stuiver and Suess (1966).

Principles and Application of Microanalysis

If one assumes that the organic portion of bone is the only material that can be dated with acceptable accuracy using radiocarbon (unless extenuating circumstances, such as unusual contamination, are present), one must first deal with the question of the amount needed for dating. This question can easily be solved. Microanalysis, or FUN-dating (a relative dating technique, actually one part of R-dating) can be used for such computation (Oakley, 1968).

FUN-dating (fluorine, uranium and nitrogen) was used in the last century (Middleton, 1844) and in the 1950s by Oakley to obtain some idea of the relative time periods involved in the burial of osteologic materials. Fluorine can be found in: the mineral portion of bone fluorapatite, the result of conversion from hydroxyapatite (calcium hydroxyphosphate); to some extent, living bone material; and to a large extent, buried osteologic materials. The amount of fluorine found in adult human individuals is about 20g. Of that, over 95 per cent is contained in the skeletal system as fluorapatite. Of this, the dentition contains 10–70mgF/100g dry substance. A healthy organism ingesting a daily minimum of fluorine will achieve a protective effect in his teeth against caries by drinking water containing $1mgF^-$/litre. The fluorine balance is at equilibrium up to a daily intake of about 10mg ($6mgF^-$/litre water). Larger amounts are normally eliminated through the kidney.

Deviations in the accumulation of fluoride do occur, however. In such cases, just as extreme accumulation during dental development (eight to ten years) obstructs the development of enamel (dental fluorosis), an extreme uptake in bones can lead to certain pathologic changes. A probable example is seen in the femur of *Homo erectus* (*Pithecanthropus erectus*), found in 1981 by Dubois in Java. It was suggested that in this locality, *Homo erectus* drank water with an unusually high content of fluorine, resulting in a large bony tumor mass (Soriano, 1970). Bones displaying such unusual antemortem histories can, of course, hardly be used as relative dating tools.

Under normal circumstances, it is the irreversible substitution of fluorine, as well as uranium, in bones that makes them suitable as a relative dating tool. With the passing of time, both elements accumulate in greater amounts. When bones are buried in different levels at the same location, older bones positioned in lower levels show greater amounts of fluorine and uranium than do those positioned above them. The accumulation of both elements is dependent on time and water action present at that location.

On the other hand, the presence of uranium in bones immediately after combustion of a sample for radiocarbon dating, could interfere somewhat with the count-rate of the sample itself. This interference is due to the fact that combustion converts uranium to radon (Rn^{222}). One could theoretically calculate the total original amount of uranium present in the bone immediately after combustion. In fact, in order to avoid counting interference in C^{14}, it is necessary to store the CO_2 for a few weeks to allow Rn^{222} to decay to sufficiently low amounts so that only the C^{14} activity is present (Rn^{222} half-life is 3,825 days).

In addition to fluorine and uranium, nitrogen can also be a useful tool for dating purposes. The concentration of nitrogen decreases with time and is directly related to the amount of total collagen present in bones, and thus also to the total carbon content of the organic portion. Nitrogen decreases with increasing bone age as protein is removed from the bones. All amino acids containing the element are removed with the protein as well. In order to compute the total amount of a sample needed for radiocarbon dating, one simply performs a microanalysis of the N-content.

Studies on modern bones have shown that the amount of collagen, and therefore of nitrogen, differs only insignificantly in bones of the same individual, of individuals of different species, and of individuals of different ages of the same species. Comparative readings on bones in the same and in different strata of the same location supply a fairly accurate estimation of their time association relative to each other. Since collagen constitutes 20–25 per cent of modern bone, which contains about 50 per cent carbon, one can easily calculate the total amount of bone needed for radiocarbon dating. Nitrogen estimates in modern bone have ranged from 4.7 per cent (Berger, Horney and Libby, 1964) to 5.36 per cent (Eastoe and Courts, 1963), and 4.63 to 5.41 per cent in a series of experiments using young to very old sheep bones.

A 30mg sample of bone, from whatever locality or environmental condition, submitted for radiocarbon dating, and no matter what the collagenous carbon content, is simply subjected to a nitrogen microanalysis. The result of the latter is multiplied by 3.0 ± 0.5. Bones can usually be dated unless they stem from a very old geological and unfavourable environmental location in which nitrogen has decreased to unusually low amounts. This will indicate that practically no collagen, and therefore no carbon, is present. Total samples used for radiocarbon collagen dating have been as small as 8g and as large as 6kg.

The nitrogen content of bones can be determined by two different methods — Dumas or Kjerdahl. The Dumas method is based on the fact that organic compounds containing nitrogen yield nitrogen and oxides of nitrogen when decomposed at red heat (925°C) with copper oxide. A section of copper causes the oxides of nitrogen to be reduced to nitrogen. The procedure is accomplished in an atmosphere of pure carbon dioxide. Liberated nitrogen is carried through the decomposition chamber by the carbon dioxide and then absorbed in potassium hydroxide. The weight of the remaining nitrogen is then calculated at atmospheric pressure according to volume and temperature.

Even with counters having a volume of less than 1 litre, which need little collagenous material (a smaller, 200ml, counter is, for example, used by Berger at the University of California, Los Angeles), it is not feasible to date and process bone samples with microanalytical readings below 0.10 per cent. Theoretically, a counter could still be filled to capacity using several hundred grams of bone with an extremely low nitrogen reading but, during the process of hydrochloric acid treatment, and no matter how weak that solution, most of the extremely small collagen portion in the form of tropocollagen and free amino acids would be hydrolysed. Decalcification, with the sole purpose of saving the organic materials, can no longer be controlled. In the last 20 years about one-third of all samples submitted to the Frankfurt Radiocarbon Laboratory, dating relatively to between 20,000 and 50,000 years, have had readings below 0.10 per cent, particularly in samples from temperate environments, which were therefore unsuitable for radiocarbon dating. Such samples could only be used for amino acid dating, providing the temperature palaeohistory is known.

A number of such samples from well-known archaeologic sites, and with great significance for palaeoanthropology, stem from Africa: for example, the Klaasies River, Otjiseva, Cave of Hearth, Tuinplaas, and Cape Flats. All are samples belonging to a geographical subspecies of *Homo sapiens*, 'anatomically modern man', as distinguished from other hominids such as Neanderthal man, *Homo erectus*, and so on. However, such small samples can now be dated by radiocarbon accelerator dating, which can be used effectively to date extremely small bone samples.

Even in the case of decisively low nitrogen readings, the chances are that a radiocarbon date on a hominid skull could be obtained by conventional dating. In such a case, one might be forced to use all of the osteologic material present. This is obviously unacceptable. Accelerator C^{14} dating now provides a solution in these cases. Direct association of hominid remains and abundant faunal remains, with the same or very similar microanalytical readings, would prove both to be contemporaneous. If that is the case, then the more abundant faunal material could be used for radiocarbon dating. That absolute date could, in turn, be applied to the hominid material. Such a procedure is called A_2-dating by Oakley (1968). Examples are absolute dating methods such as protactinium/thorium dating or all of the uranium series dating techniques applied to materials associated with australopithecines in East Africa.

It makes no difference whether the 'associated' materials are vulcanics or other associated bones. In osteologic association there is at least the assurance that no intrusion from other levels has occurred. Substantiated by the additional support of relative microanalytical dating, it becomes a reliable dating tool. It might then be proper to term bone/bone association A_1–A_2 dating, since it is neither direct absolute dating (A_1) nor purely associated (A_2) dating. Examples of such A_1–A_2 dating application have been done in Africa on hominids and fauna from sites like Florisbad, Saldanha, Naivasha, Lukenya Hill (Grm–22) Fish Hoek, Border Cave and Bushman Rock Shelter.

Several of these hominids and their associated fauna could be dated directly by radiocarbon, yielding in all cases nearly the same or at least comparable results on both materials. This type of dating procedure was used at Florisbad (A_1–A_2 date 38,680 ± 2,000, UCLA 1745B, on fauna; A_2 date on charcoal/wood 38,550 ± 3,800, UCLA 1745C). They are, undoubtedly, quite accurate. Unfortunately, even in more recent articles (Rightmire, 1979), older, contaminated C^{14} dates on the peat are still referred to as 'accurate' dates. It is particularly odd that these peat dates are accepted as being accurate, since the peat itself has been proven to yield highly erroneous dates, as has been pointed out repeatedly over the years in all radiocarbon articles dealing with such plant materials.

This error is due to humic acid concentration, which cannot be completely eliminated. There is also serious doubt at Florisbad as to the direct association of the peat with the hominid finds. Florisbad is a typical example of a site where only bone-dating, be it by radiocarbon or amino acid, can be applied. In this site, other organic materials are contaminated beyond control. The associated fauna at Florisbad, mainly *Hippopotamus amphibius*, illustrate that a C^{14} date can be obtained on associated materials. In this case it was necessary to use unfortunately large amounts of bone (746g) due to low collagen content.

Dating the Inorganic Portion of Bone (Apatite)

Until a few years ago, most radiocarbon specialists and archaeologists claimed that radiocarbon dating of bone materials, both organic and inorganic, was in most cases unreliable. This caused specialists to run a series of dates on identical samples, using both the inorganic apatite portion and the organic collagen portion, and then to compare the results.

Haynes (1968) obtained different results using both portions of osteologic materials recovered from exactly the same horizon at the Lehner Site. The archaeologic age of the bone was estimated relatively to be of Clovis age. Using the apatite from mammoth bone he obtained an age of 9,980 ± 220 years BP (A-874C), whereas the organic portion yielded a much more recent date of 5,610 ± 350 years BP (A-806A). The apatite of the mammoth dental material yielded yet another age, 7,780 ± 150 years BP (A-876C). The dates, estimated relatively to be the same, in fact showed a discrepancy among bone collagen, and tooth and bone apatite of as much as 2,000 years.

The various inorganic fractions had obviously suffered from different external exchange phenomena under changing geochemical and hydrological conditions at the site. What was not taken into account at the time was that all the fractions were essentially of different materials. Exchange of CO_2 in the two inorganic portions cannot be compared to organic materials under any circumstances, as shall be seen below. There is also a great microanalytical difference in the makeup of bone apatite and of teeth. Even though they are essentially the same chemically, their crystal packing differs substantially, that of compact bone being much less condensed than the enamelic structure of teeth.

Collagen, the main organic component of bone, undergoes a completely different exchange phenomenon, and this will be dealt with below. Collagen can never yield dates older than its true age if treated correctly in the laboratory: contamination can only result in younger dates. However, apatite can be contaminated either way, resulting in an age either too old or too young.

The two-phase system of bone consists of a bone mineral, plus minor components of other elements, surrounding the organic structure (McLean and Urist, 1968). The ultramicroscopic crystals are composed of hydroxyapatite with additional components of citrate and carbonate. Vaughan (1970) suggested, and it is now generally accepted by most specialists, that the hydroxyapatite is built up by surface exchange, internal defects and substitutions. If one converts these impurity percentages to molar ratios, the composition of bone (apatite) mineral is as follows, as determined by Neuman and Neuman (1958):

$$\left[Ca^{++}_{g}\ (H_3O^+)_2\ (PO_4^{\equiv})_6\ (OH)_2^-\right] \quad \left[Ca^{++}.\ Mg^{++}_{0\cdot 3}.\ Na^{+}_{0\cdot 3}.\ CO_3^{=}.\ Cit^{\equiv}_{0\cdot 3}\right]$$

The two major phases are calcium carbonate and calcium phosphate, the latter being amorphous and probably a second major phase of the apatite. The crystalline structure decreases with the age of the individual (Termine and Posner, 1967).

Neuman and Neuman (1958) also suggested three zones in the crystals — the crystal interior, the crystal surface and the hydration shell — and that the superficial surface ions of the crystal lattice of hydroxyapatite are hydrated. In turn, surface ions absorb ions from extracellular fluid and the latter bind another layer of water, which is called the hydration shell. As can logically be assumed, under normal conditions the exchange in the hydration shell on the surface is rapid, while that of the interior of the crystal structure can be slow.

The substitution mechanisms in the hydroxyapatite of bone are important for relative dating (F,U,N). The mechanisms are, first, an intercrystalline exchange, similar to that of surface ions but occurring at a lower rate and, second, a recrystallisation due to dissolution and reformation of crystals, with the addition of new ions into the crystal structure.

Third, thermal diffusion, analagous to movement within the structure of other solids, plays an important part.

A common substitution in fossil bone is that which turns hydroxyapatite into fluorapatite. The amount of ground water available at any particular site is responsible for this substitution. The fluorine occupies the vacant oxygen site in order to maintain the electroneutrality of the crystal (McClellan and Lehr, 1969). Two faces of the phosphate tetrahedron are parallel to the C-axis in the apatite structure. Whereas the two other faces are mutually inclined to the C-axis, the substitution carbonate takes up one of the planes inclined to the C-axis. Another substitution in the apatite structure is carbonate for phosphate.

Theoretically, primarily deposited bone apatite (*in vivo*) could be dated, if found in a geological environment free of secondary contaminants such as limestone or other calcium carbonate- and calcium phosphate-rich deposits (loess and travertine). Such secondary deposits in fossil bones could lead to a date either too recent or too old. Specifically, the formation of secondary apatite, replacing the calcium carbonate originally present in the primary makeup of the osteologic material, is responsible for this phenomenon. Exchange studies on $^{14}CO_2$ in synthetic apatite show that 60 per cent of the CO_2, possibly in the form of $CO_3^=$, is contained within the crystal lattice, and that a possible 40 per cent of HCO_3^- remains in the hydration shell. The latter portion can be substantially reduced (by half) by drying for 24 hours at 110°C.

Such treatment should be recommended if dating the inorganic portion of bone is absolutely necessary. A major portion of exchangeable CO_2 must be eliminated from the bone. However, this approach may have serious side-effects, in that it causes artificial racemisation of some of the amino acids in the organic portion, rendering the sample useless for other dating techniques.

Polach and Golson (1966) showed that a difference in assimilation of the three major isotopes of carbon exists in living organisms. The sequence of accumulation of the C^{13} concentration from lowest to highest in carbon compounds is: CO, CH_4, C_D, CO_2, $CO_3^=$. There are complications in the isotopic fractionation of fossil bone used for radiocarbon dating. In the case of sedimentary phosphorates, it was found that the oxygen and carbon in apatite CO_2 are enriched in the light isotopes of each element, as compared to the coexisting calcite. Another complication might be an isotopic exchange of the carbon present from another carbon of different energy.

Diagenetic processes may also be involved in fossil bone. Ames (1959) hinted that alkali phosphate solution might replace calcite with a carbonate-apatite of variable composition. The relative replacement rate was dependent on the solution pH, the $PO^{-3}{}_4$ content relative to HCO_3^-, and the calcite grain size. Other experiments simulated the conditions under which calcite can be converted into hydroxyapatite. Calcite can be easily converted into apatite. A fossil's environment could have been exposed to numerous processes that introduce CO_2 into the osteologic material. These are all possible sources of error when the inorganic portion is being dated, and include secondary apatite from alkali phosphate solutions, carbonate substitution in bone apatite, and carbon isotopic exchange.

In order to determine possible contaminants and their origins, investigators have analysed the carbon isotopic composition of fresh and fossil bone, and have found their various fractions to be practically identical to present-day carbon isotopic composition. There are also no significant variations in the isotopic carbon composition of animals of different age, sex, or species, or of different parts of the same animal. The isotopic carbon composition of these did show some slight differences from those of the atmosphere, but it was usually less than 1.5 per cent where fractionation or preferential assimilation occurred.

However, a series of fossil bones of mammoth (*Mammuthus primigenius*) and cave bear (*Ursus spelaeus*) has been analysed to show a great variation in chemical composition and, particularly in $CO_3^{=}$, $PO_4^{\equiv}$, and F^-, a good indication of carbonate substitution for phosphate. Here it is interesting to note that the finds originated from different environments, i.e. an open-air site as opposed to a cave. Even bones in the same location but from different layers (either above or below each other) manifested a great variation in chemical makeup. This is precisely why microanalytical studies on different layers in the same location are of such great value. They can be used as relative age indicators, unless the site is greatly disturbed. Even bones of different species but in close proximity to one another in the same layer show, in most cases, quite similar results. Electron microscopic and thin-section studies using different crystals make it possible to distinguish variations in porosity, crystallinity, and formation of secondary apatite, as well as fossil algae and fungi in the osteonic and osteocytic structure throughout the compacta.

Since there is carbonate substitution in the apatite structure, researchers have tried to remove the substituted carbonate by fractional hydrolysis. Using a method for collagen–apatite residue hydrolysis, they analysed the CO_2 for C^{13}/C^{12}, and then measured the C^{14} of the CO_2. McConnell (1962) employed a method of pyrolysis and hydrolysis. The heat separation of collagen–apatite from samples was undertaken at different temperatures (200, 400, 600 and 800°C) in a CO_2 purification system. The CO_2 was collected and analysed for C^{13}/C^{12}. The weight differences of the analysed sample were measured after pyrolysis and the X-ray pattern of every fraction examined. After that, the sample was hydrolysed and the C^{14} of the CO_2 of every fraction was measured. The radiocarbon dates were then obtained through hydrolysis and compared to radiocarbon ages on collagen and known ages.

The latter procedures probably cause a side-effect of carbon isotope fractionation due to heating and hydrolysis (Grey, Damon and Haynes, 1969). Fractional hydrolysis and pyrolysis does not make possible the elimination of substituted carbonate, and C^{14} can neither be separated nor detected in the carbonate in the interior of apatite crystals. All results using these techniques are thus unreliable when compared to known historical or, as shown later, radiocarbon collagen ages.

One of the major problems is that although secondary formation is known and can be detected by electron microscopy and luminescence studies, it cannot be removed from the primary apatite by any technique, including fractional hydrolysis. Radiocarbon dating of the apatite in fossil bone thus yields unreliable dates; only the organic portion seems to provide valid dates.

Dating of Bone by Accelerators

Conventional radiocarbon dating measures the β-decay of radiocarbon. Sample sizes vary quite drastically depending on the size of the gas counters used. In laboratories using very large counters, and based on the gas proportional CO_2 or on liquid scintillation equipment, some samples can never be processed. This problem led to a continuous improvement of the equipment, electronics as well as counters, with a specific tendency towards a reduction of the latter's volume. Today, some conventional CO_2 laboratories use counters as small as 200ml; ten years ago their research was dependent on counters as large as 7.5 litres. Regardless of how small a conventional counter is, it usually requires a relatively large sample to start with, and many grams of osteologic material may have to be sacrificed to obtain a date.

A second shortcoming of the conventional method of dating is the time required to obtain a date. This is because a long time period is required to achieve a statistically adequate number of atomic disintegrations or counts for dating, at least 10,000 or more for older dates. Usually, the smaller the counter, the longer the time needed to obtain a fairly accurate date. The time range may vary from an overnight count, to several days or even weeks. Large counters mean larger samples but shorter counting times, while small counters use small samples but require longer counting periods. Most laboratories use several counters, showing some extreme size ranges. For example, UCLA uses a 7.5 litre and a 200ml counter. After combustion of the sample, an even greater time period is required to remove the radioactive radon gas. If not properly recognised, this can severely hamper an accurate measurement. There is a technique for the immediate removal of radon (Rn^{222}), but this is also time-consuming and requires special equipment.

Depending on the equipment (electronics, counters, and so on), most conventional dating laboratories can provide dates of up to 40,000 and possibly 50,000 years. Only with very special equipment is an extension of up to 75,000 years possible (Berger, 1979, Grootes, 1978; Stuiver, Heusser and Yang, 1978). The total cost of a conventional radiocarbon laboratory, depending on the country, is anywhere from US $30,000 to $70,000.

A new method — accelerator or cyclotron radiocarbon dating — was announced in 1977 (Muller, 1977) with the aim of dating a minimal amount of material (much less than that needed in conventional radiocarbon dating), to give a longer time range and a shorter counting time. Muller reported on an analytical method of dating using an 88-inch cyclotron. Soon after, another approach using a Van de Graaf accelerator was reported by Nelson, Korteling and Scott (1977). Nuclear accelerators seem to be effective spectrometers and potentially useful for the dating of extremely small samples by radiocarbon (Pavlish and Banning, 1980). Attempts at dating by conventional mass spectrometers have failed in recent years, but accelerator dating provides a possible solution for small-sample dating. Accelerator analysis completely eliminates N^{14}, which usually interferes with the minute signals of C^{14} (these occur in only one atom in 10^{11}–10^{14} in nature; the atmosphere is composed of 78 per cent nitrogen of identical mass 14).

For accelerator dating purposes, a sample is converted into carbon dioxide, which goes into the cyclotron in the ionised state. Positively charged ions are subsequently accelerated to 30–40 meV until they emerge from the accelerator as a beam. Since nitrogen has an equal mass, it is also mixed in the beam and has to be separated. This is accomplished by passing the beam through a cell containing xenon gas. Full separation of carbon and nitrogen is allowed by the fact that the atomic number of carbon (6) and nitrogen (7) differ and thus their travelling speeds through xenon also differ, with C^{14} travelling about 30 per cent faster than N^{14}. Since C^{14} is detected completely separately, an age calculation is possible for it. A sample of known age, dated by conventional radiocarbon method by Berger (1979), was used to calibrate one of the first accelerator dates (Muller, Stephenson and Mast, 1978).

As is usually the case with any new method, there are some shortcomings to accelerator dating. First, the equipment costs 10 to 20 times as much as conventional equipment. Refinement has to proceed in such a way that statistical errors of cyclotron measurements better those presently achieved by conventional radiocarbon counters by at least one order of magnitude. Berger (1979, p. 102) points out that large high-energy cyclotrons probably contain some contaminant radiocarbon that might inadvertently be produced by physical experiments. He also points out other shortcomings of accelerators. There is also, of course, the question of mixture of radiocarbon from the inorganic portion of the bone, which might be secondary and thus either too recent or too old, as well as mixture of the inorganic with the organic portion. A clear separation of these is hard to achieve. These contamination and procedural factors might render cylotron dating useless and favour conventional dating. It is suggested, therefore, that conventional radiocarbon counters are still more accurate and more suitable, whenever enough osteologic material is available from an estimated time frame of up to 50,000 years BP, and if nitrogen microanalysis suggests a fairly high carbon/collagen content in the sample.

The following factors favour accelerator dating: first, a low sample weight of a few (15) milligrams; second, a greater time range; and, third, a shorter counting time. Each of these are true providing that the shortcomings of the method can be eliminated in the future.

Using either a cyclotron or a Van de Graaf accelerator (Berger, 1979; Bennett *et al.*, 1977; Nelson *et al.*, 1977), the new technique extends conventional dates from 50,000 to 70,000 or even up to 100,000 years, doubling the present age range reached by conventional counters. It would also provide a means to date rare samples of fossil hominids, for example, *Homo sapiens neanderthalensis*, *H. sapiens rhodesiensis* and *H. sapiens soloensis*, which existed sympatrically and allopatrically with 'anatomically modern man' during a crucial time period in human evolution. Most of these hominids cannot be dated by A_2 techniques such as K/Ar, fission track or uranium series dating, since most of them lack either associated materials suitable for such dating methods or associated faunal material. The only other method available for relative/absolute dates would be amino acid dating. The shortcomings of that technique could, however, be cross-checked by accelerator dating. Both require only minimum amounts of material, and the shortcomings and contamination of each method could be controlled, or at least limited, by their application to the same sample material.

Berger (1979) pointed out that accelerator dating is subject to sample contamination and a 100,000-year-old sample, contaminated by as little as 10 ppm of modern carbon, will be wrong by as much as 10 per cent, thus dating to only 90,000 years. In a strict sense, accelerator radiocarbon dating of bones is, at present, only a 'relative' dating technique. 'Absolute' dating applies to those methods that show a deviation of ± 5 per cent. Those over 5 per cent are considered relative techniques. Many famous hominid samples presently available for accelerator dating have undergone not only natural contamination, but were also subject to contamination caused by preservatives applied by an archaeologist, anthropologist, or museum specialist. This contamination makes impossible the accelerator or amino acid dating of well-known samples in palaeoanthropology, excavated several years ago, unless improvements are made in the future. Keeping this in mind, the next section will describe in detail the chemical preparation, combustion and dating of the organic components of bone by the conventional radiocarbon-collagen method.

Dating the Organic Portion of Bone (Collagen)

It is obvious that apatite, the inorganic portion of bones, can only be used for radiocarbon dating when circumstances prevail during the entire duration of the burial of the bones to prevent the infiltration of secondary calcium carbonate and calcium phosphate.

The organic portion of bone, collagen, does not suffer under exchange phenomena. Only under very rare circumstances does one observe an occurrence of non-osteological organics in the bone other than the collagen itself. These organics usually originate from algae, fungi and saprophytes, most of which also feed on the organic portion of fresh bone. Even if they have intruded the bone itself, they also become 'fossilised' organics. Just as the collagen itself, they present an age which, if not identical, is fairly close to that of the original burial date of the bone.

The only possible source of contamination could result from the entry of excessive humic acids from the surrounding environments at a later date. This is usually due to the excessive presence of plants, or even a bog deposit, in the immediate area associated with its location. However, this source of contamination is easily eliminated through laboratory procedures (Longin, 1971).

A second contaminant, though quite rare, could result from deposits such as tar. Berger, Horney and Libby. (1964) have shown that these can quite easily be eliminated through liquid chromatography, a procedure described below and probably standard in most laboratories dating osteological material.

Bone is composed of an interstitial substance that has a fibrillar structure similar to that of connective tissue. These fibres consist mainly of collagen and some reticular fibres. Collagen, the basic substance of bone, is characterised by its content of mucopolysaccharides. Within the organic matrix of collagen, a complex mineral substance is deposited, consisting chiefly of calcium, phosphate, carbonate and citrate. Depending on the age of the individual, there are also some minor percentage differences among species. Anywhere from 25 per cent to as much as 35 per cent of the dry, fat-free weight of all bone is made up of collagen (McLean and Budy, 1964; McLean and Urist, 1968). Because younger individuals are still in a developmental state, in which chondral ossification is in progress, they show a greater organic portion percentage. However, in older individuals, and under physiological conditions, differences also occur as a result of pathologic conditions. These deviations may occur in either direction, but the trained dating specialist can easily distinguish differences among bones of very young individuals, of different species, and also of pathologic conditions, using thin-section microscopy and other techniques. Most dating specialists are not experienced in this particular technique and may not consider it as a preliminary step in bone processing. In spite of this, it is certainly an important aid in calculating the total bone material needed for radiocarbon dating.

Of the interstitial substances, collagen, the organic matrix, has two chief components: collagenous fibres and the ground substance. Bone collagen, which yields glue or gelatine when boiled, constitutes as much as 90–96 per cent of the dry, fat-free weight of the total organic fraction of bone. When intact, collagen is present in fibrils with a double-cross banding at intervals averaging 640 A° and it produces a characteristic X-ray diffraction pattern. It is characterised chemically by a high content of pyrolidine amino acids and a low content of aromatic amino acids. Hydroxyproline is, for example, an index for a quantitative estimation of mature collagen, as is hexosamine for the ground substance. Individual collagen fibrils in fresh bone and in fossil bone found in locations protected from environmental influences, are typically 0.3–0.5μ in diameter and are often found in small bundles 3–5μ in thickness. Quite a large amount of collagen, tropocollagen or amino acids must be present in fossil bone to allow extraction of organic substances for radiocarbon dating, in contrast to the small amounts of amino acids necessary for amino acid dating.

To allow sufficient extraction of the organic portion of bone, one would typically select compacta of long bones and, in some cases, the lamina externa or interna of the neurocranium. Less useful for maximum extraction of collagen are the spongiosa and diploë, since they are quite often intruded upon by rootlets and other foreign organic materials. In addition, they are frequently the first areas to be attacked by water, algae, fungi and saprophytes due to their fragile and spongy nature. Compact bone, on the other hand, is ideal for the extraction of collagen. The dense mineral structure shields the inner organic portion from infiltration by both foreign substances and water, which would drastically diminish the organic portion. Studies on organics in one and the same bone have shown that the organic portion of compacta, as compared to the organic portion of the spongiosa, can differ by a magnitude of three.

Cement and dentin are preferentially selected over enamel when dating teeth of macrofauna. The total organic portion of enamel is much smaller than that of the compacta of bone; that of cement and dentin is somewhat less than that of spongiosa in living animals. Preservation of organics in enamel, cement and dentin is enhanced over long periods by their denser mineralogic structure. A fully developed crown enamel consists of only 3 per cent organic substance, containing a glucoprotein; the rest consists entirely of calcium salts in the form of apatite crystals.

However, dentin, by volume the largest portion of a tooth (depending again on type of tooth and species), is a fibrillar, calcified collagenous matrix traversed by odontoblastic processes composed of a specialised protein called elastin. Dentine is still much harder than compacta, but resembles bone in its structure, chemical nature and development. As in bone, the organic portion of dentine consists of about 25–28 per cent organic, with about 70–72 per cent inorganic fractions. Throughout the life of an individual, dentine is formed with a somewhat greater activity in old age, when the pulp cavity becomes reduced in size. Large teeth, for example, those of mammoth, hippopotamus and other macrofauna, are quite suitable for radiocarbon collagen dating. First, the extremely dense inorganic structure of the enamel and cement overlying the dentine provide an excellent shield against deleterious environmental influences, which could diminish the organic content of the dentine itself. Second, the enamel is thick enough in some animals to provide this shielding effect, yet it most often constitutes only 6–8 per cent of the total dental weight. It is erroneous to assume that the low organic content of teeth makes them unsuitable for dating. This is based on the misconception that enamel is the material to be dated. In fact, it is the dentine that provides the greater mass of dateable material. The organics in the cement, the layer of the tooth that continues to cover the dentin below the alveolar level, are much higher than in enamel and somewhat lower than in compacta.

Bone-processing procedures, starting with a thin-section analysis and ending with the sample count, can be seen in Figure 1.1. Processing of bone and dental material usually starts with the elimination of possible contaminants. If careful and tedious examination and treatment of materials in practically sterile laboratory conditions is undertaken, dates should ultimately be as correct as those obtained from other organic materials. The first step should be a thin-section analysis of the bones in order to determine whether contaminants such as algae, fungi and others are present. With this method, man-made substances such as shellac or other preservatives, or tar, can also easily be recognised. Recent additions of saprophytes (algae and fungi) and bacteria can also be detected easily. This is important because their photo- or chemosynthesis could cause faulty radiocarbon results.

Contamination by these substances is equally possible in charcoal, wood, or other organic materials normally considered more suitable for radiocarbon dating (Sellstedt, Engstrand and Gejvall, 1966). Naturally occurring substances are thus a serious contamination problem in all organic substances. In fact, due to the dense nature of bone, it is actually less prone to contamination than all other materials.

The second step in analysis before bone dating is an estimation of total bone material needed. Nitrogen microanalysis of a few milligrams of compacta, specifically for nitrogen, makes it possible to calculate the total bone material needed. Only dentine should be used for analysis in the case of dental material.

Figure 1.1: Bone Processing Procedures

Thin section analysis

FUN-analysis

Visual contaminant removal

Sand removal by doubly distilled H_2O

1mm compacta (shellac, etc) removal

Acetone/2-ethoxyethanol reflux

4% HCl-20 min-periosteal/shellac border treatment

Hydrolysis by HCl

Collagen strands

Renewal of solution

NaOH humic acid treatment

Repeated doubly distilled H_2O washing

Neutralisation with HCl

Washing in H_2O

Collagen strand drying

Combustion of collagen

Tropocollagen

H_2O solution—weak HCl

Humic acid removal by NaOH

Repeated doubly distilled H_2O washing

HCl treatment—slight acidity

Gelatine conversion

6N HCl-24hr treatment

Liquid chromatography

Amino acid collection

Amino acid drying

(Collagen/gelatine/amino acid)

Sample combustion

CO_2

CO_2 analysis for C

$\sigma C^{12}/\sigma C^{13}$ analysis

CO^2 cleaning by hot CuO

$KMnO_4$ and $AgNO_3$ treatment

CO_2 thermal circulation (CuO/Ag)

Sample count in 2 litre proportional counter

(2,600 min. minimum)

In about half the samples processed for radiocarbon dating, extraneous recent materials such as rootlets are present. Hasty processing after failure to inspect the samples with a microscope is probably responsible for the fact that a large number of dates differ from the relative estimation of a sample. This type of visual inspection and removal by hand often constitutes the greater part of the total processing time of a sample. Part of this removal includes the estimation of sandy deposits within the spongiosa, often necessitating the latter's total removal. Sand and rootlets often infiltrate the Haversian canals and trabeculae and, unless the spongiosa is scraped totally clean, these cannot be detected and subsequently removed.

Insects of a more recent time period are quite often mixed in with the sand and are again a cause for contamination. The removal of sandy materials at this final stage should be carried out using doubly distilled water. The bones should then be dried and again inspected visually. Thereafter, about 1mm of the compacta should be removed by scraping off the periosteal surface. This removes any remaining adherent extraneous natural contaminants that may have infiltrated the more porous periosteal surface of the compacta.

Also removed are man-made contaminants such as shellac, which may have been applied by the excavators. In the Frankfurt Radiocarbon Laboratory, preservatives have been found in every third sample of bone supplied by the excavator, even when it was indicated that the bones had not been treated. A contaminated bone would certainly falsify the date by a large margin. Bones should therefore be broken up into smaller pieces and treated on a reflux in a solution of acetone and 2-ethoxyethanol for several days to be absolutely certain that no contaminants remain. Bones treated several times over the years in museum collections will often require a much longer reflux treatment. In some cases this could last for weeks. After the bones have been dried, a mild hydrochloric acid treatment might be necessary. Usually, preservatives like shellac are easily recognised after the acetone reflux treatment, emerging as easily peeled-off, hardened, whitish deposits. In cases where several coats have been applied, a brief immersion in a 4 per cent HCl solution for 20 minutes creates a space between the periosteal–compacta surface and the hardened preservatives. Persistent preservatives can ultimately be peeled–off by this procedure. After repeated washing in doubly distilled water, the sample should again be dried.

Frequently, another type of contamination may be present, which would ultimately affect samples used for amino acid dating. The author (Protsch, 1976b), as well as Sidle (1967) and Wolman and Miller (1971), conducted a number of experiments and found considerable amino acid contamination in commercially available analytical grade HCl (37 per cent). Such contamination could also be present in ammonium hydroxide used during the liquid chromatography stage. HCl may contain up to 8,300nmol contamination, and ammonium hydroxide could have five to ten times as much. This can result in serious errors in very old samples. HCl used for processing bone samples should thus be doubly distilled; ammonium hydroxide should be made from distilled ammonia gas on a vacuum line if contamination is suspected.

Depending upon the preference of the investigator, the total bone pieces are treated in an HCl solution until only the collagen strands show up. As an alternative, the bones could be pulverised and then treated with a weaker HCl solution and doubly distilled water. In the first case, the investigator is able to see how well the collagen is preserved but has to renew the solution constantly, possibly up to seven days. The concentration of the solution is always dependent on the reaction of the inorganic portion of the bone. A solution that is too concentrated could easily hydrolyse a large amount of the collagen itself, and thus a sample must be watched constantly over several days. The second method, pulverising the bone, is certainly more expedient and might take only a few hours, but the strength of the solution is more difficult to control.

After the inorganic portion is totally dissolved, the insoluble portion is separated with a Buechner funnel. Only glass filter paper is suitable for this purpose. The remaining organic portion is then washed several times with doubly distilled water and treated with a 1 per cent solution of sodium hydroxide to remove humic acids. It is washed again repeatedly in distilled water, neutralised with HCl, and washed again repeatedly until neutral. It is usually at this stage that smaller rootlets, which might still be present in the sample, can easily be detected on the filter paper.

The collagen or tropocollagen is then converted to gelatine by heating in doubly distilled water (pH = 3) at about 90°C, using a heater with a stirring magnet (IKA Combimag RCT) for four to six hours. Usually, some gelatine can be detected by two or three hours, and nongelatinous preservative substances, if still present, would harden and show up in the solution. These can be removed by hand. Other remaining impurities are removed by centrifugation. Finally, only organic materials will be present in the gelatine formed.

The gelatine is then placed in a flask containing 6N HCl, refluxed for 24 hours, and filtered. The clear solution is then concentrated to a volume of about 100ml. At this point, two different processes for further analysis can be followed:

1. If no petroleum (tar) products are suspected in the sample, it can be neutralised, dried and combusted in a pure oxygen atmosphere.
2. If contamination is still suspected, the following procedure should be applied. A volume (5cm diameter) containing Dowex 50–WX8 resin charged with 4N HCl is used to collect pure amino acids by liquid chromatography. The filtered hydrolysate is passed through the column at

Figure 1.2: Converting Collagen (Dried Amino Acids) to Carbon Dioxide (Frankfurt Radiocarbon Dating Laboratory)

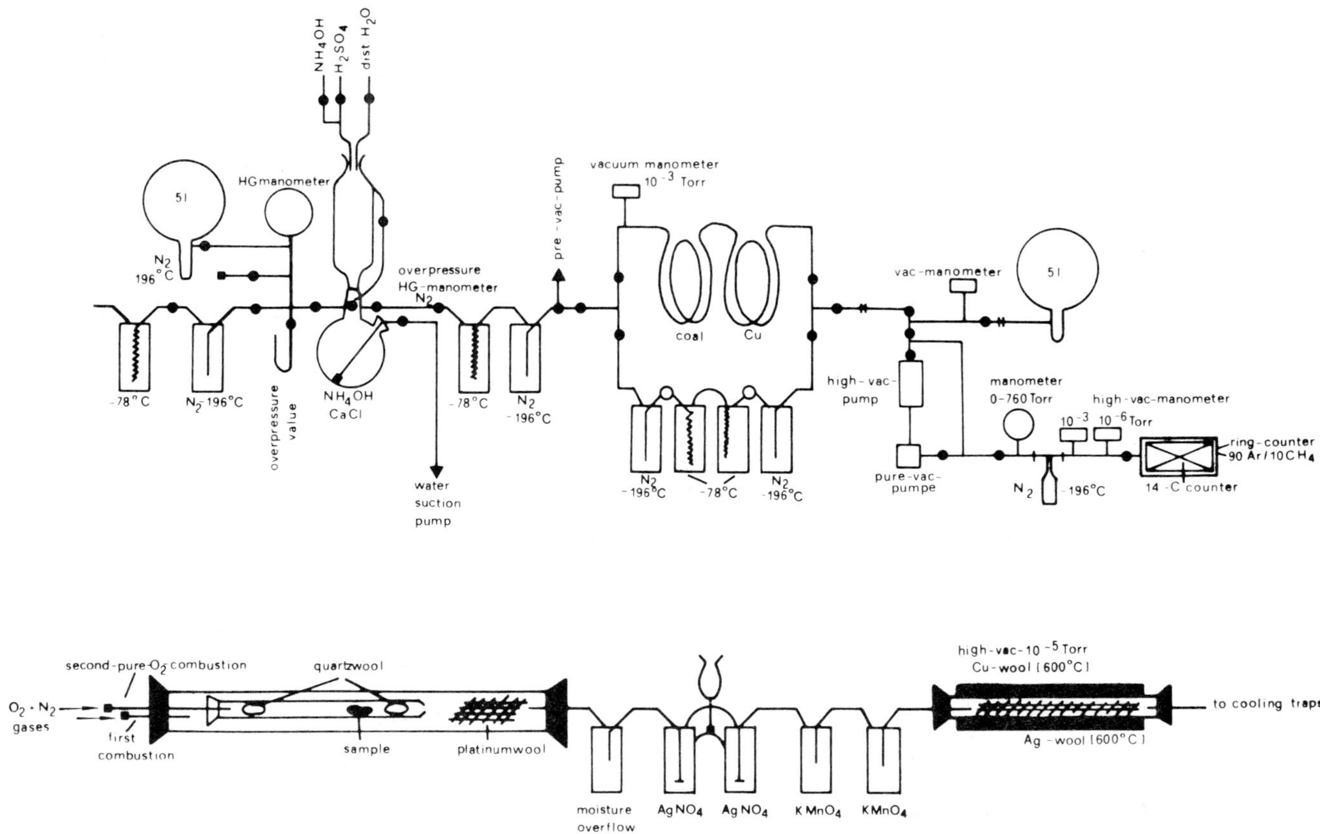

a rate of 30ml/h (Ho, Marcus and Berger, 1969). The resin is then washed repeatedly with distilled water to remove all contaminants. The column is eluted with an excess of 5N ammonium hydroxide at a flow rate of 30–60ml/h, in order to liberate the amino acids. One litre of yellow-brown affluent is collected per sample run. This affluent, containing the amino acids, is then evaporated to dryness, becoming dark brown, and with a distinctive odour.

In both cases, the procedures by which the collagen, or dried amino acids, are converted to carbon dioxide in a stream of oxygen are identical at the Radiocarbon Laboratory in Frankfurt (Figure 1.2). A two-step combustion procedure involves the burning of the sample in an inner tube and passing of nitrogen through that tube, while oxygen is passed through the outer tube. Complete combustion is achieved by a second process whereby the oxygen is again passed through the inner tube, thus burning the sample in a pure oxygen atmosphere.

The nitrogen gas simply operates as a pushing device in the first burning step. Further purification on the processing line is achieved by passing CO_2, the product of the combusted carbon of the organic sample, and the oxygen gas through five traps containing 0.1N silver nitrate, chromic acid, potassium permanganate, hot copper oxide (500°C), hot copper/silver at 450°C, and dry ice/acetone water traps. Final purification is by thermal circulation over hot copper (600°C) and silver wool in a 5 litre glass container with an external quartz convection tube. Before this, and at each counter filling, the CO_2 is routinely passed over hot Cu/Ag (450°C) at least six times in a separate system and vacuum-distilled at −78°C. The gas quality after this procedure is excellent

Counter plateaus are 500V long with slopes of 0.6 per cent/100V for muons and less than 1 per cent/100V for C^{14} + background. The standard working point is 4,300V. The absence of electronegative impurities is verified before and after each run by measuring the count rate at the steep part of the muon characteristic curve. An experimental linear correlation between guard gross and sample muon rates, valid within counting statistics, gives a post-measurement check on working point and a long-term check on instrument stability. Background is nominally 7.9 ± 0.08 cpm with slow seasonal variations in the maximum range of ± 0.2 cpm (95 per cent confidence level). No dependence on atmospheric pressure was found on the basis of 90 background measurements, each of 2,000 minutes. Four months of continuous monitoring of the laboratory aerosol radioactivity with a NaI detector and multichannel analyser (Canberra Series 30, 1,024 channels) showed an unchanged background in the range covering 200–3,000keV.

The modern standard CO_2 is prepared by wet oxidation of NBS oxalic acid. The AD 1950 C^{14} count rate is 12.10 ± 0.08 cpm at 20°C with normalisation of the $\sigma -C^{13}$ value of −19.75°/oo to the standard −19°/oo. Counting periods are two days for background, sample and standard. The samples are measured at least twice and the background at least once a week. Routine χ^2-analysis is applied to 100 minute print-outs. Age errors correspond to measured $\pm\ \sigma$ variations of sample, background, and standard. Calculated errors smaller than 100 years are increased to this figure as a minimum. C^{13}/C^{12} measurements are done only for selected samples. Thus, 80 years have been added to all bone measurements to give a theoretical correlation for C^{13} fractionation, with an additional $\pm$ 60 years squared added to the variance (Lerman, 1972). All dates are expressed in C^{14} years relative to AD 1950, using the half-life of 5,568 years. Calendar estimates and archaeologic comments are based on calibrated dates (Pearson and Baillie, 1983; Pearson, Pilcher and Baillie, 1983).

To calculate the uranium present, the CO_2 sample is passed into the counter after repeated passing over hot copper and elimination of free oxygen. Radon (Rn^{222}), the byproduct of uranium decay, allows this computation with fairly good accuracy. The radon count makes it possible to estimate the exact time at which the interference in the count due to Rn^{222} has decreased to less than one count per minute.

The 2 litre copper proportional counter is then filled to 1,013mbar with the purified CO_2 sample. The counter is protected against cosmic and surrounding radiation by a 3.5 ton lead shield and a copper multiwire anticoincidence ring-counter flushed constantly with purified 90 Ar/10 CH_4. Electronics are of the commercial NIM type.

Kelsterbach — A Case Study in C^{14} Collagen and Amino Acid Dating

In spring 1952 during bulldozing operations in the north-west corner of the Willersinn gravel pit, which is south-west of Frankfurt between Kelsterbach and Raunheim, a fossilised hominid calotte was discovered.

The original site of discovery has remained undisturbed until today because further digging was continued solely in northerly and easterly directions. These circumstances made it possible, 20 years later, for geologists to examine the precise stratigraphic relationship of the different finds in the location and stratigraphic levels. On the basis of the still undisturbed stratigraphic sequence and the graphs and drawings made at the time of discovery of the calotte, it was possible to make a relatively reliable and accurate estimation of the geological situation of the hominid and faunal remains in question (Protsch and Semmel, 1978).

Accordingly, the calotte can be attributed to a layer at a depth of 4.60m. The analysis of some sandy remains extracted from the left auditory meatus supported the accuracy of the observation that all materials originated from the stratigraphic layers as originally reported. The stratigraphic layer that yielded the hominid find belongs to the lower part of a 5m thick gravel bed and is formed by intermittent sand gravel layers

Figure 1.3: Stratigraphy of Kelsterbach

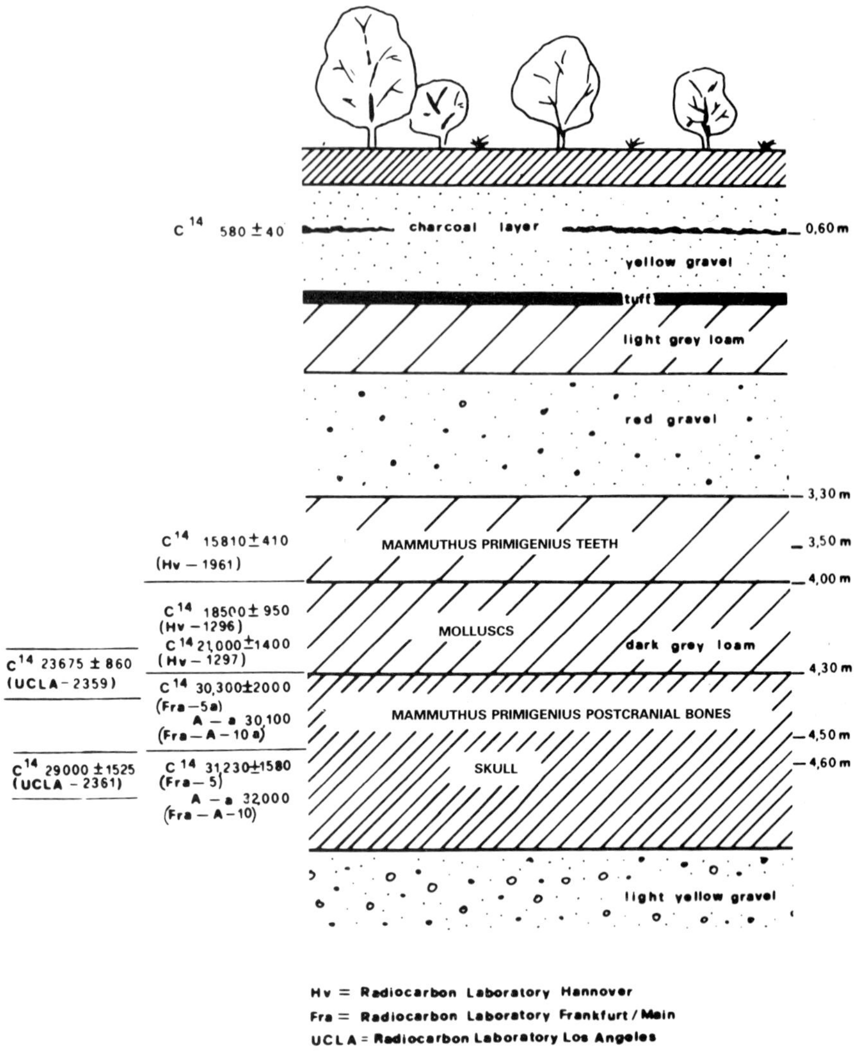

Relative and Absolute Chronology and Dating of the Kelsterbach Layers

located at the 'Obere Niederterrasse', the so-called $t_{(6)}$ terrace (Becker, 1965, 1967; Semmel, 1969, 1972, 1974). It was deposited by the River Main in that location during the Upper Pleistocene period. It consists of clay flood-plain deposits and quicksand and is covered by remains of sediments of the 'Laacher pumice-tuff' (Figure 1.3).

This gravel bed in which the fossils were found is the second youngest terrace of the River Main. Its chronological origin has been a point of frequent discussions (Becker, 1965, 1967; Semmel, 1969, 1972, 1974; Protsch and Semmel, 1978). Other faunal remains were also discovered within the terrace, but in a much higher level. One of these, a mammoth molar, was dated by radiocarbon at the radiocarbon laboratory of the Niedersächsische Landesamt für Bodenforschung in 1960 and yielded a date of 15,810 ± 410 years BP (Hv–1961). Since the mammoth fragments were recovered from a level substantially above those of the hominid, it is reasonable to believe that the hominid is substantially older. However, because this date is based on the inorganic portion of the bone (apatite) and not the organic (collagen) components, there is a probability that the apatite was contaminated with recent or old radiocarbon. However, the mineralogical makeup of the bed in the terrace makes such contamination improbable, since it was void of any calcitic material and leakage through the layers was practically nonexistent. Percolating waters with high solutions of calcium carbonate and calcium phosphate could cause secondary apatite accumulation into the bone, thus yielding C^{14} dates either too old or too recent. Another method had to be found to date the River Main terrace, and therefore the calotte, more accurately. Geomorphological analysis of the stratigraphy of the upper layer, and the results of pollen analysis of its clays, assign the terrace 'relatively' reliably within the late phase of the Würm-Glacial.

To obtain a more accurate date for the calotte, a small bone sample was removed from it in the radiocarbon laboratory of the Anthropological Institute of the University of Frankfurt. A sufficient amount of organic material could be extracted allowing both radiocarbon collagen dating as well as amino acid dating. The process of extraction and absolute dating for Kelsterbach is described in detail by Protsch and Semmel (1978) as well as in this section. According to the radiocarbon results the calotte dates absolutely to 31,230 ± 1580 years (Fra–5). By means of amino acid dating of the same sample an age of approximately 32,000 years BP could be estimated (Protsch and Semmel, 1978). The latter date should be judged as relative and should be viewed with caution, due to the difficulty of reconstruction of the palaeotemperature necessary in making an accurate amino acid dating age calculation. However, the radiocarbon results can be considered quite accurate and therefore as absolute. In this case the results of both dating methods coincide and thus also supply a good palaeotemperature reconstruction. Because both dates coincide quite well, one can assume that a date of at least 31,000 years BP is acceptable for the calotte. Blind dates on the mammoth positioned somewhat above the hominid and on the hominid itself were also processed at UCLA and yielded somewhat younger dates: 23,675 ± 860 (UCLA–2359), and 29,000 ± 1525 (UCLA–2361). These mammoth fragments were positioned below in a stratum substantially older than those dated by the Hannover laboratory to 15,810 ± 410 years (Hv-1961).

Interpretation of Microanalytical Results

The results of nitrogen analysis are unusually high for bones of such age (Table 1.1). This is easily understandable given the ecological circumstances in an early bog-type deposit. The same can be said for the increase of uranium and fluorine over time. Fluorine is unusually high but, here again, this is in keeping with expectations in a water-rich environment. The uranium is considered average and shows a continuous increase, from the higher stratigraphic levels to the lowest hominid-bearing horizon. Overall, the microanalytical results show a clear relative age succession and placement, which support the absolute dates.

Table 1.1: FUN Analysis of *Mammuthus primigenius* and Hominid from Kelsterbach

Species	Material	Location	F(%)	U(ppm)	N(%)
M. primigenius	Bone	420m	2.80	109	2.00
M. primigenius	Bone	450m	2.93	130	1.55
H. s. sapiens	Bone, os occipitale	460m	2.98	138	1.20

Radiocarbon Collagen Dating of the Kelsterbach Hominid and Mammoth

Sample pretreatment and collagen extraction as outlined in the previous section was identical in both the hominid and the mammoth. This treatment was basically the same as that described in 1964 (Berger, Horney and Libby), with a few additional pretreatment techniques as described by Protsch (1972). It was preceded by examination of the microstructure of the bone by a thin-section technique using polarised light. The organic content of bone samples was identical and provides additional proof to substantiate the assumption that the samples were nearly or very closely associated in a similar environment.

The nitrogen content of the Kelsterbach hominid was 1.2 per cent N; 70g of bone yielded about 6g of collagen and 2.60g of pure carbon, which yielded about 4.5 litres CO_2. This was more than enough for a total counter filling at the Frankfurt laboratory. Actually, only 30 to 35g of bone would have been enough for C^{14} dating purposes. This sample was processed in 1978, before the samples of the stratigraphically associated mammoth material, but several years after radiocarbon measurements on mammoth teeth and shells were processed at the Hannover laboratory. The yield of CO_2 from the collagen sample in the case of the Kelsterbach hominid was between 90 and 95 per cent. The hominid sample was counted for 36 hours (2,160 minutes) three times (two sigma-standard deviations), and yielded an age of 31,000 ± 1,600 years (Protsch and Semmel, 1978).

It was obvious from the similar microstructure of the bone that little material was needed for radiocarbon dating of the *Mammuthus primigenius* stratigraphically positioned slightly above the hominid. Microanalysis of the faunal material showed 1.55 per cent N, a slightly higher content than in the stratigraphically older hominid. A 150g sample of bone yielded 13.4g of collagen, of which about half (about 6.8g) was carbon. This amount was more than enough for two counter fillings. The sample processing was continued until 4.6 litres of CO_2 was collected for sample counting. The results are shown in Table 1.2.

Table 1.2: Hominid (Fra–5) — Radiocarbon Collagen Dating Results

Counting period	Date
1.	31,600 ± 1,800 years BP
2.	31,000 ± 1,500 years BP
3.	31,000 ± 1,450 years BP
Final age (Fra – 5)	
	31,200 ± 1,600 years BP
Final age (UCLA–2361)	
	29,000 ± 1,525 years BP

The sample was run for three counting periods, each of 2,160 minutes. Individual measurements show good reproducibility and give an average value of 31,200 years BP.

The sample was then run again for three counting periods, each of 2,000 minutes: the results are shown in Table 1.3.

Table 1.3: Faunal Sample (Part of Humerus) — *Mammuthus primigenius* (Fra–5a) at 4.60m

Counting period	Date	
1.	29,650 $^{+2,300}_{-1,800}$ years BP	
2.	29,940 $^{+2,400}_{-1,850}$ years BP	Average 30,300 $^{+2,500}_{-1,900}$ years BP
3.	31,350 $^{+2,950}_{-2,200}$ years BP	

The individual measurements in Table 1.3 show good reproducibility and give an average value of 30,300 ± 2,500 years BP for the Kelsterbach mammoth.

The final age of *M. primigenius* (UCLA-2359) at 4.30m is 23,675 ± 860 years BP.

It should be remembered that all radiocarbon dates are essentially asymmetric in accuracy and that the averaging of individual measurements should be preferred on the count rates, and not on the derived radiocarbon dates. This is because the radiocarbon date is a logarithmic function of the count rates of sample, background, and modern standard. However, the asymmetry is of importance only for very old samples, in which there is weak signal detection in the presence of noise caused by a counter background (Currie, 1968).

In the special case of the 'very old' Kelsterbach mammoth, sample measurements were performed at different times during a time period of six weeks. It was stored for three months before the first measurement to allow for complete decay of Rn^{222} in the sample CO_2. The sample count rates were compared with 20 background measurements (each of 2,000 minutes) performed during a time period of four months bracketing the sample repetitions. In the age calculations, the fluctuations of background, sample, and modern standard are accounted for on a 95 per cent confidence level (chi-squared test). The error margins thus calculated are more rigorous than the usual ± 1 standard deviation error margins used for young radiocarbon samples.

It can be noted that each of the individual sample measurements gave count rates at a 3 sigma level above background. The rigorous statistical test is necessary because the dating is not too far from the systems limits of resolution (age limit about 41,000 years BP). The procedure used in error calculation is in accordance with the recommendations of Stuiver, Minze, and Polach (1977), and Walanus and Pazdur (1980) for age reporting on very old radiocarbon dated samples.

Comparative Amino Acid Dating of the Kelsterbach Hominid and Mammoth

Bada, Protsch and Schroeder (1973) showed that comparative dating by C^{14} and amino acid dating can supply absolute dates on hominids at sites that could not be archaeometrically determined by other methods. The site of Kelsterbach provides different samples in several stratigraphic levels (bone, teeth and molluscs), which can also be dated by different methods.

Cranial hominid and postcranial faunal material of *Mammuthus primigenius* was processed by the isoleucine epimerisation method. For each sample, 5g of compact bone was used. Small pieces of bone were removed from the calotte of the hominid, specifically on the os occipitale close to the lambdoid suture, and part of a femur of *M. primigenius* was sampled. The outer surface was mechanically abraded and cleaned with a dental abrading machine and thus about 1mm of the surface removed.

Samples were subsequently treated in doubly-distilled water and ultrasonification twice (each for five minutes) and 0.2N $HC1_{aq}$ (five minutes). The pretreated bone material was then submerged in 2N $HC1_{aq}$ and the nearly totally preserved collagen strands removed and immediately neutralised. Hominid as well as faunal sample showed nearly completely preserved collagen strands.

The collagen samples were subjected to hydrolysis with 6N HCl_{aq} at a temperature of 105°C for 24 hours in order to dissolve the peptide bonds. The resulting clear fluid was evaporated at a temperature of 55°C; the residue was dissolved in double-distilled water and desalted on Dowex 5 W–X8 resin. The amino acids were eluted with 1.5N NH_4OH. The solution was reduced again with the help of a rotary evaporator at 55°C and the remainder resolved with buffer pH 2.28.

The actual analysis was done using a Multichrome B 4255 automatic amino acid analyser, by Beckman. Its column was filled with the ionic exchanger M82. The time for analysis took 210 minutes with a total sample of four.

Excellent preservation of the organic substance could be observed on the chromatograms. Qualitatively, even those amino acids that normally disintegrate rapidly (threonine, serine, and hydroxyproline) were present. Even the diastereomere type of L-hydroxyproline could be shown. Unfortunately, no constants of this amino acids are available yet, which presently makes dating by this amino acid impossible.

The isomeres of isoleucine (L- and D-allo-Ile) used for dating displayed the D/L-ratios for the samples as shown in Table 1.4.

Table 1.4: D/L Ratios of Isoleucine Isomeres

Sample	Laboratory no.	D/L-ratio
1	FRA-A-10a-(3b)	0.02548
2	FRA-A-10a-(3c)	0.02693
3	FRA-A-10a-(3d)	0.02593
4	FRA-A-10a-(4)	0.02548

The equation for the age computation is as follows:

$$1.\ \text{Age (yr)} = \frac{\ln \dfrac{1 + (D/L)}{1 - (D/L)\,K'} - C}{(1 + K')k}$$

For the equilibrium constant $K' = K^{-1}$ the value 0.725 was taken; for the reaction rate constant k the value $10^{19.41\ -\ 7,304.O/T}$; and the value of 0.028 was taken for the integration constant C (Bada, 1972).

The following equation is then applicable:

$$2.\ \text{Age (yr)} = \frac{\ln \dfrac{1 + (D/L)}{1 - 0{\cdot}725\,(D/L)} - 0{\cdot}028}{1{\cdot}725 \times 10^{19{\cdot}41\ -\ 7304{\cdot}0/T}}$$

The temperature during burial of the fossils (hominid and mammoth) was taken as 8.8°C (Protsch and Semmel, 1978). The ages of the samples are then as follows:

Mammuthus primigenius	
Sample 1	28,650 years BP
Sample 2	33,170 years BP
Sample 3	30,060 years BP
Sample 4	28,650 years BP

The average age computed for *Mammuthus primigenius* on the basis of four sample runs was 30.100 ± 2,100 years BP. The isomeres of isoleucine used for dating the hominid show the following D/L ratios for the samples:

Sample	Laboratory no.	D/L Ratio
1	Fra-10-(1)	0.02655
2	Fra-10-(2)	0.02653

Using the same equation (2) as above on the *M. primigenius* samples, the ages for the hominid are as follows:

Sample 1	32,000 years BP
Sample 2	31,900 years BP

The average age of the hominid on the basis of two sample runs is 32,000 years BP.

Since the mammoth was positioned stratigraphically slightly above the

Table 1.5: Absolute Dates and Stratigraphic Depth of Samples

Depth (m)	Method	Material	Species	Age (years BP)	Laboratory
0.50	C^{14}	Charcoal		510± 100	Fra–52
3.30 to 3.50	C^{14}	Molars	*Mammuthus primigenius*	15,810± 410	(Hv–1961)
4.00	C^{14}	Molluscs	—	18,500± 950	(Hv–1296)
4.20	C^{14}	Bone	*M. primigenius*	21,000±1,400	(Hv–1297)
4.30	C^{14}	Bone	*M. primigenius*	23,675± 860	(UCLA–2359)
4.30 to 4.50	C^{14}	Bone	*M. primigenius*	30,300±2,000	(Fra–5a)
	A–a (isoleucine)		*M. primigenius*	30,100±2,100	(FRA–A–10a)
4.60	C^{14} (collagen)	Bone	*H.s.sapiens*	29,000±1,525	(UCLA–2361)
	C^{14} (collagen)	Bone	*H.s.sapiens*	31,230±1,580	(Fra–5)
	A–a (isoleucine)	Bone	*H.s.sapiens*	32,000	(FRA–A–10)

Notes:
Hv — Radiocarbon Laboratory, Hannover;
Fra — Radiocarbon Laboratory, Frankfurt;
UCLA — Radiocarbon Laboratory, Los Angeles.

hominid (see Figure 1.3), it follows that these ages should be expected to be younger than the hominid positioned below. This is indeed the case.

All radiocarbon and amino acid dates from the different laboratories in their respective stratigraphic position are shown in Table 1.5.

As can be easily be seen in the case of Kelsterbach, a series of radiocarbon collagen dates can be supported or verified by amino acid dates. If, in addition to this, relative chronological evaluation by means of microanalysis, faunal analysis, and geomorphological/stratigraphic analysis is undertaken, one can be assured of a reasonably sound chronological placement for any fossil hominid.

Acknowledgements

To the VW-Stiftung (VW-Foundation) (Grant-I-37939) and Optical Works Ernst Leitz Wetzlar for their generous support of this research.

References

Ames, L. L. Jr (1959) 'The Genesis of Carbonate Apatites', *Econ. Geol., 54*, pp. 829–41

Bada, J.L. (1972) 'The Dating of Fossil Bones Using Racemization of Isoleucine, *Earth Planet. Sci. Lett., 15,* 223–31

—— and Protsch, R. (1980) 'Racemisation Reaction of Aspartic Acid and Its Use in Dating Fossil Bones', *Proc. Nat. Acad. Sci., 70*, 1331–4 (orig. 1973) in K. A. Kvenvolden (ed.), *Geochemistry of Organic Molecules, Part V, Molecular Application*, Dowden, Hutchinson and Ross Inc., Stroudsburg, Pennsylvania, pp. 325–8

——, Protsch, R. and Schroeder, R. A. (1973) 'The Racemisation Reaction of Isoleucine Used As A Paleotemperature Indicator', *Nature, 241* (5389), pp. 349–95

Beaumont, P.B., de Villiers, H. and Vogel, J. C. (1978) 'Modern Man in Sub-Saharan Africa Prior to 49,000 Years BP: 'A Review and Evaluation with Particular Reference to Border Cave', *S. Afr. J. Sci., 74*, pp. 409–19

Becker, E. (1965). *Stratigraphische und Bodenkundliche Untersuchungen an Jungpleistozänen und Holozänen Ablagerungen im Nördlichen Oberrheintalgraben*, Frankfurt am Main, Diss

—— (1967) 'Zur Stratigraphischen Gliederung der Jungpleistozänen Sedimente im Nördlichen Oberrheintalgraben', *Eiszeitalter und Gegenwart, 18,* 5–10

Bennet, C. L., Beukens, R. P., Clover, M. R., *et al.* (1977) 'Radiocarbon Dating Using Electrostatic Accelerators: Negative Ions Provide the Key', *Science, 198*, 508–10

Berger, R. (1979) 'Radiocarbon Dating with Accelerators', *Journal of Archaeological Science, 6*, pp. 101–4

—— (1979) 'UCLA Date List X', *Radiocarbon, 25*

—— Horney, A. G. and Libby, W. F. (1964) 'Radiocarbon Dating of Bone and Shell From Their Organic Components', *Science, 144*, pp. 999–1001

—— and Protsch, R., Rozaire, C., *et al.* (1971) 'New Radiocarbon Dates Based on Bone Collagen of Californian Palaeo-Indians', *Contributions of the University of California, Archaeological Research Facility*, UC-Berkeley, 12, pp. 43–7

—— and Protsch, R. (1973) 'Domestication of Plants and Animals in Europe and the Near East' in G. Buccelati (ed.), *Approaches to the Study of the Ancient Near East*, Orientation, Pontifical Biblical Institute, Roma, *42*, RASC. 1–2, pp. 214–27

Currie, L. A. (1968) 'Limits for Qualitative and Quantitative Determination', *Analytical Chemistry, 40*, pp. 586–92

Damon, P. E., Long, A. and Grey, D. C. (1966) 'Fluctuation of Atmospheric C^{14} During the Last Six Millenia', *Journal of Geophysical Research, 71*, 1055–63

de Vries, H. and Barendsen, G. W. (1953) 'A New Technique for Radiocarbon Dating by a Proportional Counter Filled with Carbon Dioxide', *Physica, 19*, pp. 987–1003

de Vries, H. (1958) 'Variation in Concentration of Radiocarbon with Time and Location on Earth', *Koninklijke Nederlandse Akademie van Wetenschappen, Verhandelingen, B61*, pp. 94

Eastoe, J. E. and Courts, A. (1963) *Practical Analytical Methods for Connective Tissue Proteins*, E. and F. N. Spon, London

Elsasser, W. E., Ney, P. and Winkler, J. R. (1956) 'Cosmic-ray Intensity and Geomagnetism', *Nature, 178*, pp. 1226–7

Glowatzki, G. and Protsch, R. (1973) 'Das absolute Alter der Kopfbestattungen in der Großen Ofnet-Höhle bei Nördlingen in Bayern', *Homo, 24*, pp. 1–6

Grey, D. C., Damon, P. E. and Haynes, C. V. (1969) 'Carbon-isotope Fractionation during Wet Oxidation of Oxalic Acid', *Radiocarbon, 11*, pp. 1–2

Grootes, P. M. (1978) 'Carbon-14 Time Scale Extended: Comparison of Chronologies', *Science, 200*, pp. 11–15

Haynes, V. (1968) 'Radiocarbon: Analysis of Inorganic Carbon of Fossil Bone and Enamel', *Science, 161*, pp. 687–9

Henke, W. and Protsch, R. (1978) 'Die Paderborner Calvaria — Ein Deluvialer Homo Sapiens', *Anthropologischer Anzeiger 36*, pp. 85–108

Ho, T. Y., Marcus, L. F. and Berger, R. (1969) 'Radiocarbon Dating of Petroleum- impregnated Bone from Tar Pits at Rancho La Brea, California', *Science, 164*, pp. 1051–2

Kigoshi, K. and Hasegawa, H. (1965) 'Secular Variation of Atmospheric Radiocarbon Concentration and its Dependence on Geomagnetism', *Journal of Geophysical Research, 71*, pp. 1065–72

Leakey, M. D., Hay, R. L., Thurber, D. L., Protsch, R. and Berger, R. (1972) 'Stratigraphy, Archaeology, and Age of Ndutu and Naisiusiu Beds, Olduvai Gorge, Tanzania', *World Archaeology, 3*, pp. 328–41

Leakey, L. S. B., Protsch, R. and Berger, R. (1968) 'Age of Bed V, Olduvai Gorge, Tanzania', *Science, 162*, pp. 559–60

Lerman, J. C. (1972) 'Carbon-14 dating: Origin and Correction of Isotope Fractionation Errors in Terrestrial Living Matter' in T. A. Rafter, and T. Grant-Taylor (eds.), *International Conference on Radiocarbon Dating, 8th Proc: Wellington, Royal Soc. New Zealand, 2*, H17, pp. 612–24

Libby, W. F. (1952) *Radiocarbon Dating*, University of Chicago Press, Chicago

—— (1955) *Radiocarbon Dating*, 2nd edn, University of Chicago Press, Chicago

Longin, R. (1971) 'New Method of Collagen Extraction for Radiocarbon Dating', *Nature, 230*, pp. 241–2

MacKay, C., Pandow, M. and Wolfgang, R. (1963) 'On the Chemistry of Natural Radiocarbon', *Journal of Geophysical Research, 68*, pp. 3929–31

Mann, W. B., Marlow, W. F. and Hughes, E. E. (1961) 'The Half-life of Carbon-14', *International Journal of Applied Radiation and Isotopes, 11*, pp. 57–67

McClellan, G. H. and Lehr, J. R. (1969) 'Chemical Investigation of Apatite', *Am. Min., 54*, pp. 1374–91

McConnell, D. (1962) 'The Crystal Structure of Bone', *Clin. Orth., 23*, pp. 253–68

McLean, F. C. and Budy, A. M. (1964) *Radiation, Isotopes, and Bone*, Academic Press Inc, New York

—— and Urist, M. R. (1968) *Bone Fundamentals of the Physiology of Skeletal Tissue*, 3rd ed, University of Chicago Press, Chicago, p. 314

McNeish, R. S., Berger, R. and Protsch, R. (1970) 'Megafauna and Man from Ayacucho, Highland Peru', *Science, 168*, pp. 975–7

Middleton, J. (1844) 'On Fluorine in Bones: Its Source and its Application to the Determination of the Geological Age of Fossil Bones', *Proc. Geological Society of London, 4*, pp. 431–3

Muller, R. A. (1977) 'Radioisotope Dating with a Cyclotron', *Science, 196*, pp. 489–94

—— Stevenson, E. J. and Mast, T. S. (1978) 'Radioisotope Dating with an Accelerator: A Blind Measurement', *Science, 201*, pp. 347–8

Neuman, W. F. and Neuman, M. W. (1958) *The Chemical Dynamics of Bone Mineral*, University of Chicago Press, Chicago
Nelson, D. E., Korteling, R. G. and Scott, W. F. (1977) 'Carbon-14: Direct Detection at Natural Concentrations', *Science, 198*, pp. 507–8
Oakley, K. P. (1968) *Frameworks for Dating Fossil Man*, Aldine Publishing Company, Chicago
Pandow, M., MacKay, C. and Wolfgang, R. (1960) 'The Reaction of Atomic Carbon with Oxygen: Significance for the Natural Radiocarbon Cycle', *Journal of Inorganic Nuclear Chemistry, 14*, pp. 153–8
Pavlish, L. A. and Banning, E.B. (1980) 'Revolutionary Developments in Carbon-14 Dating', *American Antiquity, 45*, pp. 2, 290–7
Pearson, G. W. and Baillie, M. G. L. (1983) 'High Precision ^{14}C Measurement of Irish Oaks to Show the Natural Atmospheric Variations of the AD Time Period' in Stuiver, Minze and Kra (eds.), *Internatl. Radiocarbon Conf. 11th, Proc., Radiocarbon, 25*, pp. 2, 179–86
——, Pilcher, J. R. and Baillie, M. G. L. (1983) 'High Precision ^{14}C Measurement of Irish Oaks to Show the Natural ^{14}C Variations from 200BC to 4000BC' in M. Stuiver and R. Kra (eds), *Internal Radiocarbon Conf., 11th Proc. Radiocarbon, 25* (2), pp. 179–86
Polach, H. A. and Golson, J. (1966) 'Collection of Specimens for Radiocarbon Dating and Interpretation of Results: Canberra, Australian Institute of Aboriginal Studies', *Man, 2*, pp. 42
Protsch, R. (1972) 'The Dating of Upper Pleistocene Subsaharan Fossil Hominids and Their Place in Human Evolution: With Morphological and Archaeological Implications'. Diss. UCLA, Anthropology.
—— (1974a) 'The Fish Hoek Hominid: Another Member of Homo Sapiens Afer, *Anthropologischer Anzeiger, 34* 3/4 pp. 241–9
—— (1974b) 'Florisbad: Its Palaeoanthropology, Chronology, and Archaeology', *Homo, 25*,pp. 2, 68–78
—— (1974c) 'The Age and Stratigraphic Position of Olduvai Hominid I', *Journal of Human Evolution, 3*,pp. 379–85
—— (1975) 'The Absolute Dating of Upper Pleistocene SubSaharan Fossil Hominids and Their Place in Human Evolution', *Journal of Human Evolution, 4*, pp. 297–322
—— (1976a) 'The Naivasha Hominid and its Confirmed Late Upper Pleistocene Age', *Anthropologischer Anzeiger, 35*, 2/3, pp. 97–102
—— (1976b) 'New Absolute Dates on Upper Pleistocene Fossil Hominids from Europe and South America', in R. Berger and W. F. Libby (eds), *Proc. 9th Intern. Rad. Conf. UCLA and UC-San Diego*, University of California Press, Los Angeles
—— (1978a) 'The Chronological Position of Gamble's Cave II and Bromhead's Site (Elmenteita) of the Rift Valley, Kenya', *Journal of Human Evolution, 7*, pp. 101–9
—— (1978b) 'Wie Alt Ist der Homo Sapiens?' *Archäologische Informationen, 4*, pp. 8–32
—— (1978c) *Catalog of Fossil Hominids of North America*, G. Fischer, New York, Stuttgart
—— (1978d) 'Ursprung und Migration der Fossilen Subspecies des "Anatomisch Modernen Menschen" des Oberen Pleistozäns', *Archäolog. Informationen, 4*,pp. 8–32
—— (1978e) 'Der Mensch Stammt Aus Afrika', *Umschau, 78*, pp. 18, 554–62
—— (1981) 'The Palaeoanthropological Finds of the Pliocene and Pleistocene; Part II Monograph Eyasi; Die Archäologischen und Anthropologischen Ergebnisse der Kohl-Larsen-Expedidionen in Nord-Tanzania 1933–1939', *Verlag Archaeologica Venatoria, Tübingen*, TMU 4/3, p. 386
—— and Berger, R. (1973) 'Earliest Radiocarbon Dates for Domesticated Animals from Europe and the Near East', *Science, 179*, pp. 235–9
—— and Glowatzki, G. (1974) 'Das absolute Alter des paläolithischen Skeletts aus der Mittleren Klause bei Neuessing, Krs. Kelheim in Bayern', *Anthropologischer Anzeiger, 34*, pp. 2, 140–4
—— and Oberholzer (1975) 'Mumbwa: Its Absolute Chronology and Archaeology', *Zeitschrift f. Morphologie u. Anthropologie, 68*, pp. 1, 1–7
—— and de Villiers, H. (1974) 'Bushman Rock Shelter, Origstad, Eastern Transvaal, South Africa', *Journal of Human Evolution, 3*, pp. 387–96
—— and Oberholzer, J. J. (1975) 'Palaeoanthropology, Chronology and Archaeology of the Matjes River Rock Shelter', *Zeitschrift f. Morphologie u. Anthropologie, 67*, pp. 1, 32–43

—— and Semmel, A. (1978) 'Zur Chronologie des Kelsterbach-Hominiden. *Eiszeitalter u. Gegenwart, 28,* pp. 200–10

Ralph, E. K. and Stuckenrath, R. Jr. (1960) 'Carbon-14 Measurements of Known Age Samples', *Nature, 188,* pp. 185–7

Rightmire, G. P. (1979) 'Implications of Border Cave Skeletal Remains for Later Pleistocene Human Evolution', *Current Anthropology, 20*, pp. 1, 23–35

Semmel, A. (1969) 'Quartär'. In: Erl. geol. Kt. Hessen 1:25 000, Bl. 5916 Hochheim (eds.: Kümmerle und Semmel), Wiesbaden, *3*: pp. 51–99

—— (1972) 'Untersuchungen Zur Jungpleistozänen Talentwicklung in Deutschen Mittelgebirgen', *Z. Geomorph., N.F., Suppl., 14*,pp. 105–12

—— (1974) 'Der Stand der Eiszeitforschung im Rhein-Main-Gebiet', *Rhein-Main-Forsch., 78*, pp. 9–56

Sellstedt, H., Engstrand, L. and Gejvall, N.-G. (1966) 'New Application of Radiocarbon Dating to Collagen Residue in Bones', *Nature, 5*, 1966

Sidle, S., (1967) 'Amino Acid and Peptide Synthesis from Hydrogen Cyanide', *Nature, 216*, pp. 408

Soriano, M. (1970) 'The Fluoric Origin of the Bone Lesion in the *Pithecanthropus erectus* Femur', *Amer. J. Phys. Anthrop., 32*, pp. 49–58

Suess, H. E. (1965) 'Secular Variations of the Cosmic-ray-produced Carbon-14 in the Atmosphere and Their Interpretations', *Journal of Geomorphological Research, 70*, pp. 5937–52

Stuiver, M. (1961) 'Variations in Radiocarbon Concentration and Sunspot Activity', *Journal of Geophysical Research, 66*, pp. 273–6

—— (1965) 'Carbon-14 content of 18th and 19th Century Wood: Variations Correlated with Sunspot Activity', *Science, 149*, pp. 533–5

—— Heusser, C. J. and Yang, T. C. (1978). 'North American glacial history extended to 75,000 years'. *Science, 200*, pp. 16–21

—— and Polach, H. A. (1977) 'Discussion: Reporting of 14-C Data', *Radiocarbon, 19*, pp. 355–63

—— and Suess, H. E. (1966) 'On the Relationship Between Radiocarbon Dates and True Sample Ages', *Radiocarbon, 8*, pp. 534–40

Termine, J. D. and Posner, A. S. (1967) 'Amorphous/Crystalline Interrelationship in Bone Minerals', *Calc. Tiss. Res., 1*, pp. 8–23

Vaughan, J. M. (1970). *The Small Physiology of Bone*, Clarendon Press, Oxford p. 325

Vogel, J. and Beaumont, P. (1972) 'Revised Radiocarbon Chronology for the Stone Age in South Africa', *Nature, 237*, pp. 50–51

Walanus, A. and Pazdur, M. F. (1980) 'Age Reporting of Very Old Samples', *Radiocarbon, 22*, pp. 1021–7

Willis, E. H., Tauber, H. and Münnich, K. D. (1960) 'Variations in the Atmospheric Radiocarbon Concentration over the Past 1,300 Years', *Radiocarbon, 2*, pp. 1–4

Wolman, Y. and Miller, S. L. (1971) 'Amino Acid Contamination of Aqueous Hydrochloric Acid', *Nature, 234*, pp. 548–9

2 AMINO ACID RACEMISATION DATING — A REVIEW

Patricia M. Masters

Since 1970 a new method of measuring the ages of archaeological and geological materials of biological origin has been developed. The technique, amino acid racemisation dating, is applicable to samples such as bone, teeth, ivory, mollusc shell, foraminifera, and calcareous sediments. What these materials have in common originally is a mineral lattice encasing structural proteins. Remnant indigenous protein components survive after burial, diagenesis, and in some cases fossilisation, although these proteinaceous components will have undergone various chemical reactions in the process. Among those reactions is amino acid racemisation.

Twenty common amino acids can be found in protein chains (Figure 2.1). All protein-bound amino acids have a three-dimensional configuration called, by convention, the left-handed or L-enantiomer. When amino acids are synthesised chemically, rather than by biological organisms, the synthetic mixture consists of equal proportions of L- and D-enantiomers and is referred to as a racemic mixture. Glycine, having two hydrogens attached to its α-carbon, is the exception in that it cannot have L- and D-forms. L- and D-enantiomers of the same amino acids are identical in atomic composition, mass and charge, but spatially they are mirror images of each other. The utilisation of L-amino acids exclusively by animals and plants during biosynthesis of proteins is guaranteed by the stereoselectivity of the synthesising enzymes. (In some lower organisms, D-amino acids occur in peptide linkage, but either they are generated by racemases or are incorporated by mechanisms other than the RNA-ribosomal system of protein synthesis.) This asymmetry is vital to biological function and, apparently, has characterised all life on earth since its origin. The chemical reaction which converts one stereochemical form of amino acid into its mirror-image form is known as racemisation. In the case of amino acids having two centres of asymmetry, such as isoleucine and threonine, the reaction is more properly called epimerisation.

Mechanism and Kinetics of Racemisation

The mechanism of racemisation is illustrated in Figure 2.2 with the amino acid alanine. Abstraction of the proton from the asymmetric α-carbon results in formation of a planar carbanion intermediate. This negatively

Figure 2.1: Stoichiometric Formulae of the 20 Amino Acids Commonly Found in Proteins from Living Organisms, Two Additional Amino Acids Present in Collagen, and Some 'Non-protein' Amino Acids Found Only in Minor Amounts in Organisms. The molecular structure common to most amino acids is shown in the stippled areas; the individual R-groups are shown in the clear areas. The 'non-protein' amino acids are produced by decomposition of various protein amino acids (Schroeder and Bada, 1976). Glycine, β-alanine and γ-aminobutyric acid have no centre of optical asymmetry; isoleucine, threonine, hydroxyproline and hydroxylysine have two centres; and all others have a single asymmetric centre (at the α-carbon position)

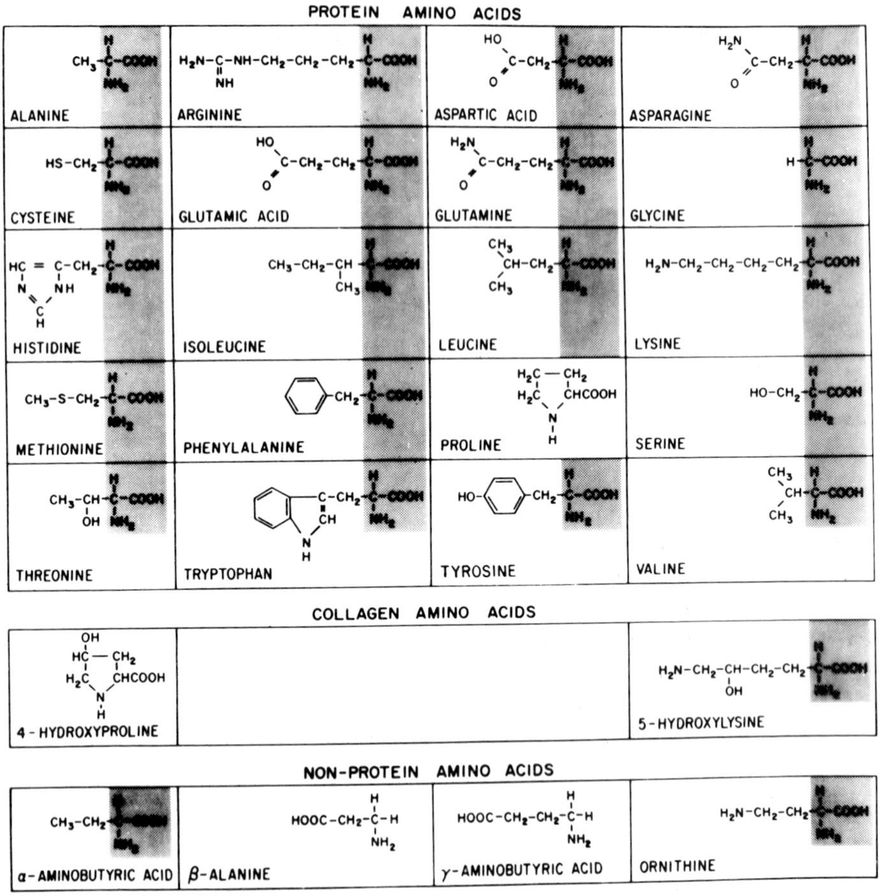

Source: Schroeder and Bada, 1976.

charged intermediate is unstable and will rapidly add back a proton. Depending on the side to which the proton adds, either the L-amino acid reforms or the D-enantiomer is generated. Both highly acidic or basic solutions will facilitate the loss of the α-proton, but water is an absolute requirement for carbanion formation.

Figure 2.2: Mechanism of Amino Acid Racemisation Showing Formation of Carbanion Intermediate

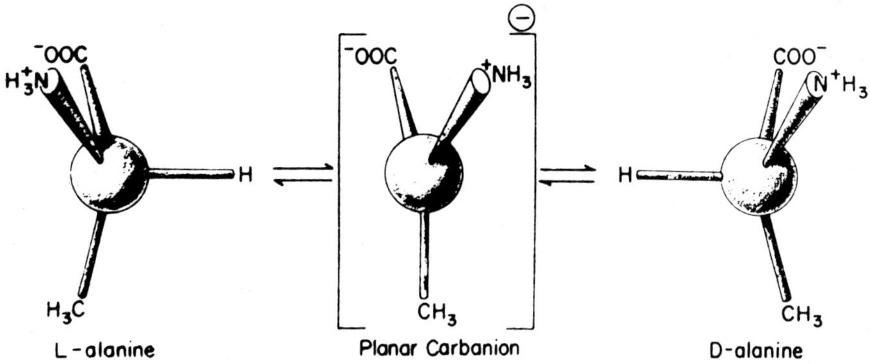

Source: Bada and Schroeder, 1975.

When looking at materials of biological origin, such as bone or mollusc shell, the initial D- to L-enantiomeric ratio is essentially zero (D/L = 0) since all proteinaceous amino acids start out as the L-form. Over a short amount of time at high temperatures (discussed further in Chapter 9), or over a long period of time at low environmental temperatures (most burial or depositional environments), the D/L ratio will increase due to spontaneous racemisation. The interconversion reaction will continue as long as there are amino acids remaining in the sample to react, but the D/L ratio does not move beyond the equilibrium value which is usually 1.0. The extent of amino acid racemisation can be determined by relatively simple laboratory techniques (Masters and Bada, 1978), and the D/L ratio measured is then used to calculate the age of the fossil specimen.

The racemisation reaction can be written as:

$$\text{L-amino acid} \underset{k_i}{\overset{k_i}{\rightleftharpoons}} \text{D-amino acid}$$

k_i = first-order rate constant for interconversion of the two enantiomeric forms.

The integrated rate expression for the reversible racemisation reaction is

$$\ln\left[\frac{1 + D/L}{1 - D/L}\right] - \ln\left[\frac{1 + D/L}{1 - D/L}\right]_{t=0} = 2k_i t$$

This equation (equation 2.1) applies to amino acids which attain an enantiomeric ratio of D/L = 1 at equilibrium. The rate equation can be modified to apply to amino acids such as isoleucine which have two asymmetric centres and reach equilibrium at ratios different from 1. The t=0 term is used to correct for racemisation during sample processing and, in some cases, for the fact that initial D-enantiomeric content may not be 0.

As with any chemical reaction, racemisation rates can be influenced by temperature and pH. This is in contrast to the decay 'reaction' of a radioactive nuclide such as carbon–14 that has a decay rate unaffected by chemical or physical factors. Applications of amino acid racemisation reactions to chronometric problems, then, have certain limitations to which radiometric dating methods are not subject. These will be discussed later.

The racemisation rate increases with higher temperatures. This means that the warmer the burial environment of a bone, for example, the less time is required for the reaction to reach equilibrium. Once a racemic mixture of any amino acid is achieved, its usefulness as a chronological indicator has ended. Conversely, the cooler the depositional environment, the longer the effective dating range is for amino acid racemisation.

The temperature dependence of rates in four different amino acids, as measured by their half-lives at 0° and 25°, is seen in Table 2.1. When using the aspartic acid racemisation reaction in bone, the maximum dating limit at mid-latitude continental locales such as central Europe is about 500,000 years; in equatorial regions like East Africa (Olduvai Gorge), it is about 60,000 years (Bada, 1981). The dating ranges can be expanded by measuring the rate in a more slowly reacting amino acid such as isoleucine.

Table 2.1: Racemisation Half-lives in Years for Several Amino Acids at 0°C and 25°C, pH 7.6

Amino acid	0°C (Years × 10^4)	25°C (Years × 10^4)
Serine	5(?)	0.04(?)
Aspartic acid	43	0.30
Alanine	140	1.20
Isoleucine	~600	~5.00

Source: Bada, 1982.

It is also apparent from Table 2.1 that different amino acids racemise at different rates. Kinetic studies at elevated temperatures have been carried out on free amino acids in aqueous solution, on various proteins, and on bone. The relative order of racemisation rates in these samples is shown in Table 2.2, along with rates in fossil bone and shell. Bone closely resembles the generalised protein pattern, while shell looks quite different. Enantiomeric analysis of several amino acids in a fossil can be used to determine its degree of contamination, since intrusive amino acids would disrupt this pattern of racemisation rates.

Table 2.2: Order of Amino Acid Racemisation Rates Relative to Leucine or Isoleucine

	k_{asp}	: k_{phe}	: k_{glu}	: k_{ala}	: k_{val}	: k_{leu}	: k_{iso}
Free amino acids[a] pH 7.6, 135°C	8.6	4.4		2.2	0.8		1.0
Proteins[b] pH 7.6, 135°C	8–10			2–3			1.0
Proteins[c] 0·1N NaOH, 65°C	8–10	5–7	3–4	2–3	0.4–0.9	1.0	
Modern bone[d] 135°C	10		3	2.5		1.0	
Fossil bones[e] (uncontaminated) total amino acids	12–14	3.3	3–4	2–3	0.9	1.0	~1
Fossil shell[e] *Chione* total amino acids	~2	1.2	1	2	0.6	1.0	~1

Notes:
a. Bada, 1972;
b. Schroeder and Bada, 1976;
c. Masters and Friedman, 1979;
d. Bada, Kvenvolden and Peterson, 1973;
e. Bada and Shou, 1980.

The mechanism of the amino acid racemisation reaction (Figure 2.2) can explain the variability in rates among the different amino acids. The alacrity of the α-proton to leave the amino acid molecule is influenced by the electronegative micro-environment of the molecule. Disregarding the charge state of the α-amino and α-carboxyl groups, since they are similar for all free amino acids, and irrelevant for peptide-bound amino acids, the inductive strength of the R-group becomes the principal determinant of racemisation rate. The more electron-withdrawing is the R-group, the more it can stabilise the carbanion intermediate in the reaction, and the faster the racemisation reaction can proceed. The electron-withdrawing or inductive strengths of the R-substituents, known as σ^*, were first correlated with racemisation rates of free amino acids (Bada, 1972). For protein-bound amino acids, racemisation also occurs via the carbanion-intermediate mechanism. Again the σ^* values can account quite satisfactorily for the rate differences among amino acids (Figure 2.3).

Figure 2.3: Relationship of Amino Acid Racemisation Rate Constants in Casein, 0.1N NaOH, 65°C, to the Inductive Strengths of Their R-substituents as Measured by σ^*. The rate constants are plotted as the log of the individual amino acid's rate constant relative to k_{ala} since the R-group of alanine is the reference value of $\sigma^* = 0$. The line is the least squares fit to the data points (•) using the σ^* values for asparagine (asn) and glutamine (gln). The weighted averages between aspartic acid and asparagine and between glutamic acid and glutamine in the protein's composition are indicated by (x). The correlation coefficient for the regression line is r = 0.995

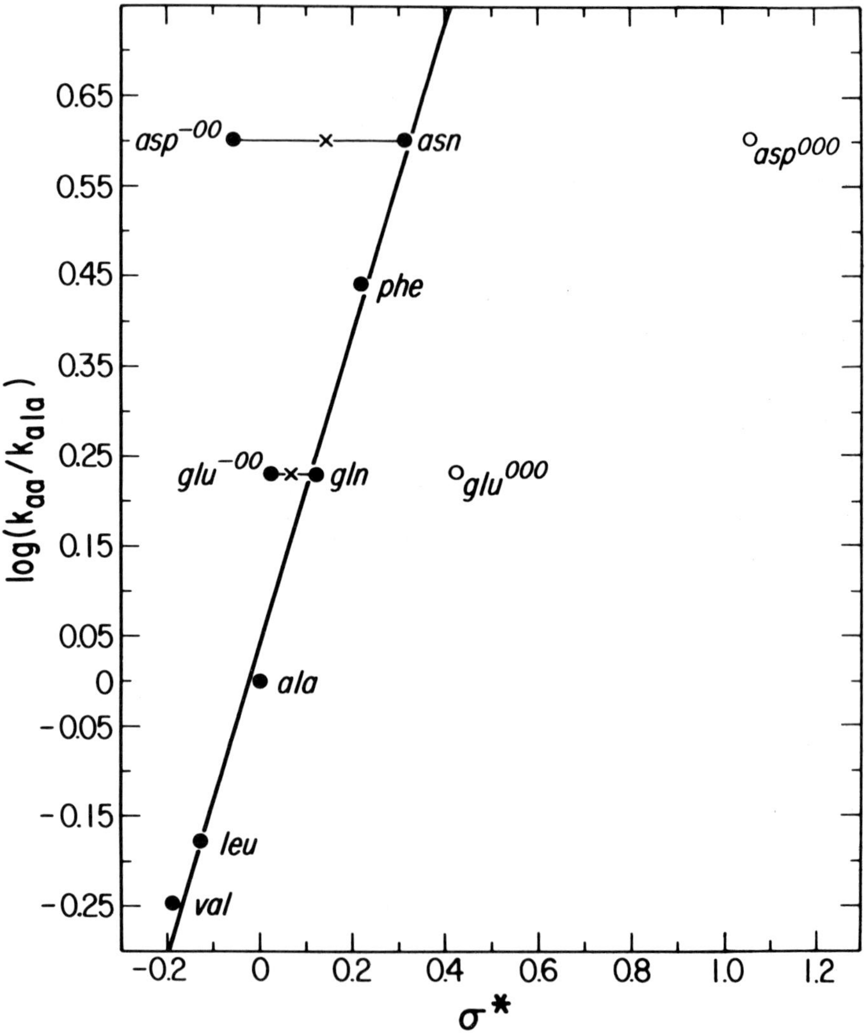

Source: Masters, 1982a.

Rate dependence on pH is somewhat more complex. Figure 2.4 shows the rate of free alanine as a function of pH. The pH-rate plot for alanine (and other free monocarboxylic amino acids) reveals three regions: an acid-catalysed region at pH values less than 1, a plateau I region over the pH range 3 to 6.5 where the rate is independent of pH, and a plateau II between pH 9 and 12, where k is also independent of pH. For dicarboxylic amino acids (e.g. aspartic acid), there is an additional rate acceleration at a pH value of about 3. Of importance to biological and geochemical applications of the racemisation reaction is the width of the plateau I region.

Figure 2.4: Plot of log k_i Against pH for Alanine at 142°C. Short-dashed lines indicate k_i for the acid-catalysed reaction; long-dashed lines, k_i of the ionic species having protonated amino and carboxyl groups; and short- and long-dashed line, k_i of the ionic species with positively charged amino group and negatively charged carboxyl group

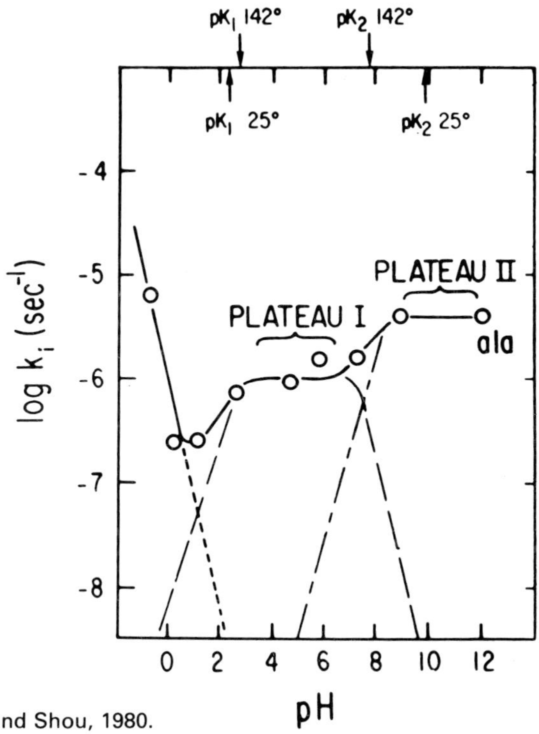

Source: Bada and Shou, 1980.

Here, the reaction rate is independent of pH over the range pH 3 to 6 (mildly acidic) at high experimental temperatures (142°C). As temperature decreases, the width of plateau I increases (see Figure 2.4) due to the different values of pK_1 and pK_2. At temperatures of 25°C or less, the pH independent region extends from pH 3 to about 8.5–9 (slightly acidic to slightly basic). Since most environmental temperatures average less than

25°C and most natural and biological systems maintain pH values between 3 and 9, the rate of free amino acid racemisation in these contexts can be considered essentially independent of pH.

Dating of Fossil Materials

The first recognition of the presence of D-amino acids in fossils was by Hare and Mitterer (1967) of the Carnegie Institution in Washington. They reported the accumulation with age of D-allo-isoleucine relative to L-isoleucine in *Mercenaria* shell from geological deposits dating from recent to Miocene. Hare and Abelson (1968) expanded the *Mercenaria* observations to seven other amino acids that were also undergoing racemisation, but it was not until 1969 that the potential usefulness of the racemisation reaction as a dating method was proposed (Hare and Mitterer). Within a year, the first paper appeared that applied the L-isoleucine $\rightleftharpoons$ D-allo-isoleucine reaction to dating of calcareous sediments from deep sea cores (Bada, Luyendyk and Maynard, 1970).

It was also recognised that foraminifera, the organisms contributing carbonates to the calcareous marine sediments, and molluscs in general, exhibit complex racemisation kinetics (Wehmiller and Hare, 1971; Bada and Schroeder, 1972; Mitterer, 1975; Masters and Bada, 1977). In carbonaceous fossils, the racemisation reaction does not follow reversible first-order kinetics to equilibrium. This situation is illustrated in Figure 2.5. The non-linear plot means that racemisation is occurring at one (relatively faster) rate early in diagenesis of a carbonaceous fossil and at a second, slower rate later in its diagenetic history. Such rate changes have been attributed to the fact that mollusc shells are composed of a large number of different proteins and different proteins appear to have different rates of racemisation (Masters and Bada, 1977). Diagenetic hydrolysis, releasing free amino acids from the polypeptides, could also influence the kinetics because free amino acids can racemise very rapidly, particularly in the presence of metal ions (Bada and Schroeder, 1972).

The implication of the complex racemisation kinetics in shells is that dates must be calculated from the kinetic equation (equation 1) using a rate constant (k_i) that applies only to the initial section of the curve. Dates can be derived from enantiomeric ratios if the ratios do not exceed the D/L value at the curve's point of inflection. Higher ratios fall beyond the region of the initial rate constant, and only minimum ages can be calculated for such samples.

Figure 2.5 also illustrates the racemisation kinetics of isoleucine in bone at elevated temperature. The complexities seen in calcareous marine sediments (inset Figure 2.5) and other carbonaceous fossils do not occur in bone, and the reaction proceeds linearly to equilibrium. Consequently, a

Figure 2.5: Kinetics of Racemisation of Isoleucine in Bone at 148°C. The straight line is a least squares fit of the data. The inset shows the kinetics of racemisation of isoleucine in a calcareous marine sediment at 148°C. In sediments, reversible first-order kinetics are approximated up to alleu/iso ≟ 0.35, as indicated by the straight line in the inset

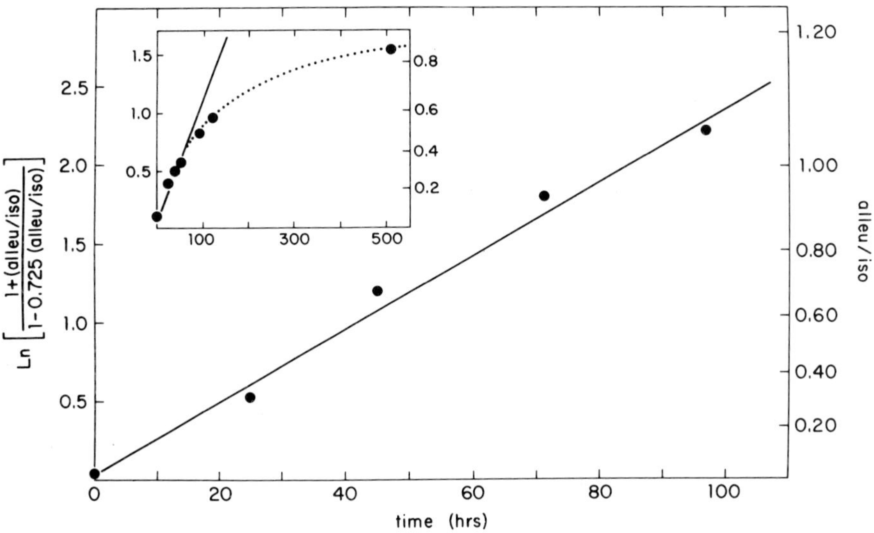

Source: Bada, 1972.

single rate constant can be used to calculate an age for a fossil bone no matter how high its enantiomeric ratio, as long as it falls below the equilibrium value. The same linear kinetics behaviour appears to apply to teeth as well (Bada *et al.*, 1979). It has been suggested that the relatively simpler protein composition of bone and teeth, approximately 90 per cent collagen, may account for the reversible first-order kinetics (Masters and Bada, 1977).

Another difference between shell and bone matrices in fossils is that molluscs and foraminifera exhibit species-specific rates of racemisation (Miller and Hare, 1976; Masters and Bada, 1977; King and Neville, 1977), while bone and teeth appear to have similar enantiomeric ratios, regardless of species origin. Therefore, to date a series of strata within a geological sequence using shells, it is necessary to select well-preserved valves representing a single species from each stratum. On the other hand, any type of *in situ* bone or tooth specimen, whether hominid or animal, can be analysed for an amino acid date. Teeth and bones racemise at identical rates (Bada *et al.*, 1979).

One difficulty which may arise with dating animal remains from human living sites is that they may have been cooked, thus yielding erroneously high racemisation ratios and ages. This phenomenon has been observed at two Upper Palaeolithic sites in Italy (Belluomini, 1981) and at several Middle Palaeolithic cave sites in Israel (Masters, 1982b, and unpublished observations). With human cremations, the evidence of charring is usually recognisable and such bones can be avoided. Another, and more insidious problem is the bone's state of preservation. This problem faces both radiocarbon and amino acid dating laboratories. As yet, no method of assessing the preservational status of a bone sample has been successfully applied.

Calibration of the Rate Constant

The final point regarding the application of amino acid racemisation reactions to dating of fossil materials is that the *in situ* rate constant must be determined. Reference to equation 2.1 reveals that once the enantiomeric ratio for a fossil has been measured, there remain two unknowns: k_i and t. To solve for t, a value must be independently determined for k_i. It has been shown that temperature is the primary influence on racemisation rate in the natural environment (Bada and Helfman, 1975; Bada *et al.*, 1979). Therefore, a means of accounting for the temperature history of the fossil is essential to the accuracy of the k_i. This has been accomplished with both bone and shell by the 'calibration' technique (Bada and Protsch, 1973).

To 'calibrate' a rate constant for a particular site requires a bone or shell from a stratum that has already been dated by some other means. Dates can be derived from historical evidence (e.g. Egyptian dynasties), geological association (last interglacial terrace or beach deposit), or an independent dating technique such as radiocarbon. A fossil sample of known age is then analysed for its enantiomeric ratio (usually of aspartic acid, leucine or isoleucine). This ratio (D/L) and the age (t) are substituted into equation 1, and the rate constant k_i can be calculated. The k_i from 'calibration' samples up to 10,000 or 11,000 years old is the appropriate rate constant for dating any other materials from the site that are Holocene in age. Beyond 11,000 years, fossils have been exposed to temperatures cooler on the average than Holocene temperatures. Pleistocene samples, therefore, need to have a Pleistocene 'calibration'. It has been demonstrated at a number of sites (Bada, Schroeder and Carter, 1974; Bada, *et al.*, 1974; Bada and Deems, 1975) that bones 17,000 to 20,000 years old make reliable 'calibration' samples for much older bones. It has also been shown that mollusc shell from marine terraces of last interglacial age (oxygen isotope substage 5e, approximately 120,000 years) can 'calibrate' for two to three glacial–interglacial cycles further back into the past (Masters and Bada, 1977; Karrow and Bada, 1980).

There is palaeoclimatological evidence to explain why these two calibration procedures work. In the case of bones, an approximately 18,000-year-old specimen will have experienced the cold of the last full glacial maximum (or its equivalent at lower latitudes; see CLIMAP, 1976), as well as warming temperatures of the end-Pleistocene and the full interglacial temperatures of the Holocene. Thus, such bone will have been exposed to an abbreviated glacial–interglacial temperature history, which on average approximates the mean temperature over the last 100,000 years or so. Shells dated by other means to the last interglacial high seastand have undergone depositional temperatures of a complete glacial cycle; approximately 110,000 years based on oxygen isotope curves from deep sea cores (Shackleton and Opdyke, 1973). The curves exhibit a periodicity of about 100,000 years for glacial–interglacial intervals extending beyond 700,000 years ago. Given this regularity and the similar ranges between temperature maxima and minima, it is reasonable to expect a 'calibration' derived from the last cycle to apply to the temperature histories of earlier cycles.

A small sample of the more than 300 aspartic acid racemisation dates on fossil bone produced by the Amino Acid Dating Laboratory at Scripps Institution of Oceanography is summarised in Table 2.3. The samples range from historically 'calibrated' dates of the Egyptian dynasties to Middle Palaeolithic cave deposits of the last interglacial. The majority of 'calibrations' have been based on radiocarbon dates (Bada *et al.*, 1974; Bada and Deems, 1975; Bada and Helfman, 1975).

The bulk of the amino acid dates in Table 2.3 agree nicely with dates derived by other methods. The only disputed racemisation dates are from North America where the calibration was provided by a UCLA radiocarbon date on the Laguna Woman skeleton. A radiocarbon date determined by accelerator mass spectrometry (AMS) at the University of Oxford in England (Bada *et al.*, 1984) has revised the Upper Pleistocene age downward to the mid-Holocene. The authors conclude that in poorly preserved bone having a low amino acid content, the extent of racemisation is not correlated with radiocarbon age.

Table 2.4 summarises D-alloisoleucine/L-isoleucine dates for uplifted marine terraces and bay-filling deposits in the vicinity of San Diego, California. These fossil deposits represent high seastands of the Pleistocene period. The faulting and uplifting that characterise this section of coast have complicated while preserving the record of past seastands. Relative ages cannot be estimated on heights above sea level because of faults and differential rates of uplift. Yet the general trend to uplifted coastal cliffs has contributed to the survival of the series of ancient beaches, since subsequent terracing has taken place below the previous, tectonically raised shoreline. Terraces are assigned to three different high seastands corresponding to oxygen isotope stages 5, 7, and 9, i.e. the three

Table 2.3: Comparisons between Aspartic Acid Racemisation Dates and Either Radiocarbon, Historical, or Geologically Inferred Ages

Site	D/L aspartic acid[a]	Radiocarbon, historical, or geologically inferred age (yr)	Aspartic acid age (yr)
Europe			
Muleta Cave, Mallorca, Spain[j]	0.273	16,850 ± 200 UCLA–1704D	$k_{asp} = 1.25 \times 10^{-5} yr^{-1}$
	0.293	18,980 ± 200 UCLA–1704E	18,600
	0.455	28,600 ± 600 UCLA–1704A	33,700
Abri Pataud, Les Eyzies, France[j]	0.148	23,010 ± 170 GrN–4721	$k_{asp} = 3.41 \times 10^{-6} yr^{-1}$
	0.178	33,260 ± 425 GrN–4719	32,100
Stránská skála, near Brno, Czechoslovakia			
layer no. 7, 5.0–5.1m deep	0.512 }	690,000[d]	$k_{asp} = 7.7 \times 10^{-7} yr^{-1}$
layer no. 5, 5.5–5.6m deep	0.574 }		
Külna Cave, near Brno, Czechoslovakia			
layer no. 7a, 3.8–4.0m deep	0.072[b]	45,600 ± 2850 or −2200 GrN–6060	50,000
layer no. 11, 7.0–7.2m deep	0.092[b]	Last Interglacial (?)[e]	80,000
Middle East[k]			
Cayönu, Turkey	0.169	8,340 UCLA–1703B	$k_{asp} = 1.21 \times 10^{-5} yr^{-1}$
Sarab, Iran	0.173	7,620 UCLA–1703C	8,700
Asiab, Iran (near Sarab)	0.155	7,620 ± 70 UCLA–1714A	$k_{asp} = 1.14 \times 10^{-5} yr^{-1}$
	0.167	8,700 ± 100 UCLA–1714C	8,600
Tarkhan, near Cairo			
F 1516	0.450[c]		$k_{asp} = 9.31 \times 10^{-5} yr^{-1}$
F 1691	0.475[c], 0.465[c]	1st and 2nd Dynasty	3180BC
F 1556	0.453	3200–2700BC	2910BC
Sakkara, near Cairo	0.137[b]	4th Dynasty 2500–2400BC	760AD(?)
Tura, near Cairo	0.416	Middle Kingdom 2100–1600BC	2030BC

South African Coast			
Nelson Bay Cave[l]	0.151	16,700 ± 240 I–6516	$k_{asp} = 4.92 \times 10^{-6} yr^{-1}$
	0.167	18,100 ± 500 UW–185	20,000
		18,660 ± 100 GrN–5889	
Klasies River Mouth Cave[l]			
level 13[f]	0.370		65,000
level 16	0.467		89,000
level 18	0.474	~80,000 to 120,000[g]	90,000
level 19	0.548		110,000
Swartklip I[m]	0.55	beginning of last glacial? (i.e. ~80,000)[h]	~110,000
North America			
Laguna Woman, Laguna Beach, CA[n]	0.25	17,150 ± 1,470 UCLA–1233A	$k_{asp} = 1.08 \times 10^{-5} yr^{-1}$
Los Angeles Man, near Baldwin Hills, CA[n]	0.35	>23,600 UCLA–1430	26,000
Laguna Woman[o]	—	5,100 ± 500 Oxford AMS[i]	$k_{asp} = 3.6 \times 10^{-5} yr^{-1}$
Los Angeles Man[p]	—	3,560 ± ? Arizona AMS	~8,000 (ibid)
Murray Springs, AZ[j]	0.33	5,640 ± 60 A–905A,B	$k_{asp} = 4.84 \times 10^{-5} yr^{-1}$
	0.52	11,230 ± 340 A–805	10,500
Double Adobe, AZ[k]	0.50	10,420 ± 100 A–1152	9,500

Notes:

a. Except as indicated, samples hydrolysed 24hr;
b. Sample hydrolysed 6hr. D/L aspartic acid ratio for modern bone hydrolysed 6hr = 0.026; thus, t = 0 term used in equation 1 for these samples is 0.052;
c. Sample hydrolysed 4hr. D/L aspartic acid ratio for modern bone hydrolysed 4hr = 0.023; thus, t = 0 term used in equation 1 for these samples is 0.046;
d. Brunhes/Matuyama palaeomagnetic reversal (690,000 years) occurs between layers 5 and 7;
e. Valoch, K., personal communication;
f. The larger the level number, the lower the sample stratigraphically;
g. Klein, 1974;
h. Klein, 1975;
i. Radiocarbon analysis by accelerator mass spectrometry;
j. Bada, *et al.*, 1974;
k. Bada and Helfman, 1975;
l. Bada and Deems, 1975;
m. Bada, *et al.*, 1979;
n. Bada, Schroeder and Carter, 1974;
o. Bada, *et al.*, 1984;
p. Taylor, *et al.*, 1985.

previous interglacial episodes. Since the racemisation dates fall on the interglacial stages rather than in between them, the results support the validity of the last interglacial 'calibration'. Mitterer (1975) has also used a last interglacial 'calibration' to calculate isoleucine ages for Pleistocene deposits in Florida. His dates are consistent with known interglacial stages as well as stratigraphic evidence of age.

Table 2.4: Fossiliferous Marine Deposits of San Diego County, California, Dated by Amino Acid Racemisation in *Chione*

Site	Elevation (m)	Alleu/Iso	Date	Oxygen isotope stage
No. 14	6	0.30	100,000	5
SDSU 318	13	0.35	120,000	
UCLA 3458	18	0.34	110,000	
SDSU 1854	12	0.39	130,000	
SDSU 2577	18	0.36	Calibration[a]	
Del Mar NE	18	0.58[b]	210,000	7
Border Field	9	0.50[b]	170,000	
Tecolote Creek	12	0.51[b]	180,000	
No. 1	6	0.55	200,000	
No. 130	18	0.59	210,000	
Morena Blvd.	16	0.68[b]	~300,000 (?)	9

Notes:

a. Calibration site date by uranium series measurements on fossil coral, 120,000 ± 10,000 years, by Ku and Kern (1974);

b. Data from Karrow and Bada (1980); all other results from Masters and Bada (1977).

Without a 'calibration', a relative 'aminostratigraphy' can be determined for a series of marine deposits (Mitterer, 1974; Andrews, Bowen and Kidson, 1979; Miller *et al.*, 1979; Atwater, Ross and Wehmiller, 1981). The principle here is that deposits can be ordered by D/ L ratios into a sequence of relative, not absolute, ages.

Efforts to date Indian shell middens by degree of racemisation in the mollusc shell remains (Masters and Bada, 1977) have yielded variable results. Compared with a suite of nine radiocarbon dates on shell at one site, only one racemisation date on shell fell within 5 per cent of the radiocarbon date for samples from the same excavated level. At that site, the average difference between racemisation and radiocarbon dates was 88 per cent, although other sites exhibited less discrepancy. Problems with midden shell dating may be due to a number of factors. Lack of stratigraphic consistency in the C^{14} dates as well as the amino acid dates indicates reworking of the midden either by human actions or rodent disturbance. Errors in C^{14} dates also are known to occur, and shell carbonates are subject to recrystallisation which can affect isotopic composition. Finally, the amino acid ratios, relative to the C^{14} results, err on the high side indicating an elevated temperature history, perhaps due to cooking. Consequently, midden shell appears to make a poor choice of sample for racemisation dating.

Amino acid racemisation reactions as age measures have also been studied in other organic materials. Lee, Bada and Peterson (1976) looked at fossil woods from Africa. Petit (1974) investigated the degree of racemisation in plant remains from woodrat middens, an important source of palaeoclimatic information in the American South-west. Engel *et al.*, (1978) analysed remains of various plant taxa from packrat middens. Attempts have also been made to date coral from fossil formations in the Caribbean (Wehmiller, Hare and Kujala, 1976), but the results are not particularly promising. Overall, the most successful applications of racemisation reactions to chronometric problems have been with bone, mollusc shell from geological deposits, and foraminiferal deep-sea sediments.

The accuracy of a racemisation date is difficult to assess. Known sources of error are the measurement techniques themselves (amino acid analyses, 5–10 per cent; gas chromatography, 10–20 per cent; high performance liquid chromatography, 15–20 per cent). The calibration procedure also incorporates the uncertainty of the calibrating date. A radiocarbon determination, for example, always has a statistical counting error assigned to it. But in addition, the contamination of the sample and its actual temperature history are not known and cannot be evaluated in an error analysis. For this reason, errors are not ordinarily listed with amino acid dates.

Use of Racemisation as a Palaeotemperature Indicator

Due to the temperature dependence of the racemisation reaction, k_i can be utilised to calculate an average temperature for the depositional history of the fossil specimen if the age is known (see equation 1). This provides a means of cross-checking the accuracy of racemisation dates (Bada and Helfman, 1975; Belluomini, 1981; Masters, 1982b). It also has been used to develop palaeotemperature models of Pleistocene climates (Schroeder and Bada, 1973; Mitterer, 1975; McCullough and Smith, 1976), and heat flow values or temperature histories for deep ocean sediments (Bada and Man, 1980).

Temperatures can be calculated from D/L ratios in the following manner. First, the analytically determined enantiomeric ratio is substituted into equation 1, with the independently known age of the fossil. The k_i is calculated and is called k_1. The unknown being sought is T_1. The k_1 is next substituted into the Arrhenius equation, as follows in equation 2.2:

$$\ln\left[\frac{k_2}{k_1}\right] = \frac{E_a}{R}\left[\frac{T_2 - T_1}{T_1\ T_2}\right]$$

Reference values k_2 and T_2 are taken from another well-calibrated locality where T_2 represents the average annual temperature (in degrees Kelvin) and is usually taken from meteorological records. E_a is the activation energy of the reaction, which is 33.4kcal mol $^{-1}$ for aspartic acid racemisation in bone. R is the gas constant 1.987cal degree^{-1} mol^{-1}. The T_1 computed from equation 2 is the average temperature to which the fossil has been exposed since the time of its burial or deposition.

Twenty comparisons between racemisation-based temperatures for fossil bones of Holocene age and temperatures from nearby meteorological stations are listed in Table 2.5. The reference values for T_2 and k_2 are derived from the k_{asp} in human teeth at 37°C, which is discussed more thoroughly in Chapter 9. Agreement is generally very close despite a variety of burial conditions, soil types, degree of water saturation, climates, and distance from meteorological stations. The concordance provides evidence that the 'calibration' procedure is reasonably accurate even under highly divergent environmental conditions. The results in Table 2.5 also support the conclusions from experimental kinetics studies discussed above that temperature is the principal variable affecting racemisation rates.

There is slightly better agreement between racemisation temperatures (T_{asp}) and exponential temperatures (T_e), $r = 0.945$, than between T_{asp} and the arithmetic mean temperatures (T_a), $r = 0.937$. This would be expected because climatological records are compiled by taking the annual average of the daily maximum and minimum temperatures (T_a), which is a linear approximation to the actual integrated temperature over a year's time. Since the racemisation rate is an exponential function of temperature changes, the exponential temperature should correspond more closely with T_{asp} values than the arithmetic mean temperature (T_a). The exponential temperatures are calculated from equation 2.3:

$$T_e = \frac{T_a + 1{\cdot}232 + 0{\cdot}161\ R_T}{1{\cdot}065}$$

R_T = the annual range in temperature (Lee, 1969).

Table 2.5: Comparison of Temperatures Calculated from Aspartic Acid Racemisation Rate Constants with Mean Annual Temperatures (Ta) Taken from Meteorological Data and with Exponential Temperatures (Te). All temperatures are in °C

Site	Age	k_{asp}	Ta	Te	T_{asp}
Buhen Horse, Sudan[a]	1675BC	1.2×10^{-4}	26	28	26
Tarkhan, Egypt[b]	Old Kingdom 3200–2700BC	9.5×10^{-5}	22	24	25
Tura, Egypt[b]	Middle Kingdom 2100–1600BC	9.7×10^{-5}	22	24	25
Olduvai Gorge, Tanzania[a]	LJ 3330 1,350 ± 100	7.1×10^{-5}	23	23	24
170 Little Salt Spring, FL[c]	Tx–2636 12,030 ± 200	5.05×10^{-5}	24	24	22
Murray Springs, AZ[d]	A–805 11,230 ± 340	4.5×10^{-5}	18	20	21
Lukenya Hill, Kenya[d]	UCLA 1709C 2,120 ± 60	4.0×10^{-5}	21	21	21
Clovis, NM[e]	A–490 11,040 ± 500	3.7×10^{-5}	14	18	20
Valley of Mexico[d]	UCLA 1773 10,120 ± 600	3.3×10^{-5}	16	17	20
Domebo, OK[d]	SM–695 11,045 ± 647	2.5×10^{-5}	17	21	18
Tranquillity, CA[d]	UCLA 1623B 2,550 ± 60	2.3×10^{-5}	17	20	18
Muleta Cave, Majorca[d]	UCLA 1704C 8,570 ± 350	1.7×10^{-5}	17	17	16
Clear Lake, CA[f]	UCLA 1795C 7,750 ± 400	1.6×10^{-5}	14	17	16
Skeleton 16709, W–12[c] La Jolla, CA	Pta 1725 8,360 ± 75	1.45×10^{-5}	17	18	16
Stanford Man, CA[d]	UCLA 1861 5,130 ± 70	1.1×10^{-5}	14	17	14
Marin 152, CA[f]	UCLA 1891A 3,270 ± 70	9.2×10^{-6}	16	14	13
Lindenmeier, CO[d]	I–141 10,780 ± 135	8.6×10^{-6}	8	13	13
Dent, CO[d]	I–622 11,200 ± 500	6×10^{-6}	10	14	11
Lindoe Bluff, Alberta[e]	GSC–612 11,200 ± 20	2.1×10^{-6}	5	11	7
Cochrane, Calgary[e]	GSC–612 10,760 ± 160	2.1×10^{-6}	4	9	6

Notes:
a. King and Bada, 1979;
b. Bada, *et al.*, 1979;
c. P.M. Masters, unpublished results;
d. Bada and Helfman, 1975;
e. J.L. Bada, unpublished results;
f. Bada and Masters, 1978.

Geochemical and Cosmochemical Implications of Racemisation

Enantiomeric ratios and amino acid compositions have been studied in meteorites, Precambrian rocks, and in simulation experiments of primitive earth conditions. These and related topics are summarised in reviews by Dungworth (1976), Schroeder and Bada (1976), and Bada (1982).

Summary

During the past 12 years, amino acid racemisation dating of fossil materials has emerged as a promising new technique with applications in oceanography, geology, palaeontology, and archaeology. The method is based on the fact that the L-amino acids found in the proteins of living organisms undergo a slow process of spontaneous chemical racemisation following burial. The rate of the racemisation reaction depends on the average temperature to which the fossil has been exposed since deposition. For most lower- and mid-latitude continental environments, the rate of racemisation is sufficiently fast to date materials which are Holocene to Middle Pleistocene in age.

Although the limit of radiocarbon dating is effectively ~40,000 years, amino acid racemisation can provide the means of dating much older fossil bone. Mid-Pleistocene to recent times mark an especially critical period for the evolution of hominids biologically and culturally, a period for which no other bone dating technique presently exists. Furthermore, racemisation analyses require 5–10g of bone, whereas standard radiocarbon techniques may need 100–200 times that amount. When fossil remains are very scanty or are anthropologically too valuable to be totally sacrificed for a date, and until accelerator radiocarbon dating has demonstrated accuracy beyond about 40,000 years, the racemisation technique could usefully be applied, once a 'calibration' has been determined for the site.

The most insidious problem facing amino acid racemisation dating, and radiocarbon dating as well, is the preservational state of the sample. It is clear now how a cascade of errors originates with an incorrect calibration (see Bada *et al.*, 1984; Bada, 1985). The challenge to the racemisation method in the coming years is to develop criteria for assessing the preservational status of a bone (or shell) in order to apply the chronometric analysis successfully.

References

Andrews, J. T., Bowen, D. Q. and Kidson, C. (1979) 'Amino Acid Ratios and the Correlation of Raised Beach Deposits in South-west England and Wales', *Nature, 281*, pp. 556–58

Atwater, B. F., Ross, B. E. and Wehmiller, J. F. (1981) 'Stratigraphy of Late Quaternary Estuarine Deposits and Amino Acid Stereochemistry of Oyster Shells beneath San Francisco Bay, California', *Quaternary Res., 16*, pp. 181–200

Bada, J. L. (1972) 'Kinetics of Racemization of Amino Acids as a Function of pH', *J. Am. Chem. Soc., 95*, pp. 1371–73

—— (1981) 'Racemization of Amino Acids in Fossil Bones and Teeth from the Olduvai Gorge Region, Tanzania, East Africa', *Earth Planet. Sci. Lett., 55*, pp. 292–98

—— (1982) 'Racemization of Amino Acids in Nature', *Interdisc. Sci. Rev., 7*, pp. 30–46

—— (1985) 'Amino Acid Racemization Dating of Fossil Bones', *Ann. Rev. Earth Planet. Sci. 13*, pp. 241–68

—— and L. Deems (1975) 'Accuracy of Dates Beyond the ^{14}C Dating Limit Using the Aspartic Acid Racemisation Reaction', *Nature, 255*, pp. 218–19

——, Gillespie, R. Gowlett, J. A. J. and Hedges, R. E. M. (1984) 'Accelerator Mass Spectroscopy Radiocarbon Ages of Amino Acid Extracts from California Paleoindian Skeletons', *Nature, 312*, pp. 442–44

—— and Helfman, P. M. (1975) 'Amino Acid Racemization Dating of Fossil Bones', *World Arch., 7*, pp. 160–73

—— Kvenvolden, K. A. and Peterson, E. (1973) 'Racemization of Amino Acids in Bones', *Nature, 245*, pp. 308–10

——, Luyendyk, B. P. and Maynard, J. B. (1970) 'Marine Sediments: Dating by the Racemization of Amino Acids', *Science, 170*, pp. 730–32

—— and Man, E. H. (1980) 'Amino Acid Diagenesis in Deep Sea Drilling Project Cores: Kinetics and Mechanisms of Some Reactions and Their Applications in Geochronology and in Paleotemperature and Heat Flow Determinations', *Earth Sci. Rev., 16*, pp. 21–55

—— and Masters, P. M. (1978) 'The Antiquity of Human Beings in the Americas: Evidence Derived from Amino Acid Racemization Dating of Paleoindian Skeletons', *Occ. Papers in Method and Theory in California Arch., 2*, pp. 15–24

——, Masters, P. M., Hoopes, E. and Darling, D. (1979) 'The Dating of Fossil Bones using Amino Acid Racemization', in R. Berger and H. E. Suess. (eds), *Radiocarbon Dating*, University of California Press, Berkeley and Los Angeles, pp. 741–56

—— and Protsch, R. (1973) 'Racemisation Reaction of Aspartic Acid and Its Use in Dating Fossil Bones', *Proc. Natl. Acad. Sci., 70*, pp. 1331–34

—— and Schroeder, R. A. (1972) 'Racemization of Isoleucine in Calcareous Marine Sediments: Kinetics and Mechanism', *Earth Planet. Sci. Lett., 15*, pp. 1–11

—— (1975) 'Amino Acid Racemization Reactions and Their Geochemical Implications', *Naturwissenschaften, 62*, pp. 71–9

—— and Carter, G. F. (1974) 'New Evidence for the Antiquity of Man in North America Deduced from Aspartic Acid Racemization', *Science, 184*, pp. 791–93

——, Protsch, R. and Berger, R. (1974) 'Concordance of Collagen-based Radiocarbon and Aspartic-acid Racemization Ages', *Proc. Natl. Acad. Sci., 71*, pp. 914–17

—— and Shou, M. -Y. (1980) 'Kinetics and Mechanisms of Amino Acid Racemization in Aqueous Solution and Bones' in P. E. Hare (ed.), *Biogeochemistry of Amino Acids*, John Wiley and Sons, New York, pp. 235–55

Belluomini,G. (1981) 'Direct Aspartic Acid Racemization Dating of Human Bones from Archaeological Sites of Central Southern Italy', *Archaeometry, 23*, pp. 125–37

CLIMAP Project Members (1976) 'The Surface of the Ice-Age Earth', *Science, 191*, pp. 1131–6

Dungworth, G. (1976) 'Optical Configuration and the Racemization of Amino Acids in Sediments and in Fossils — A Review', *Chemical Geology, 17*, pp. 135–53

Engel, M. H., Zumberge, J. E., Nagy, B. and Van Devender, T. R. (1978) 'Variations in Aspartic Acid Racemization in Uniformly Preserved Plants about 11,000 Years Old', *Phytochemistry, 17*, pp. 1559–62

Hare, P. E. and Abelson, P. H. (1968) 'Racemization of Amino Acids in Fossil Shells', *Carnegie Inst. Wash. Yearb., 66*, pp. 526–28

—— and Mitterer, R. M. (1967) 'Nonprotein Amino Acids in Fossil Shells', *Carnegie Inst. Wash. Yearb., 65*, pp. 362–664

—— (1969) 'Laboratory Simulation of Amino-acid Diagenesis in Fossils', *Carnegie Inst. Wash. Yearb., 67*, pp. 205–8

Karrow, P. F. and Bada, J. L. (1980) 'Amino Acid Racemization Dating of Quaternary Raised Marine Terraces in San Diego County, California', *Geology, 8*, pp. 200–4

King, K., Jr and Bada, J. L. (1979) 'Effect of *in situ* teaching on amino acid racemization rates in fossil bone', *Nature, 281*, pp. 135–7

—— and Neville, C. (1977) 'Isoleucine Epimerization for Dating Marine Sediments: Importance for Analysing Monospecific Foraminiferal Samples', *Science, 195*, pp. 1333–35

Klein, R. G. (1974) 'Environment and Subsistence of Prehistoric Man in the Southern Cape Province, South Africa', *World Arch., 5*, pp. 249–84

—— (1975) 'Paleoanthropological Implications of the Non-archaeological Bone Assemblage from Swartklip I, Southwestern Cape Province, South Africa', *Quaternary Res., 5*, pp. 275–88

Ku, T. -L. and Kern, J. P. (1974) 'Uranium-series Age of the Upper Pleistocene Nestor terrace, San Diego, California', *Geol. Soc. Am. Bull., 85*, pp. 1713–16

Lee, C., Bada, J.L. and Peterson, E. (1976) 'Amino Acids in Modern and Fossil Woods', *Nature, 259*, pp. 183–86

Lee, R. (1969) 'Chemical Temperature Integration', *J. Applied Meteorol., 8*, pp. 423–30

Masters, P. M. (1982a) 'Amino Acid Racemization in Structural Proteins' in M. E. Reff and E. L. Schneider (eds.), *Biological Markers of Aging*, NIH Pub. No. 82–2221, pp. 129–37

—— (1982b) 'An Amino Acid Racemization Chronology for Tabun' in A. Ronen (ed.), *The Transition from Lower to Middle Palaeolithic and the Origin of Modern Man*, B.A.R. Press, London pp. 43–56

—— and Bada, J. L. (1977) 'Racemization of Isoleucine in Fossil Molluscs from Indian Middens and Interglacial Terraces in Southern California', *Earth Planet. Sci. Lett., 37*, pp. 173–83

—— (1978) 'Amino Acid Racemization Dating of Bone and Shell' in G. F. Carter (ed.), *Archaeological Chemistry II, Advances in Chemistry Series*, No. 171, American Chemical Society, Washington DC, pp. 117–38

—— and Friedman, M. (1979) 'Racemization of Amino Acids in Alkali-treated Food Proteins', *J. Agric. Food Chem., 27* pp. 507–11

McCullough, E. A. and Smith, G.G. (1976) 'Correction in the Glacial-postglacial Temperature Difference Computed from Amino Acid Racemization', *Science, 191*, pp. 102–3

Miller, G. H. and Hare, P.E. (1976) 'Use of Amino Acid Reactions in some Arctic Marine Fossils as Stratigraphic and Geochronological Indicators', *Carnegie Inst. Wash. Yearb., 74*, pp. 612–17

——, Hollin, J. T. and Andrews, J. T. (1979) 'Aminostratigraphy of UK Pleistocene Deposits', *Nature, 281*, pp. 539–43

Mitterer, R. M. (1974) 'Pleistocene Stratigraphy in Southern Florida Based on Amino Acid Diagenesis of Fossil *Mercenaria', Geology, 2,* pp. 425–8

—— (1975) 'Ages and Diagenetic Temperatures of Pleistocene Deposits of Florida Based on Isoleucine Epimerization in *Mercenaria', Earth Planet. Sci. Lett., 28*, pp. 275–82

Petit, M. G. (1974) 'The Racemization Rate Constant for Protein-bound Aspartic Acid in Woodrat Middens', *Quaternary Res., 4*, pp. 340–45

Schroeder, R. A. and Bada, J. L. (1973) 'Glacial-postglacial Temperature Difference Deduced from Aspartic Acid Racemisation in Fossil Bones', *Science, 182*, pp. 479–82

—— (1976) 'A Review of the Geochemical Applications of the Amino Acid Racemization Reaction', *Earth Sci. Rev., 12*, pp. 347–91

Shackleton, N. J. and Opdyke, N. D. (1973) 'Oxygen Isotope and Paleomagnetic Stratigraphy of Equatorial Pacific Core, V28–238: Oxygen Isotope Temperatures and Ice Volumes on a 10^5 Year and a 10^6 Year Scale', *Quaternary Res., 3*, pp. 39–55

Taylor, R. E., Payen, L. A., Prior, C. A., Slota, P. J., Jr, Gillespie, R., Gowlett, J. A. J., Hedges, R. E. M., Jull, A. J. T., Zabel, T. H., Donahue, D. J. and Berger, R. (1985) 'Major Revisions in the Pleistocene Age Assignments for North American Human Skeletons by C-14 Mass Spectrometry: None Older than 11,000 C-14 Years BP', *Amer. Antiq.*, 50, pp. 136–40

Wehmiller, J. and P. E. Hare (1971) 'Racemization of Amino Acids in Marine Sediments', *Science, 173*, pp. 907–11

—— and Kujala, G. A. (1976) 'Amino Acids in Fossil Corals: Racemization (Epimerization) Reactions and their Implications for Diagenetic Models and Geochronological Studies', *Geochim. Cosmochim. Acta, 40*, pp. 763–76

3 ELECTRON SPIN RESONANCE

Motoji Ikeya

Electron spin resonance (ESR) is a microwave absorption spectroscopy that can detect the magnetic moment induced by the spin (self-rotation) of negatively charged electrons. ESR has recently become a Quaternary dating technique spanning the interval between the upper limit of C^{14} dating and the lower end of K–Ar dating for biological and geological materials by detecting electron spins created by natural radiation. The principle and procedure of ESR dating are described with examples for carbonate fossils such as shells, corals and deep sea core planktonic foraminifera and for hydroxyapatite fossils, bones and teeth related with palaeoanthropology. Concordance of ESR dates with C^{14} and uranium-thorium dates is demonstrated. It is hoped that this chapter, the first description of ESR dating in a book of dating techniques, will stimulate further studies on application and refinement of ESR dating in archaeology and geology.

Principles of ESR and ESR Dating

Introduction

Methods of dating geological and archaeological materials have developed considerably in the last decade (Aitken, 1974; Fleming, 1976; Michels, 1973). Recently, a new method of dating using electron spin resonance (ESR) has been developed for minerals and biological materials from the Quaternary Era.

ESR is a physical method used to observe magnetic resonance absorption of microwave power by unpaired electrons that have the magnetic moment due to spin (self-rotation). The principle of ESR dating utilises electron spins created by natural radiation resulting from the decay of uranium, thorium and potassium in the material and also from the environment. Some of this radiation damage has been accumulated since the formation of the material. The total dose of natural radiation, which is related to the age, can be determined from the concentration of electron spins. This chapter is thus the first description of ESR as a dating method in a book of chronology science. It aims to encourage chronology scientists, archaeologists and geologists to undertake collaborative studies with physicists and chemists.

The term electron paramagnetic resonance (EPR) is sometimes used instead of ESR when the object of the measurements involves iron group ions and rare earth ions. The method of ESR has become as standardised as infrared and visible absorption measurements with a commercial spectrometer. A large number of books on ESR or EPR for physicists and chemists (Pake, 1962; Poole and Farach, 1972; Slichter, 1963) contain the quantum physics necessary for the analysis of ESR spectra. These are not appropriate for archaeologists, anthropologists and geologists who simply want to apply or familiarise themselves with ESR as a dating method in their fields. Therefore, ESR has been described for those whose major interest is not physics but chronology science.

The earliest ESR study on biological materials can be traced back to free radical detection in coals (Uebersfeld, Etienne and Combrisson, 1954). Radicals in coals, lignites, peats and petroleum were detected and ascribed to the effect of natural radioactivity and geothermal decomposition. Artificial gamma irradiation was used to estimate the radical production efficiency and the age of formation in an early work on ESR dating (Duchesne, Depireux and van der Kaa, 1961). Meanwhile, thermoluminescence (TL) was proposed as a dating technique for rocks and minerals (Daniels, Boyd and Saunders, 1953) and developed as a dating technique in archaeology (Aitken, 1974). The trails of radiation damage created by nuclear fissions were discovered (Silk and Barnes, 1959) and developed as fission track (FT) dating (Fleischer, Price and Walker, 1975). The use of ESR, which also detects natural radiation damage (O'Brien, 1955) as do TL and FT methods, was suggested as a dating technique in geology (Zeller, Levy and Mattern, 1967; Zeller, 1968; McMorris, 1969, 1970; Morency, Emond and Bitter, 1969). Meaningful ages that satisfy geologists have not been obtained for old geological materials.

The ESR of biological materials has been studied extensively in biophysics and medicine. The technique has only recently been used for dating fossil shells and corals (Ikeya and Ohmura, 1981, 1983) and bones (Ikeya, 1978a; Ikeya and Miki, 1980b) as well as cave deposits (Ikeya, 1975, 1977, 1978a, 1978b; Miki and Ikeya, 1978). An assessment of natural radioactivity is made for an absolute dating technique. Samples of distinctly different ages lead to a relative dating method. ESR dating of anthropologically important bones and stalagmites at Petralona (Hennig, *et al.,* 1981, 1982; Ikeya, 1977–1982; Poulianos, 1982) and Tautavel (Yokoyama *et al.,* 1980; Yokoyama, *et al.,* 1982) are now attracting attention, as are recent studies on deep-sea core planktonic foraminifera (Sato, 1982), geological faults (Ikeya, Miki and Tanaka, 1982) and geothermal fields (Ikeya, 1983b). There has been an attempt to include all related works on ESR dating at this stage in the references.

Principles of ESR and ESR dating are given first in this chapter, with the procedures of radiation assessment. Typical examples of ESR dating for carbonate fossils, shells, corals and deep-sea core foraminifera follow.

ESR dating of fossil bones related to palaeoanthropology is described in detail later. Some preliminary results on other biological materials are briefly introduced for further studies. Most of the examples in this chapter are taken from the work of Ikeya, since not much literature is available elsewhere at the moment. Some need more refinement before being used without confusion by archaeologists and geologists. Therefore, the content should be taken as a progress report for further development in the future (Ikeya, 1983a). This field is developing drastically in the 1980s.

What is Electron Spin Resonance (ESR)?

An electron can be considered as a negatively charged sphere rotating by itself. This rotation produces a circulating current in the direction opposite to the rotation of the sphere, since the charge of an electron is negative. The circular current produces a magnetic field in the vicinity of the electron. Therefore, the electron spin (self-rotation) can be considered equivalent to a tiny magnet, as is shown in Figure 3.1. These types of tiny magnets (electron spins) form magnetically neutral pairs in atoms and molecules. No net magnetic moment is present due to the pairing of electron spins in materials, except for some specific atoms and ions.

Radiation ionises atoms or molecules. When the ionised electron is trapped at some other atoms, an electron-excess atom and an electron-deficient atom are formed; these are called 'trapped electron' and 'trapped hole' centres, respectively. The trapped electrons and holes now have net magnetic moments due to the unpaired electron spins and may be considered as tiny magnets.

When the electron spin is placed in an external static magnetic field, the direction of the small magnet, i.e. the direction of the spin rotation, may be either the same as the direction of the external magnetic field or opposed to it, as shown in Figure 3.1. These two situations are energetically different and are called 'up spin' and 'down spin' states, respectively. Only two states with a spin quantum number $+\frac{1}{2}$ and $-\frac{1}{2}$ are allowed according to quantum physics. The energy splitting for different spin states is known as the 'Zeeman effect' (Carrington and McLachlan, 1967; Pake, 1962; Slichter, 1963).

The energy difference is proportional to the magnetic field and to the magnetic moment of the unpaired electron, and is written as follows in equation 3.1:

$$\Delta E = g\beta H$$

where β represents the Bohr magneton (the basic unit of the small magnet

Figure 3.1: An Unpaired Electron Spin is Equivalent to a Tiny Magnet due to the Circular Current Caused by the Self-rotation (Spin) of the Negatively Charged Sphere of the Electron. Electromagnetic wave or microwave energy is absorbed when the tiny magnet or the self-rotation (spin) changes direction under the external magnetic field. The energy levels are also shown

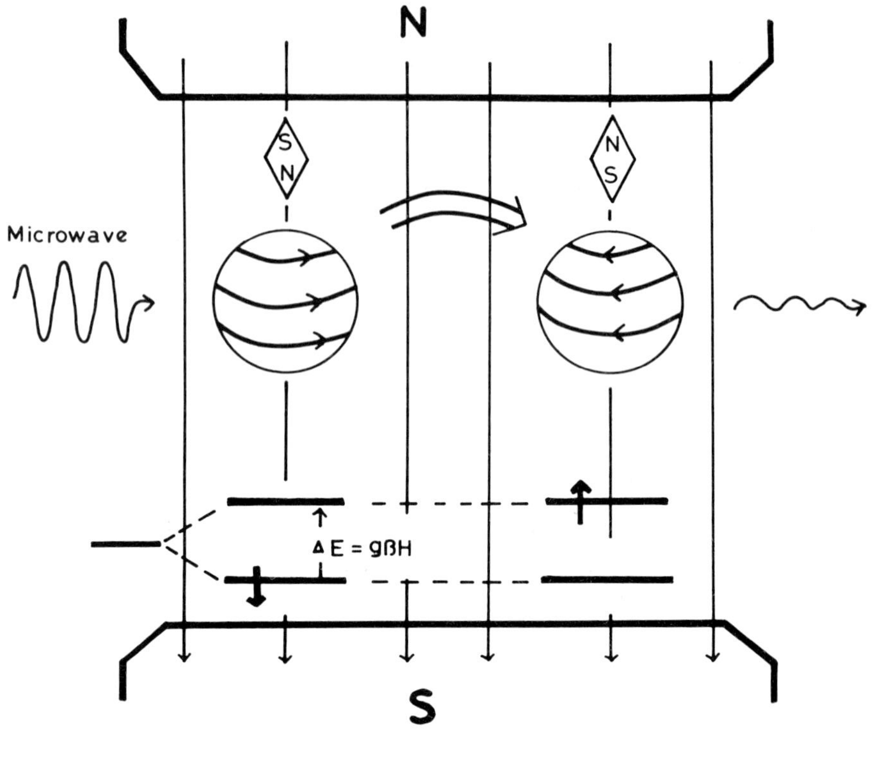

for the electron spin); g is the spectroscopic splitting factor (g value) representing the speed or frequency of the self-rotation in a classical picture; H is the strength of the external static magnetic field. The g value is about 2.0, depending on crystallographic fields near the unpaired spin. The accurate determination of g is the basis of modern analytical ESR spectroscopy.

Figure 3.2 indicates the energies of the electron spin states in the external magnetic field as a function of the strength of the magnetic field. The direction of the magnetic moment (or the spin direction) is changed when the energy difference due to the Zeeman effect is made equal to the quantum energy of the microwave with a constant frequency by varying the strength of the magnetic field. The microwave is absorbed due to the transition from one state to another. This absorption of the electromagnetic radiation (microwave) by the unpaired electrons in a magnetic field is called electron spin resonance (ESR). The condition of the resonance, or the flipping of spins from one spin direction to another, is represented by equation 3.2:

$$g\beta H = h\nu$$

where H is Planck's constant and ν is the microwave frequency.

The transition from the low-energy state to the high-energy state occurs following the absorption of the microwave quanta. Physically, the direction

Figure 3.2: Energies of Electron Spin States as a Function of the Strength of the External Magnetic Field (H). The transition from the low-energy state to the high-energy state occurs for a certain magnetic field following the absorption of microwave at a fixed frequency

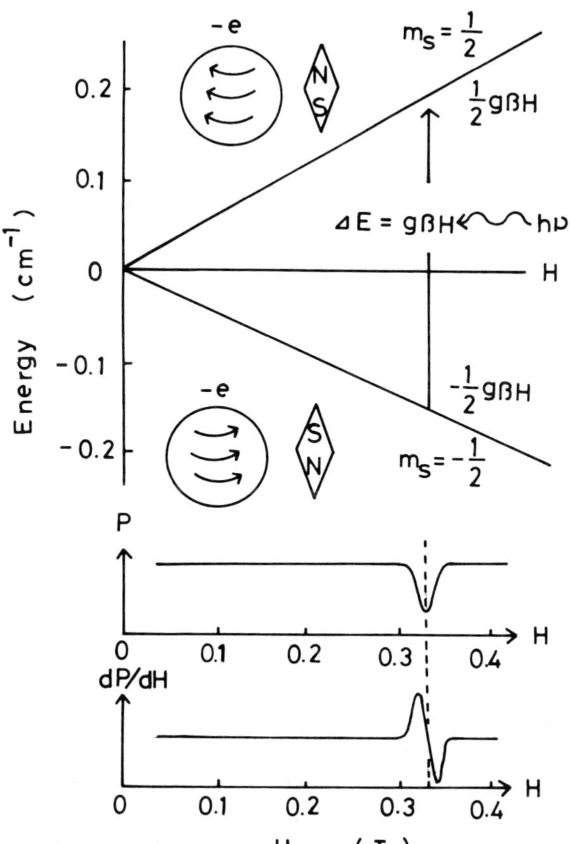

of the spin magnetic moment or the direction of spins is changed under the magnetic field by the absorption of a microwave, as shown in Figure 3.1. It must be noted again that the magnetic field and, therefore, the Zeeman energy, is changed to match the energy separation for the constant energy (frequency) of microwave quanta. The procedure for ESR is different from the similar nuclear magnetic resonance (NMR), because of the technical difficulty in obtaining a microwave oscillator that can change frequency continuously with a sufficient precision.

The resonance magnetic field of an unpaired free electron with $g = g_e = 2.0023$ is around 3,400 gauss (G) or 0.340T (tesla) for a typical X-band microwave frequency (9GHz = 9×10^9Hz). As the magnetic field where the absorption of the microwave occurs or the ESR signal appears depends on the frequency, an ESR signal is described by means of a g value (equation 3.3):

$$g = \frac{h\nu}{\underset{\cdot}{\nu}H} = 0.07144\frac{\nu}{H}$$

where ν is in GHz and H is in tesla. Hereafter, ESR signals are described by the g-values.

ESR Spectrometer

Figure 3.3 shows the block diagram of an ESR spectrometer. The microwave from an oscillator, called a Klystron or gun-diode, is led by a wave-guide to the resonant cavity in the static magnetic field produced by the electron magnet. The sample is inserted into the cavity. The strength of the magnetic field is swept in ESR spectroscopy by slowly changing the current in the magnet coils and the absorption of the microwave is measured. Generally, the magnetic field is modulated by an additional 100kHz sinusoidal magnetic field of small amplitude. Therefore, the absorption of the microwave is also modulated by 100kHz and is synchronously amplified electronically (lock-in amplifier system).

The rectified output on the recorder is, therefore, a differential line of the microwave absorption (dP/dH) as is shown in Figure 3.3 (see Pake 1962, pp. 39–47). Moreover, one may take the maximum-minimum magnitude as an index of the concentration of the unpaired electrons. If the width of the ESR signal is 1 gauss, one can detect a minimum number of 10^{10} to 10^{11} spins in the sample with an ordinary spectrometer. This sensitivity is very high if one considers that the number of atoms in $1cm^3$ of solids is on the order of 10^{21}. About 10^{15} to 10^{16} atoms per cm^3 is needed for the measurement of optical absorption. The signal-to-noise ratio or the sensitivity is further enhanced by a digital signal-averaging technique with repeated measurement.

Figure 3.3: Block Diagram of an ESR Spectrometer Which Records the First Derivative of the Microwave Absorption

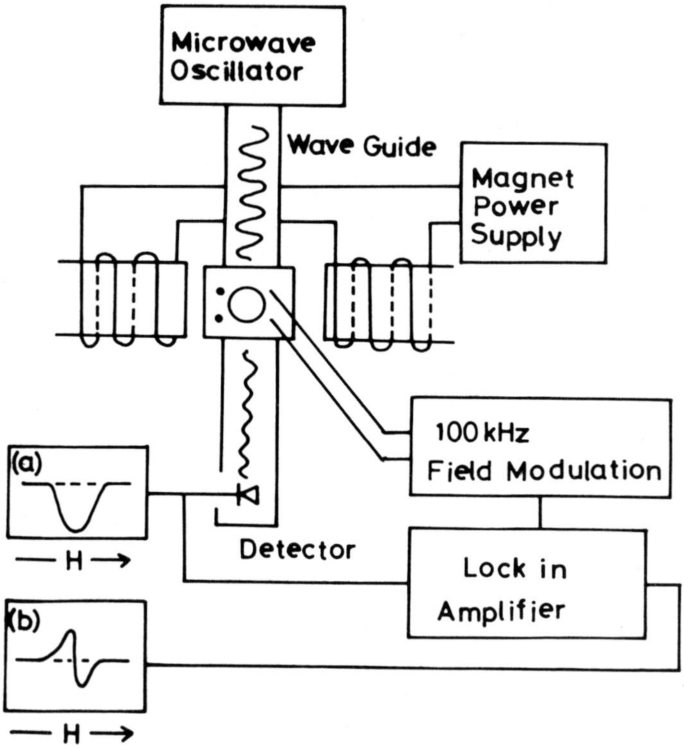

ESR Spectrum

Figure 3.4 shows a simple ESR spectrum (the first derivative of the microwave absorption) of a horse molar from Choukoutien cave which was inhabited by *Homo erectus pekinensis* around 200,000 to 500,000 years ago. The relative signal intensities of one to two for two signals, with $g_{\parallel}$ = 1.9973 and with $g_{\perp}$ = 2.0019, can be seen from the spectrum (indicated by arrows). The parallel and perpendicular g values ($g_{\parallel}$ and $g_{\perp}$) are due to the anisotropy of the g value along the crystalline axis of hydroxyapatite. The actual signal may be simulated with all angles integrated over the randomly orientated microcrystalline hydroxyapatite. Analysis of the ESR spectrum of polycrystalline or amorphous materials is described in standard textbooks.

The magnitude or amplitude of the signal is proportional to the concentration of spins or defects produced by natural radiation or, in some cases,

Figure 3.4: ESR Derivative Line of a Horse Molar at Choukoutien, China. The signals of presumably CO_3^{---} type defects in hydroxyapatite are indicated by $g_{\parallel}$ and $g_{\perp}$. The quintet (five lines) with the intensity ratio 1:4:6:4:1 is associated with alanine radicals ($CH_3CHCOOH\cdot$)

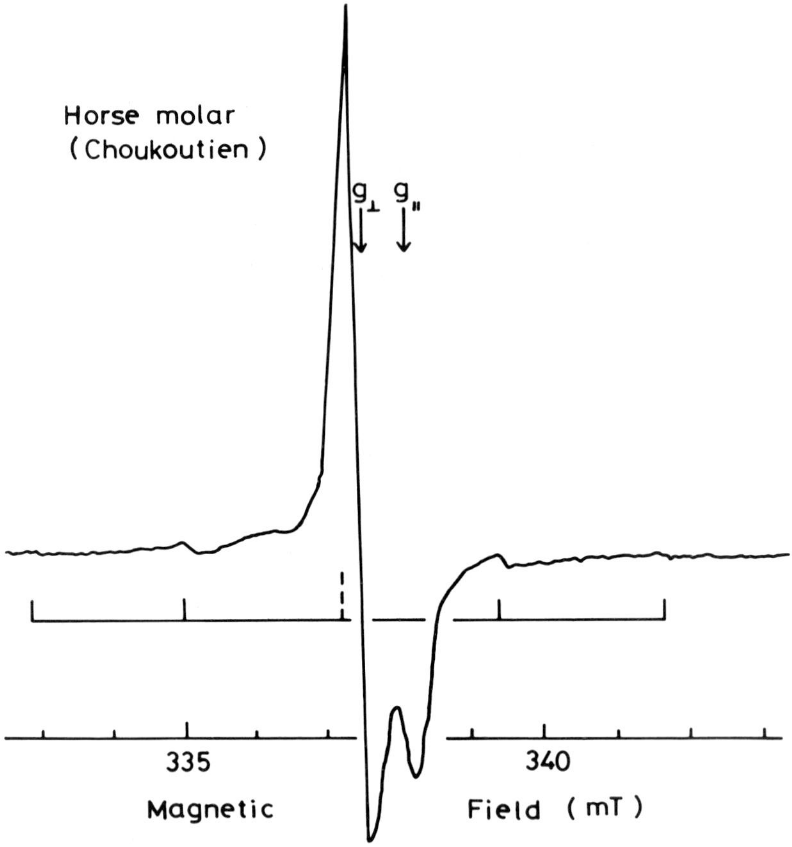

by chemical reactions. Therefore, it is related to total radiation dose (TD) and to the age for a constant annual dose rate. Additional signals due to impurities or defects in the organic part of a fossil are observable in some cases. The quintet (five-line) spectrum indicated in Figure 3.4 is associated with an electron in organic alanine radicals ($CH_3CHCOOH\cdot$). The signal is split by an internal magnetic field called 'hyperfine (hf) field' due to the small magnets of nuclear spins of four nearly equivalent protons. This is described for organic radicals in the section on fossil woods (Ikeya, 1982a).

Methods of ESR Dating

Total Dose of Natural Radiation (TD)

The basis of ESR dating is that the concentration of defects or trapped electrons created by natural radiation is related to the age. The ESR signal intensity gives the relative concentration of the radiation-induced defects or organic radicals. If one could know the concentration of the defects now and at a certain time in the future, one could estimate the time at which the defects began to occur. The current technique of ESR dating uses artificial irradiation as a 'time machine' to let the concentration of defects proceed to a future state. The past exposure is estimated from the production efficiency and present concentration.

The ESR spectrum of a sample is measured before and after an artificial irradiation dose Q. The total dose of natural radiation (TD) is determined from the enhancement of the ESR signal, assuming a linear relation between the defect concentration and the total radiation dose (equation 3.4):

$$R(Q)/R_o = (TD + Q)/TD$$

where R_o and R(Q) are the ESR signal intensities before and after irradiation, respectively.

The TD is sometimes called the 'archaeological dose (AD)', 'accumulated dose of natural radiation (AD)', 'equivalent dose (ED)', 'palaeo dose (PD)' or 'geological dose (GD)'.

Artificial irradiation is applied several times: the enhancement is plotted in Figure 3.5 for fossil bones from Choukoutien (Peking). The results are fitted to equation 3.5:

$$R(Q) = R_o (1 + Q/TD)$$

TD is obtained by linear extrapolation of the data points. This experimentally determined total dose (TD) is expressed as the dose equivalent to the Co^{60} γ-rays or β-rays used for the artificial irradiation.

In practice, TD is calibrated by a known age in a relative dating method or converted to the age by the assessment of the natural dose rate in an absolute dating. The age in thousands of years (ky) is expressed as follows (equation 3.6):

$$T(ky) = TD\ (Gy)/D(mGy/y)$$

The recommended SI unit of absorbed radiation dose, the gray (Gy), is the energy of 1 joule dissipated in 1kg of the material. The unit of 'rad' has been used instead of 'gray (Gy)' (1Gy = 100 rad). As the conventional

Figure 3.5: The Growth of the ESR Signal Intensity of Choukoutien Bone as a Function of Artificial γ Ray Irradiation from a Source of Co^{60}. The total dose of natural radiation (TD) is obtained by extrapolating the linear increase or by fitting to equation 3.5

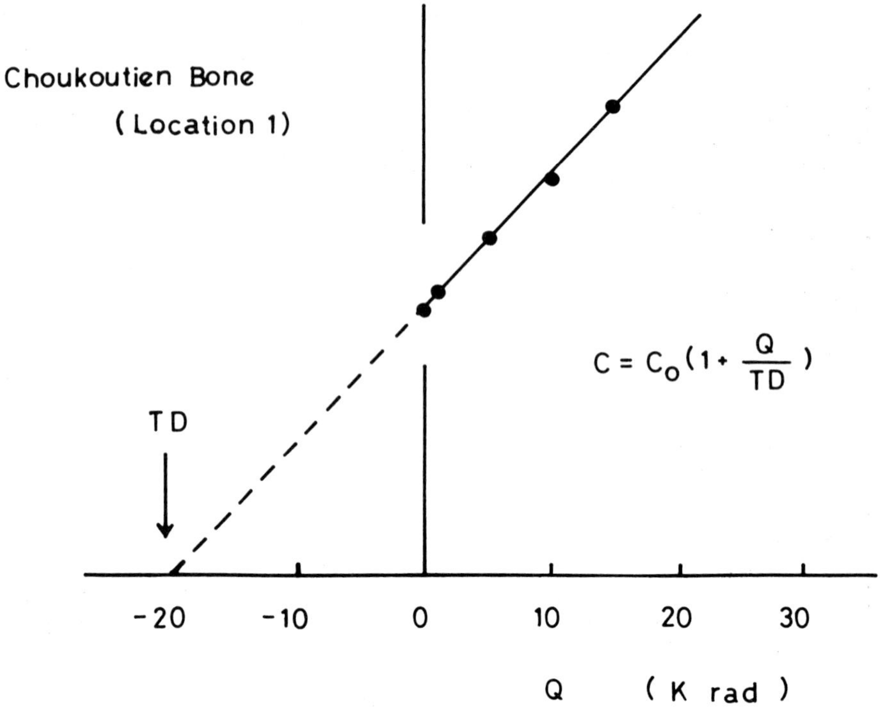

unit of 'rad' will continue in use for a while and Gy is often used for giga years (10^9y) by chronologists, then the term 'rad' is used in this book.

Thermoluminescence

The principle and procedure of thermoluminescence (TL) dating are the same as those of ESR dating. The emission of light by heating minerals is proportional to trapped electrons (holes) and to the exposure to natural radiation that the material has experienced. A TL peak does not necessarily represent the presence of a specific ESR signal. The TL light intensity must be measured in one sample before, and in other samples after, artificial irradiation of Q rad (I_n and $I_{n'+\gamma}$). Note that a TL measurement releases the trapped electrons but erases the information by reading it. The TD is obtained by the following equation 3.7:

$$I_{n'+\gamma} = I_n\,(I+Q/TD)$$

A fairly good agreement on TD by TL and ESR is obtained for cave deposits like stalactites (Miki and Ikeya, 1980), though both need not necessarily coincide (Ikeya and Miki, 1980c; Nambi, 1979).

Little work has been done on biological materials by the TL method. Chemical reactions, such as oxidation combustion due to the absorbed oxygen or decomposition of biological materials during TL measurements, produce spurious luminescence called chemiluminescence (CL). Mechanical grinding also produces defects responsible for light emission, called tribothermoluminescence (TTL). The preparation of fossil bones for TL measurement was attempted by removing the organic part (Christodoulides and Fremlin, 1971).

Annual Radiation Dose: Radiation Equilibrium

The age in years must be deduced from the TD or the equivalent dose (ED) in rad using the annual dose rate of natural radiation (D) estimated by other methods. The age is expressed using the conventional radiation dose 'mrad/y' as shown in equation 3.8:

$$T(ky) = TD(rad)/D(mrad/yr)$$

Natural radiation consists of α, β and γ rays from radioactive elements. The effect of α rays is discounted by a factor of k ($k < 1.0$) as the high defect density due to the large linear energy transfer (LET) along the α track results in a high probability of recombination of defects. The dose rate is shown in equation 3.9:

$$D = kD_\alpha + D_\beta + D_\gamma$$

where D_α, D_β and D_γ are the annual dose rate by α, β and γ rays, respectively.

Cosmic rays of about 25 mrad/y also contribute to the dose rate unless the object is shielded by sediments. Experimental k-value for carbonates was found to be 0.1 – 0.2 from the results of α ray irradiation.

Annual radiation dose rates have been estimated in TL dating from the contents of U^{238}, Th^{232} and K. The energies of α, β and γ radiation for 1 ppm of U^{238} and Th^{232} and 1 per cent of K are converted into annual dose rates of mrads per year, assuming radiation equilibrium. The dose rate for the loss of gaseous radon in the U^{238} series and of thoron in the Th^{232} series disintegration are also given in parentheses, relating to an infinite medium (Table 3.1) (Bell, 1976, 1979).

Table 3.1: Annual Radiation Dose Rates

	D	D	D
U^{238} (1 ppm)	278.3 (126.2)	14.62 (6.09)	11.48 (0.56)
Th^{232} (1 ppm)	73.8 (30.9)	2.86 (1.03)	5.14 (2.08)
K_2O(1 per cent)	0	68.93	20.69

It must be noted that the dose rate for a finite medium, e.g. for a fossil sample embedded in a sediment, must be evaluated considering the internal dose of α and β rays, the external dose rate of γ rays and, in some cases, partial β rays (Yokoyama *et al*, 1982). The external radiation (D_e) must be measured experimentally with a thermoluminescence dosimeter (TLD) or must be calculated by assuming radiation equilibrium, this time using the U^{238}, Th^{232} and K content of the sediment. It is also necessary to check the rate of loss of gaseous radon (^{222}Rn) as well as leaching or solution of α recoil radioactive elements. Confirmation of radiation equilibrium is needed if the dose rate is calculated from the contents of U^{238}, Th^{232} and K_2O using the above values.

The simplest method to determine the content of U^{238}, Th^{232} and K is chemical photometric analysis. Results of α ray and γ ray spectroscopy for Th^{230} /U^{234} dating method are also utilised. Recent development of high sensitive Ge(Li) or pure Ge radiation detectors allows the simple measurements of γ rays without chemical preparation. Another method is to count the fission tracks on a mica plate attached to the sample irradiated by neutrons for uranium content determination.

Uranium-series Disequilibrium: TD vs T Calculation

Infinite Media

If only uranium and not its daughters is included in a material like carbonate travertines, shells and corals at its formation, then Th^{230} and its daughters in U^{238}-series elements are built up close to the equilibrium after quite a long time (five half-lives) in the materials, since the half-lives of U^{234} and Th^{230} are 2.48×10^5 and 7.52×10^4 years, respectively. The assumption of radiation equilibrium is no longer valid as is known from the dating methods of uranium–uranium (U^{234}/U^{238}) and uranium–thorium (Th^{230}/U^{234} (Ivanovich and Harmon, 1982). The importance of this disequilibrium in ESR dating was noted in the TL dating study of Quaternary period carbonate deposits (Wintle, 1978).

The annual dose rate at the time t years after U^{238} incorporation is then written as in equation 3.10:

$$D(t) = D + D_{34}(r_o - 1)e^{-\lambda_{34}t} + D_{30}(Th^{230}/U^{238} - 1)$$

where D_{34} and D_{30} are the dose rates of radioactive elements for U^{234}, and from Th^{230} to Pb^{206} corresponding to the equilibrium disintegration of 1 ppm of U^{238}; r_o is the initial ratio of U^{234}/U^{238}.

The Th^{230}/U^{238} activity ratio is shown in equation 3.11:

$$Th^{230}/U^{238} = 1 - e^{-\lambda_{30}t} + (r_o - 1)\frac{\lambda_{30}}{\lambda_{30} - \lambda_{34}}(e^{-\lambda_{34}t} - e^{-\lambda_{30}t}),$$

where λ_{30} and λ_{34} are the decay constants of Th^{230} and U^{234}, respectively. The total dose of natural radiation can be considered by integrating equation 3.10 from $t = 0$ to $t = T$ (equation 3.12):

$$TD = \int_o^T D(t)\,dt$$

$$= DT + D_{34}\frac{(r_o - 1)}{\lambda_{34}}(1 - e^{-\lambda_{34}T}) - D\left\{\frac{1}{\lambda_{30}}(1 - e^{-\lambda_{30}T}) - \frac{(r_o - 1)}{\lambda_{34}}\left[1 - \frac{\lambda_{30}e^{-\lambda_{34}T} - \lambda_{34}e^{-\lambda_{30}T}}{\lambda_{30} - \lambda_{34}}\right]\right\}$$

$$D_{30} \simeq KD_\alpha'' + D_\gamma''$$

where $D_\alpha'' = 210.9$mrad/yr: $D\beta'' = 8.35$mrad/yr; $D\gamma'' = 10.97$mrad/yr for 1 ppm of U^{238} without loss of radon.

D_α'', D_β'' and D_γ'' for 100 per cent radon loss are 58.8, 0 and 0.5mrad/yr, respectively. D_{34} is 29.62, 0 and 0.029mrad/yr for α, β and γ rays, respectively. If a partial loss of radon can be determined from the activity ratio of Ra^{226}, Rn^{222} and Pb^{210}, then D and D″ should be modified for the loss ratio of radon (Ikeya, 1982c; Ikeya and Ohmura, 1983).

Figure 3.6 shows the result of the TD calculation for U^{238} equilibrium and for disequilibrium, taking the Th^{230} build-up into account using equation 3.12 for an infinite medium. In order to derive the age from the TD obtained by ESR, the solution must be done graphically or with a small computer. The dashed lines with arrows indicate the procedures to get the age T. First, the age is calculated from TD and the dose rate (D), using the equation $T_1 = TD/D$, assuming radiation equilibrium. Using this age, equation 3.12 is then solved iteratively for the true age.

Finite Media

Calculation for a finite medium for U^{238}-series disequilibrium may be made using equation 3.12 for the internal total dose of α and β rays and the average external γ ray total dose, $D_{ex}T$ as schematically shown for a fossil shell considering the range of each radiation. To avoid the effect of the

Figure 3.6: Total Dose of Natural Radiation (TD) versus Age (T) for the Equilibrium and Disequilibrium of U^{238}-series Disintegration. The annual dose rate (D), assuming equilibrium, gives a low age T_1. Using equation 3.12, the TD_1 for the disequilibrium is calculated. The age T_2 is obtained by drawing a line with the slope D at the intersection with TD (ESR). The age, T_d, is obtained by iteration procedure with a computer

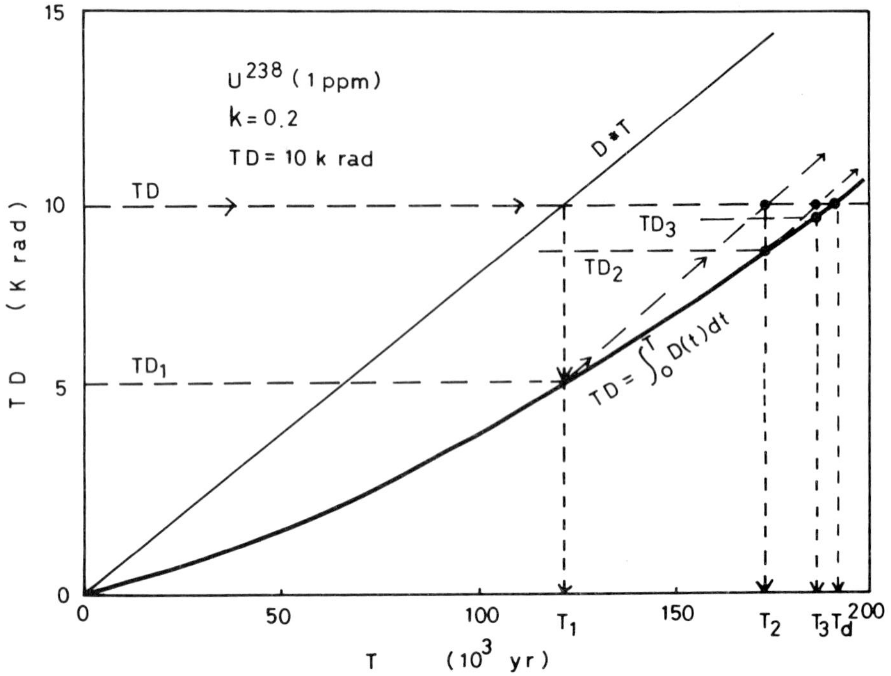

external α and β rays out of the sediments, it is wise to remove the surface by chemical etching with acids. One can use D_{ex} measured experimentally or calculated from the content of uranium in sediments. Experimental TD may then be fitted to equation 3.13 (for a finite medium):

$$TD = kTD_\alpha + TD_\beta + D_{ex}T$$

The TD versus T relation can be calculated as in an infinite media; the age for the TD obtained by ESR can be determined (Figure 3.7) in a similar way as in Figure 3.6 (Koberle *et al.*, 1973).

If radiation disequilibrium occurs due to leaching of radioactive elements by alpha recoil-induced solution, then radiation assessment must be made considering the activity of each element (Yokoyama *et al.*, 1982). An open-system model for a fossil bone that accumulates uranium will be dealt with later (Ikeya, 1982c).

Figure 3.7: Assessment of TD for a Finite Medium, a Fossil Shell in a Sediment. The internal α and β rays and the external γ rays (D_e) contribute to the TD due to the ranges of α rays (−10μm), β rays (−1mm) and γ rays (−30cm), as is schematically shown

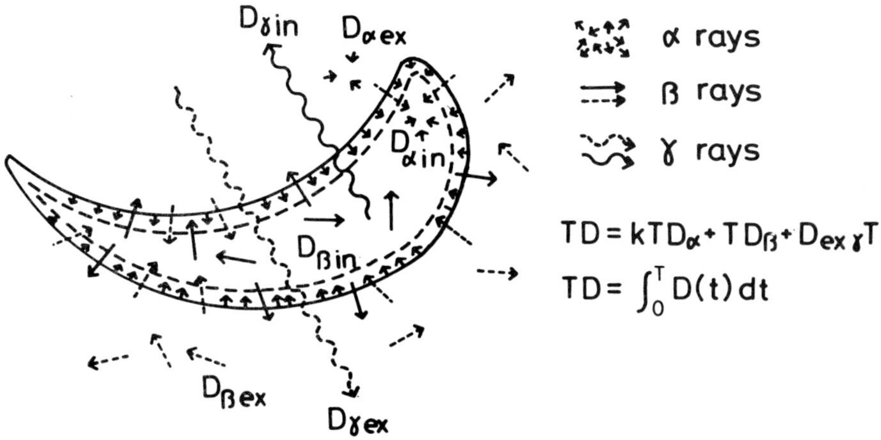

A New Dating Method: Digital ESR

In conventional ESR and thermoluminescence (TL) dating, the conversion from the obtained TD to the age in years must be made by estimating the annual radiation dose using another method. One can directly determine the age by measuring both accumulated and accumulating defects in the materials without knowing the annual radiation dose (Ikeya and Miki, 1980a).

In this new method of digital ESR dating, the computer-averaged signal intensity obtained by repeated measurements is given as a digital number together with the signal intensity of Mn^{++} in a standard sample, or in the sample itself (N_T and N_{Mn}). The ESR measurement is repeated after the passage of time, t, of a few days to a year. The signal intensity of the defect produced by natural radiation is enhanced from N_T to N_{T+t} in t years. The ratio of $N_{T+t}/N_{Mn} = R_{T+t}$ is always taken to avoid errors due to drift or fluctuation of the ESR spectrometer. If the rate of defect production is nearly constant, the concentration of defects is proportional to the elapsed time. Thus, equation 3.14 gives:

$$R_{T+t}/R_T = (T + t) / T$$

One can determine the age, T, directly by fitting the ratio R_{T+t} to equation 3.15:

$$R_{T+t} = R_T(1 + t/T)$$

The assumption that the increase in the defect concentration is linear from the beginning, and that the natural radiation dose rate is constant with time, must be studied carefully. Low-dose artificial irradiation would be necessary to confirm the linear growth behaviour of the defects. Samples should be kept at the site of the excavation between measurements. Unfortunately with this apparatus, a waiting time (t) of a few years is necessary to date an archaeological sample of 1,000 years.

This method obviously yields more precision for defects that accumulate faster, which is the case in chemically induced radicals in foods. Figure 3.8 shows an example of the accumulating free radicals (at g = 2.005) produced by the oxidation of food oil in fried potato chips, measured by a

Figure 3.8: Demonstration of Digital ESR Dating Using the Oil Decomposition of Potato Chips. The age, T, is obtained from the time-dependent ESR signal intensity by fitting the data to equation 3.13 or $I_T(1 + t/T)$

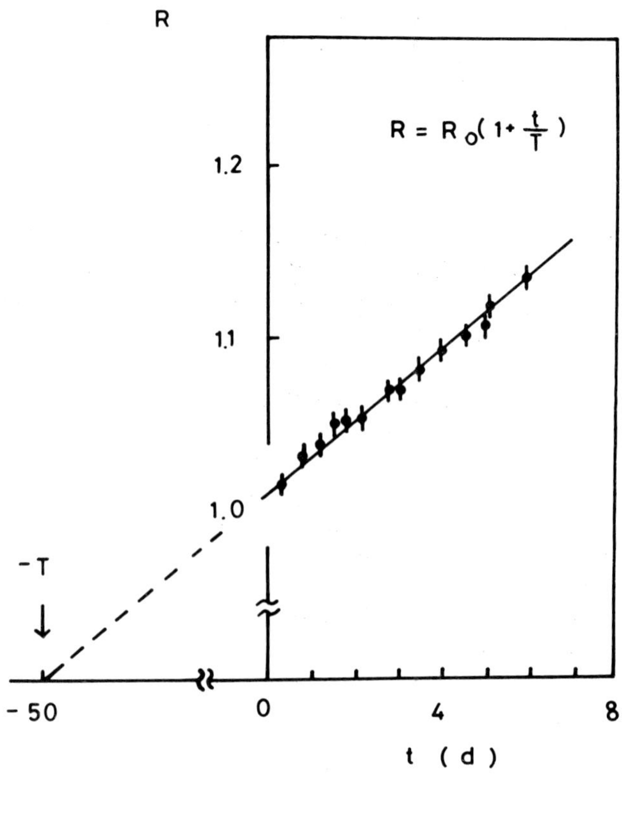

digitalised ESR spectrometer with a microcomputer (Radiosensor JES–3000). The signal intensity of the accumulating radical relative to that of Mn^{++}, also present in potato chips, is printed out as a three-digit number. A standard error of about $0.004/(2^{-8})$ is obtained using this apparatus with an 8-bit analogue to digital (A-D) converter.

The age of the potato chips can be roughly determined from the accumulating radical concentration as the food oil oxidises, assuming linear

Figure 3.9: Dating Procedures with Defects in Archaeological Materials

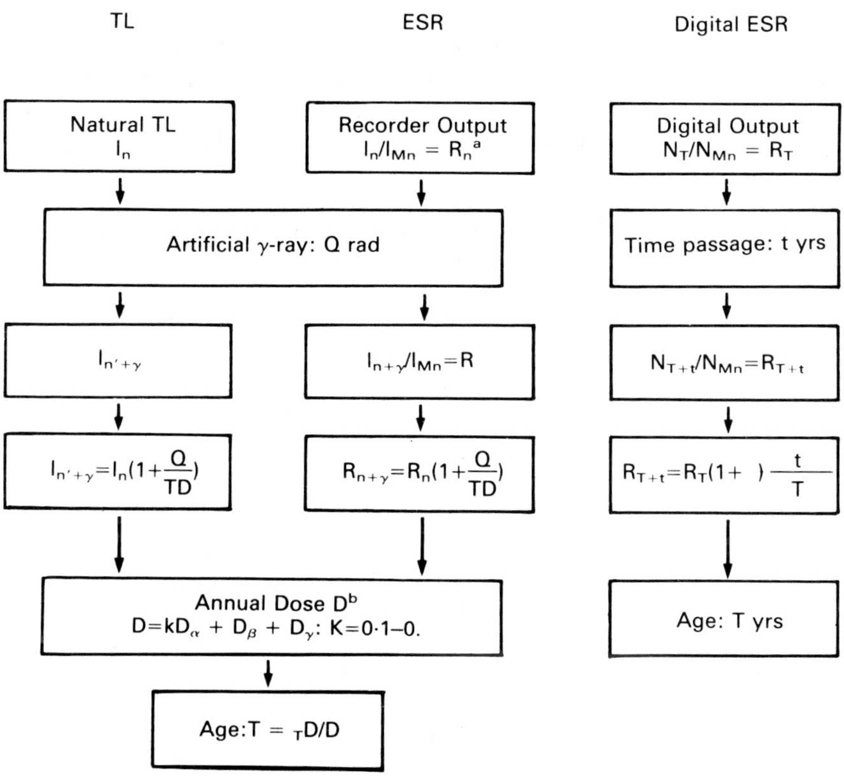

Notes:

a. Artificial radiation reduces internal I_{Mn} if the content of Mn^{++} is low or the radiation dose is high. In that case, I_n should be taken rather than R_n;

b. Annual dose rate is time-dependent in some cases due to the disequilibrium in U^{238}-series disintegration. The age T is deduced using the TD vs age relationship shown in equation 3.12.

assuming linear growth. The age obtained (about 50 days) is somewhat greater than the 35 days estimated from the date of manufacture printed on the envelope. This discrepancy may be due to a change in environment (high temperature beforehand) and to original oil oxidation at the time of production (zero age).

The stabilities of the magnetic field and microwave power of an ESR spectrometer are 10^{-7} and 10^{-6}, respectively, for an instrument in good condition. The stability and the linearity of the amplifier system is not well known at present. However, dependence on sensitivity and stability changes may be reduced considerably since the ratio of the time-enhanced signal to that of Mn^{++} is measured repeatedly and is computer-averaged. Thus, the increase of radical intensity to a precision of 10^{-5} to 10^{-4} might be detected at present with an appropriate A-D converter (Ikeya and his colleagues are developing an apparatus with a microcomputer system for use in archaeology). The dating of food or other materials of relatively young age will be helpful for legal medicine or forgery detection, even with the 8-bit precision spectrometer.

The recent progress in the digital integrated circuit (IC) and LSI technique would make this new dating method practical and compatible with C^{14}-dating (Ikeya and Miki, 1980a, 1980c) (Figure 3.9).

Time Range of ESR Dating

Stability of Defects

The stability of defects or radicals places limits on the age which can be measured by ESR dating. The stability depends on the properties of the defect and the temperature of the sample. Studies on the thermal properties or the stability of defects used for age determination are needed in order to estimate the decay time of the defect at the appropriate temperature. The fading time for some defects in carbonate is around one million years, far beyond the limit of radiocarbon dating.

It is assumed that a thermal activation process is responsible for the decay of defects or for the release of electrons from traps at the temperature T. The reciprocal decay time is given as follows in equation 3.16:

$$1/\tau = \nu_o \exp(-E/kT)$$

where E is the activation energy for the destruction of a defect; ν_o and k are the pre-exponential frequency factor and the Boltzman constant, respectively.

The analysis of the thermoluminescence curve gives the appropriate values for the parameters. The pre-exponential frequency factor is of the order of 10^{12} to 10^{13} sec^{-1} for a solid, though sometimes it is as small as

Figure 3.10: Fading Time at 20°C for a Defect Stable up to the Temperature T_p. The straight line is the theoretical one of equation 3.17; the dots are the experimental data. The stability of defects in carbonates and fossil bones is around 10^6 years, and 10^7 to 10^8 years, respectively; this is the old limit of ESR or TL dating. The time ranges of other dating methods are also shown

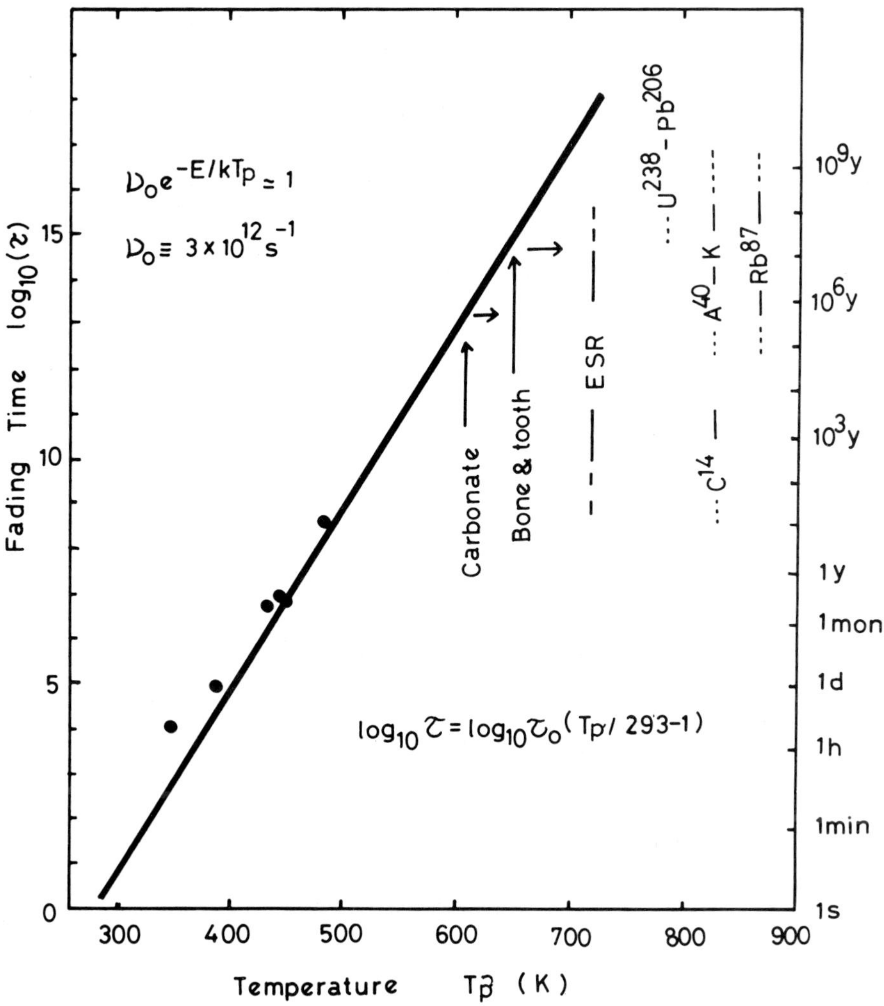

10^6 for some unknown reason. The lifetime of a defect at the peak temperature of the thermoluminescence is generally of the order of one second. The decay time at T may be written as in equation 3.17:

$$1/\tau = \nu_o \exp(-E/kT) = \nu_o [\exp(-E/kT_p)]^{T_p/T}$$

$$\text{where } \tau = 1, \text{ at } T=T_p \text{ or } \nu_o = \exp(E/kT_p)$$

A simple calculation gives equation 3.18:

$$\log_{10}\tau = (\log_{10}\nu_o)(T_p/T - 1)$$

Thus, the fading time or the stability of the defect can be roughly estimated as a function of the peak temperature of the thermoluminescence. Equation 3.18 gives a straight line on a semilogarithmic scale, as shown in Figure 3.10. The fading times of defects in several materials obtained experimentally are also shown in Figure 3.10. The typical fading times of one type of defect in carbonates and bones are shown, together with the fading time of some thermoluminescence dosimeters (TLD) to demonstrate the validity of equation 3.18. The results agree fairly well for materials used as TLD. Thus, the result might be extrapolated to a long decay time for defects with a high destruction temperature or high TL peak temperature.

The fading time in bones is from a few million years to more than 10 million years, as the defect gives a roughly estimated TL peak of around 400°C (670K). The actual stability may be one order of magnitude larger or smaller than this simple estimation. The range of applicability of other dating methods are also shown in Figure 3.10 for a rough comparison. The time range of dating methods depends on the samples. The old limit of C^{14} dating would be around 50,000 years, although recent developments with the tandem accelerator have extended the old limit.

The lower limit of ESR dating depends on the signal-to-noise ratio (S/N ratio) of the ESR signal. The most recent specimen dated in Ikeya's laboratory is a human bone of about 600 years old. If one improves the S/N ratio by a signal averaging technique, the lower limit might be improved to tens of years. Thus, it might be possible to extend ESR dating to quite recent events, which might prove useful in forensic science. The general limit with commercial equipment is about 1,000 years at present.

Saturation of Growth Due to Fading

Saturation of defect production occurs due to equilibrium between production by natural radiation and thermal fading. The growth of stable defects is described as in equation 3.19:

$$\frac{dN}{dt} = \alpha D - \frac{N}{\tau}$$

where N = concentration of defects or spins per cm^3 [cm^{-3}];
α = concentration of defects per cm^3 produced by 1 rad radiation [cm^{-3} rad^{-1}];
D = annual dose rate [rad yr^{-1}];
τ = fading time of defects at room temperature (yr).

A simple saturation curve is obtained using the condition of N = 0 at t = 0 (equation 3.20):

Figure 3.11: Saturation of the Growth of Defect Concentration Due to Equilibrium Between Fading and Production by Natural Radiation. Enhancement by artificial irradiation has the same slope as the initial production. The TD obtained by extrapolating the linear part is smaller than Dτ and the real TD

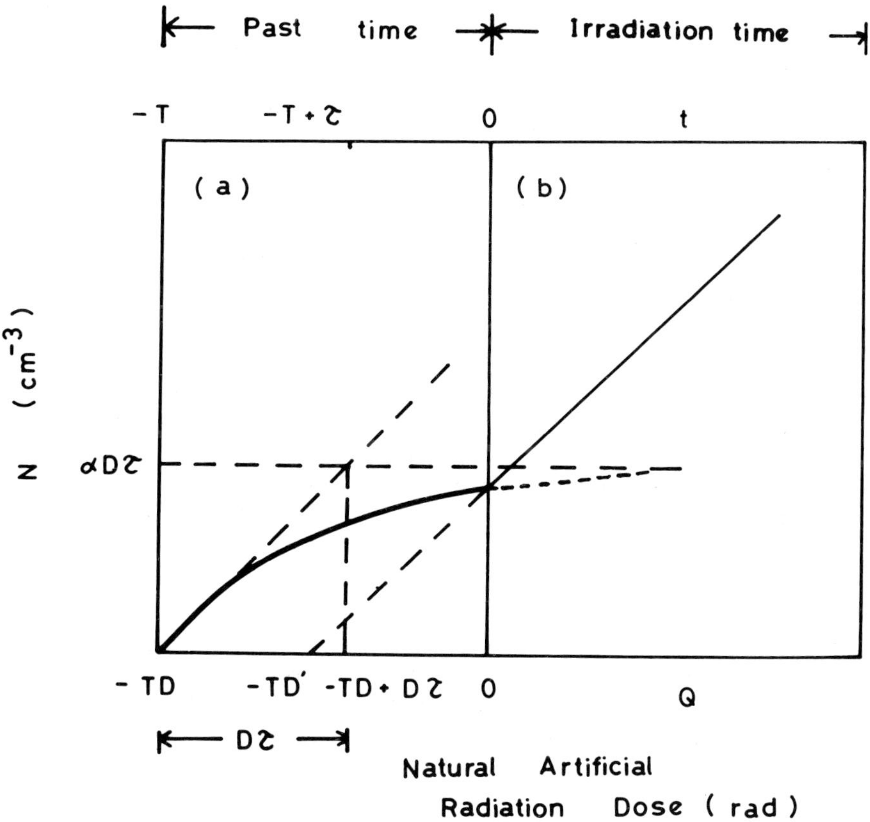

$$N = \alpha D\tau\,[1 - \exp(-\,t/\tau)]$$

Figure 3.11 is the growth curve of the defect. The growth is linear at the beginning (equation 3.21):

$$N = \alpha Dt \qquad (t<<\tau)$$

and is saturated at the following value (equation 3.22):

$$N_s = \alpha D\tau$$

It should be noted that the saturation value depends on the annual dose rate and the fading time of the defects, as is shown schematically in Figure 3.11.

The defect production per rad, α, depends on the material. The number of defects or radicals produced by a radiation energy of 100eV is expressed by the G value in radiation chemistry. G values are around 10 for organic materials such as ethanol and other alcohols at low temperatures, and around 0.1 for aromatic hydrocarbons such as benzene and naphthalene. The radiation energy of 10^{-5}J/g is the definition of the radiation energy 'rad'. The number of defects produced by 1 rad of radiation energy is 6.3×10^{11}/g for a material with G = 1.0. The density is around 2.7g/cm^3 rad. The annual radiation dose is in general 0.1 rad/yr and the fading time, τ, is in the order of 10^6 years. Thus, saturation value in the order of 10^{17} to 10^{18}/cm^3 is obtained as N_s in equation 3.22.

The additive artificial radiation dose rate is much higher than that of natural radiation in general. The exposure rate of 10^4 R/hr corresponds to a rough annual rate of 10^8 rad/y. This is 10^9 times larger than the natural radiation dose. If the concentration reached is close to the saturation value of $\alpha D\tau \simeq 10^{17}$-$10^{18}$/cm^3, then artificial irradiation produces little further enhancement. Therefore, the total dose of natural radiation (TD) obtained from a linear extrapolation is smaller than the real TD. Thus, if TD is measured as close to the calculated $D\tau$, it should be examined carefully in this respect.

Saturation by Defect Interaction

The saturation of growth by irradiation also occurs due to the interaction or recombination of defects. If the defects are electron or hole traps, the rate of the recombination of electrons and holes is increased by the presence of such traps. One can consider that a defect produced in the vicinity of another defect within a certain distance is not stabilised. A volume with the number of atoms, b, becomes inefficient for defect production. The growth rate of defects is then described using the lattice site number N_o in

cm^3 and radiation dose Q = Dt, for a schematic model shown in Figure 3.12 (equation 3.23):

$$\frac{dN}{dQ} = \alpha \left(\frac{N_o - bN}{N_o}\right)$$

Assuming N = 0 at Q = −TD, then equation 3.24 follows:

$$N = \frac{N_o}{b} \left[1 - \exp\left(- \frac{\alpha b}{N_o} (Q + TD)\right)\right]$$

For $Q << N_o/b\alpha$, then equation 3.25 follows:

$$N = \alpha (Q + TD) = \alpha TD \left(1 + \frac{Q}{TD}\right)$$

This is equal to the form in equation 3.5.

Figure 3.12: Schematic Model of the Non-sensitive Volume Around a Defect and an Impurity in a Crystalline Material. A volume of approximately b atoms around the defect and b_I atoms around an impurity are insensitive to radiation

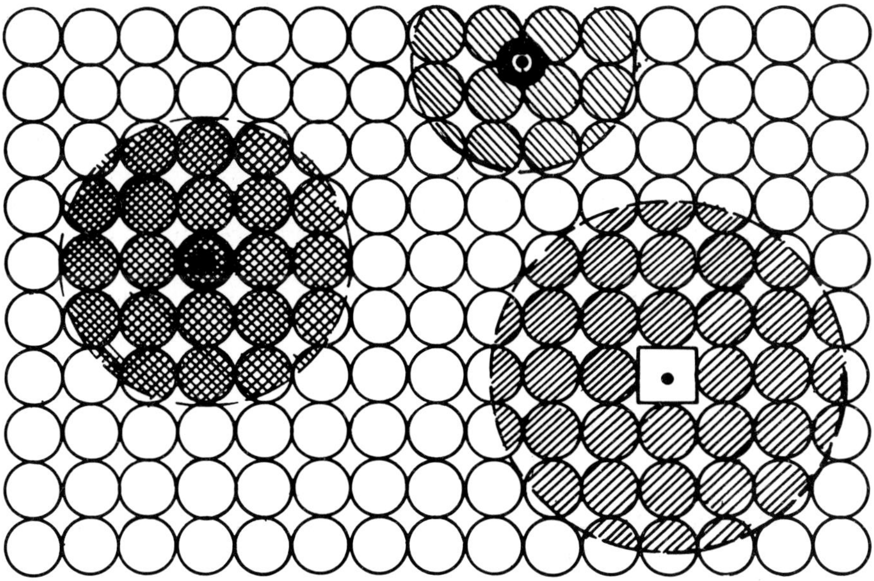

In the case of calcite, the saturation behaviour appears for a radiation dose of about 10^6 rad. Substituting the value $N_o = 2 \times 10^{22}/cm^{-3}$, and $\alpha = 2 \times 10^{12}/cm^3 rad$, and $Q = 10^6$ rad, b is obtained from equation 3.26 to give 10^3 to 10^4:

$$\alpha bQ/N_o \simeq 1$$

This corresponds to about a 10-lattice distance. This is actually the interaction volume obtained in solid state physics. The saturation value N_o/b is in the order of 10^{18} to $10^{19}/cm^3$. If the defect is saturating due to the defect interactions, the additive artificial irradiation by the high-dose rate does not give any effect. Thus, the TD obtained is larger than the real one, if one simply extrapolates the linear part as is shown in Figure 3.13. Best fitting to a saturation curve is required for TD determination.

Figure 3.13: Growth of Defect Concentration by Natural and Artificial Irradiation for a Sample Age Much Smaller Than the Fading Time. The TD obtained is larger than the real age unless a best fit to equation 3.24 is made

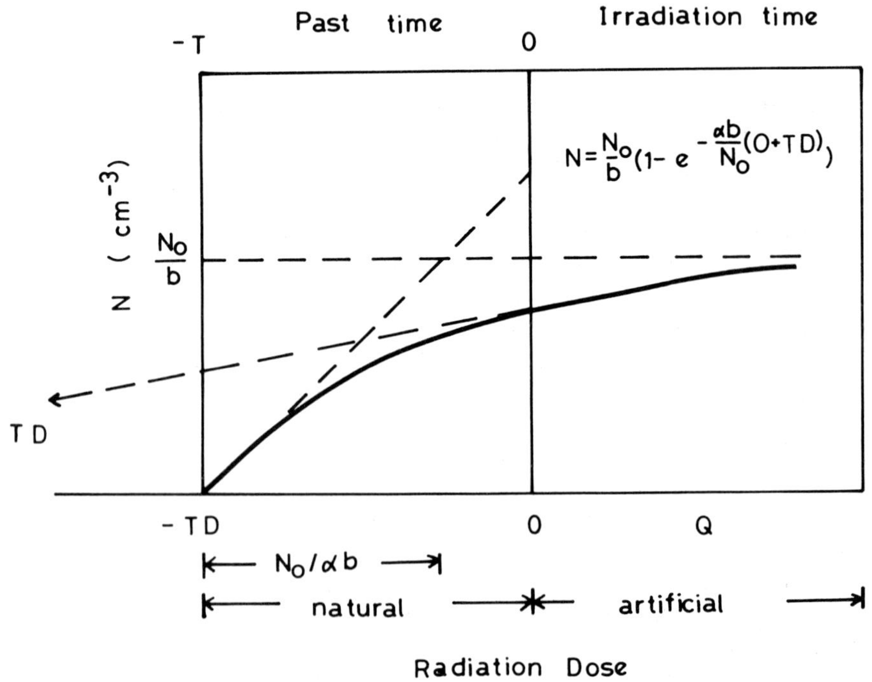

The saturation of growth due to the presence of an impurity can be explained by equation 3.27:

$$\frac{dN}{dQ} = \alpha \left(\frac{N_o - b_I N_I - bN}{N_o}\right)$$

where N_I is the concentration of impurity.

If N_I/N_o is on the order of 10^{-3}, the defect cannot be produced in a stable form. This phenomenon is often observed for carbonates (fossil shells and limestone) with a high concentration of Mn^{++}.

The same argument can be made for α ray production yield (k-value). If the content of radioactivity is high for old fossils, an overlap of α tracks occurs. Considering the α ray range of about 3μm and the diameter of 5Å, the effective volume b= 5×10^4 is obtained for 1MeV α rays. Then equation 3.24 may be divided by $\alpha = 2 \times 10^{11}$ for G = 0.1 to give the TD. Hence the k-value of about 0.2 due to the interaction in an α track is further reduced by the overlap of α tracks for TD by as much as hundreds of krad or Mrad. The ESR ages for fossil bones using k = 0.15 is thus an underestimation, as is shown in a later section. Thus, ESR dating is best applied to younger materials than Miocene and not to geological materials as old as Carboniferous or Cambrian.

Carbonate Fossils

Fossil shells are vitally important for sediment age determination in geology and age estimation of shell mounds. Detailed classification of shell species has been done in palaeontology and geology. This section describes typical ESR spectra of shells and corals whose ages have been determined from the shell classification of sediments as well as from the radioisotope dating methods. Ages deduced from TD, taking into account the disequilibrium mentioned in the previous section, are compared with Th/U, Pa/U and C^{14} ages. Calculation of TD versus age is made for the deep-sea core foraminifera, taking the initial incorporation of Th^{230} and the rate of Rn^{222} loss into account. The very good agreement obtained indicates the accuracy of the ESR dating method.

Model of Defects in Carbonate

Shells consist of mineral carbonate ($CaCO_3$) and organic conchiolin. Natural radiation produces defects in both constituent structures. A model of defects in the mineral part of shells is described briefly in this section. Carbonate is a typical constituent of cave deposits, which have been dated most successfully by ESR (Ikeya, 1975). Figure 3.14a shows the crystal structure of calcite ($CaCO_3$). An electron ionised from a planar molecular ion of $CO_3^{=}$ is trapped by a similar ion to be $CO_3^{\equiv}$ in the lattice of $CaCO_3$. The excess electron of $CO_3^{\equiv}$ enters the P_z orbital in the molecule as indicated in Figure 3.14b.

Figure 3.14: Crystal Structure of Calcite ($CaCO_3$) (a) and the Molecular Orbital Picture of a Simple Defect, CO_3^{---} Radical (b). The black circles and open circes are carbon and oxygen atoms, respectively. The shaded circles are divalent calcium atoms. The planar triangle molecule of CO_3^{---}, consisting of sp^3 hybrid orbitals, can trap an additional electron to form CO_3^{---}, as shown in the shaded orbital in (b)

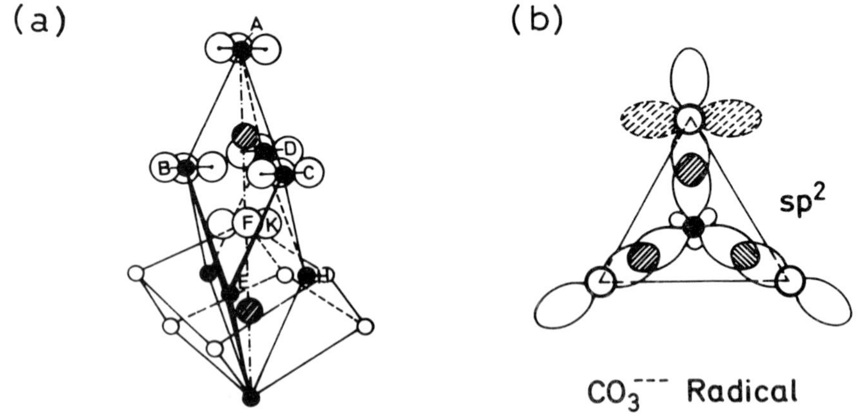

These simple defects of CO_3^- ($g_{\parallel}$ = 2.005, $g_{\perp}$ = 2.016) and CO_3^{---} ($g_{\parallel}$ = 2.001, $g_{\perp}$ = 2.003) molecular ions are not stable in $CaCO_3$ at room temperature (Serway and Marshall, 1967). The CO_3^- moves to some impurity and is stabilised. A detailed picture of the stable hole is not clear at present. The CO^{---} is stable for a few hours at room temperature, resulting in a long-lasting phosphorescence due to recombination. The electron moves to some unknown hole and emits light. Thermoluminescence peaks are observed around 50°C, 280°C and 350°C. The first peak at 50°C is associated with the migration of self-trapped CO_3^{---} ions. The fading time or the decay time of defects responsible for the second and third peaks is $(2-10) \times 10^5$ and $(2-10) \times 10^6$ years, respectively, at 20°C, according to a rough estimation using equation 3.15.

The peak at 280°C may be associated with the destruction of the complex defect CO_3^{---} associated with a trivalent ion (Serway *et al.*, 1968). Defects at high temperature are associated with aggregated defects, presumably along the path of α tracks (Ikeya, 1978b). The enhancement of the 350°C peak is observed by neutron irradiation. This is somewhat similar to the recent ESR dating work on ancient flints (Robins *et al.*, 1978) by α recoils produced by neutron irradiation (Garrison *et al.*, 1981). Stability of TL peaks at 280°C was recently discussed to solve the conflict between the ESR and TL ages of stalagmite calcite at the early hominid site of Caune de l'Arago, France (Debenham, 1983). Models and stability have not yet been well established.

ESR Spectra of Shells

Sea shells of known ages ranging from thousands to tens of million years old have been measured at room temperature by ESR. Usually the field modulation of 100kHz is about 0.1mT, sometimes reducing to 0.05mT for better resolution of overlapping radical signals. A piece of shell or powdered shell is placed in a standard quartz cell for ESR measurement. Shells are taken out of the cell and artificially irradiated to estimate the total dose (TD), using γ rays from a Co^{60} source. Dose rates of about 10^4 rad/hr have been used to avoid the high dose-rate effect discussed below (Groom *et al.*, 1978).

Figure 3.15 shows a typical ESR derivative spectrum of a shell. Only broad lines and a sharp signal around 0.34T have been observed. The signal intensity at 0.34T, as indicated by an arrow in Figure 3.15, is enhanced by artificial γ ray irradiation. One can obtain a TD of 130Gy or 13krad by fitting the linear enhancement of the signal to equation 3.5. The age estimated from the stratigraphy is about 250,000 years (middle Pleistocene). It is simply assumed that the shell has no signal at the age of zero and that natural ionising radiation has produced the signal at the central part. Actually, no signal is found to be associated with CO_3^{---} in modern shells unless irradiated artificially.

Figure 3.15: Typical ESR Spectrum (Derivative Absorption as a Function of the Magnetic Field) for a Pleistocene shell, *Anadara pseudosubcrenate.* The sharp signal at 0.34T (3,400G) is enlarged in the inserted signal

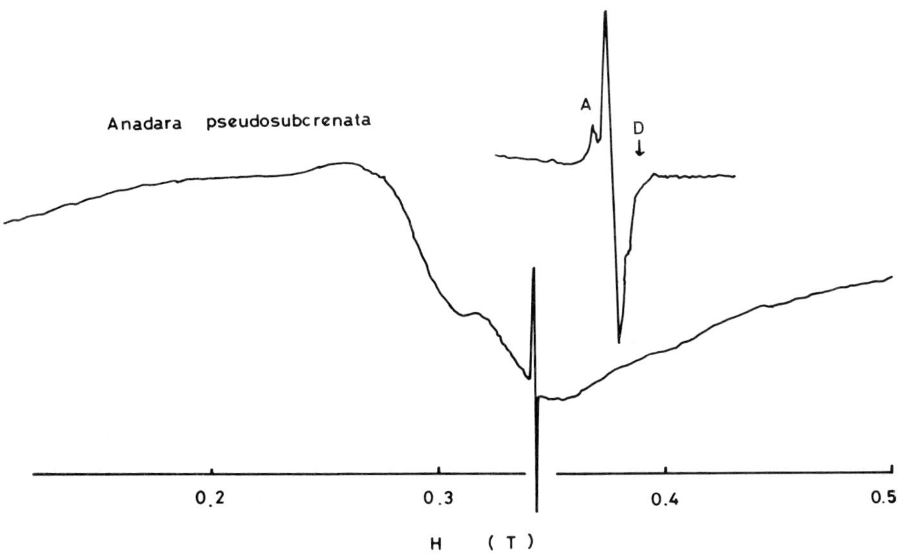

Figure 3.16: ESR Spectrum of the Sea Shell *Baryspira rubiginosa labocallosa*. Signals associated with Mn^{++} in $CaCO_3$ are also indicated. Signal A intensity is not enhanced by γ rays; signal B is associated with organic radicals. Signals C and D are associated with $g_{\parallel}$ and $g_{\perp}$ of CO_3^{---} impurity defects in the mineral part of the shell

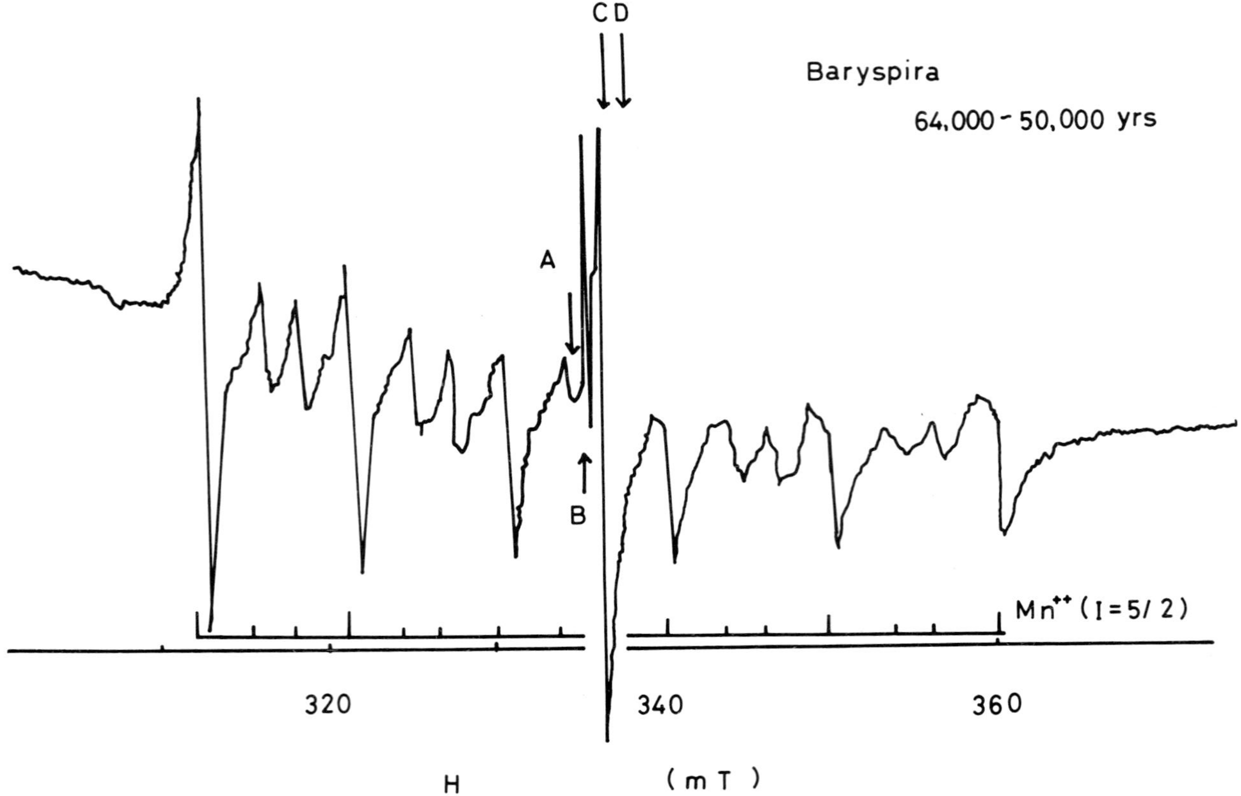

Figure 3.16 shows the other typical spectrum of a sea shell (*Baryspira*), whose age had been determined by radioisotope dating of Th^{230} and Pa^{231} as 65,000 and 60,000 years, respectively. The intensity of the characteristic hyperfine lines of Mn^{++} and its forbidden transition lines are observed as indicated by the lines in Figure 3.16. The intensity of the signal indicated by C and D is small for a shell from a shell mound, approximately one tenth of the one shown here. Difficulties are present for shells with a large quantity of Mn^{++}, as the huge signal of the impurity masks the presence of the weak signal of radicals produced by natural radiation.

The ESR spectrum of radiation-induced radicals consists of four types of signals. This is more clearly resolved in a reduced modulation field. Figure 3.17a shows ESR derivative lines of a shell (*Dantalium weinkauffi dunker*) before and after artificial γ irradiation. Four types of defect signals are observed as indicated by A, B, C and D in the figure from the low field. The signal intensities of B, C, and D are clearly enhanced by the γ irradiation. Figure 3.17b shows the growth of the signal as a function of the radiation dose. The intensity of signal A is not affected by γ ray irradiation, in this case.

The TD of about 13Krad is obtained from the signal enhancement for B and C. This coincidence is not necessarily seen in all shells. Generally, growth of the intensity of the signal B has a tendency to saturate soon upon γ irradiation. Nearly the same TD as signal C is obtained for the signal D. This is natural as signal D is associated with $g_{\parallel}$ of the CO_3^{---} type radical while the signal C is for $g_{\perp}$ 1 of the same centre. If the annual rate of natural radiation is estimated by some means, the age of the shell can be deduced from the geological dose of 13krad.

The g values of the signal A and B are 2.0060 ± 0.0005 and 2.0037 ± 0.005, respectively. The signal B may overlap the quintet (five-line) signal of organic radicals with hyperfine splitting of four protons with the coupling of 2.17mT. This is presumably associated with alanine molecular radicals as described in Figure 3.4. The intensity of signal C is nearly twice that of signal D if one integrates the derivative spectrum and compares the area. The signals C and D might be ascribed to a defect with anisotropic g values, i.e. $g_{\perp} = 2.0009$ and $g_{\parallel} = 1.9950$ in the axial field of the polycrystalline materials of the carbonate.

Figure 3.18 shows the ESR spectra of three types of shells belonging to different ages. The conditions of the measurement are nearly the same except for the relative magnitude as indicated. The signal intensity increases as the age increases from about 6×10^4 to $(5\text{–}6) \times 10^5$ years old, estimated from uranium dating and from the stratigraphy. In the case of old shells, the intensity of the signal A is even decreased by γ irradiation. Thus, only the intensity of the signal C, or half of the signal intensity if there is an overlap with other signals, is used. It should be noted that signals C and D increase linearly, while A and B seem to be saturated. The

Figure 3.17: ESR Derivative Line of a Shell (*Dantalium weinkauffi dunker*) Before and After Artificial γ Ray Irradiation (a). Signal intensities are shown as a function of the artificial radiation dose. TD is obtained from the enhancement of the signal C. The intensity of the signals A or Mn^{++} is nearly constant at this radiation dose range. The TD of signal B coincides with that of signal C only in this case

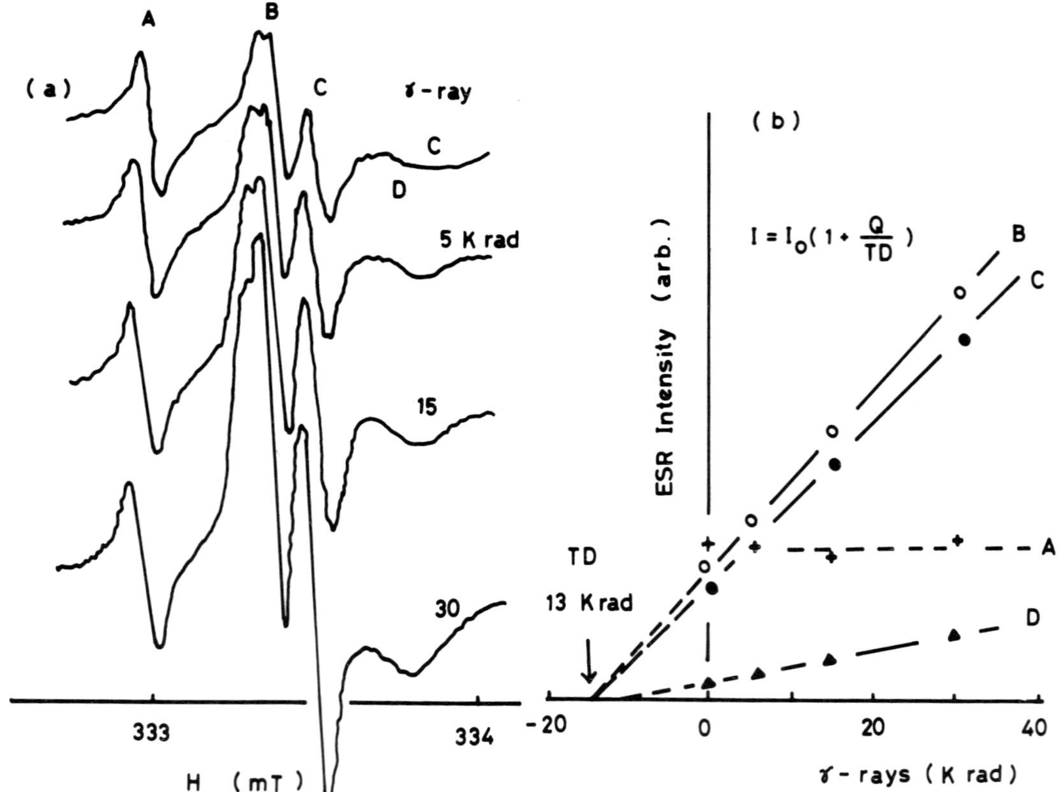

Figure 3.18: ESR Spectra of Shells Belonging to Three Different Eras: (a) *Barispira rubiginosa labocallosa Lisch*, aged 50,000–59,000 years by Th^{230}-dating, and 53,000–64,000 years by Pa^{231}-dating; (b) *Nerierita didyma Röding Ca.* aged 120,000 years; and (c) *Astarte hakodatenis Yokoyama*, Pleistocene

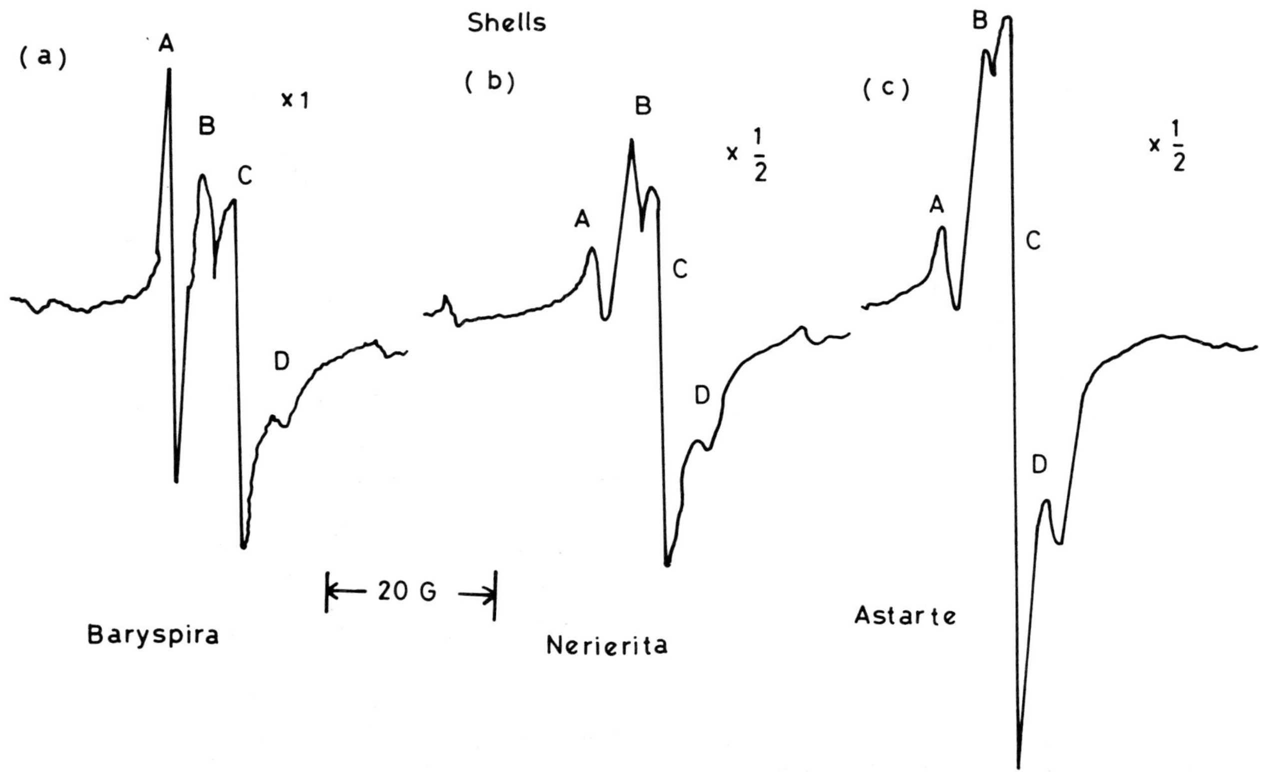

TDs obtained from signal C for these three samples are 3krad, 6krad and 13krad, respectively.

It is interesting to note that one can estimate the age of the sea shells from the magnitude of signal C and the relative magnitude of signal A to other signals. The pattern of the spectrum is different for different ages. One sea shell whose age is in dispute in the field of palaeontology has been measured: stratigraphic dating suggests that the age is about 2.5×10^4 years, while palaeontological estimation gives an age of 1.25×10^5 years. ESR dating gives a TD of about 6krad: this is about $(1.5–1.2) \times 10^5$ years for an annual dose rate of 0.04–0.05rad/yr and agrees well with the palaeontological age estimation (Omura, 1980).

Age and Total Dose

Figure 3.19 shows the relationship between the age estimated by a palaeontologist and the TD of the shells, both on a logarithmic scale. The ages determined by the radioisotope method are also indicated. The other ages, with some uncertainty, are by the sedimentary shell classification. A clear proportionality relation exists up to above one million years with a constant of about 0.05rad/yr. In this case, all the shells were buried in sands in northern Japan (Tohoku area). The content of uranium in the environment might be assumed to be roughly the same for sea shells. Moreover, the rigid shell would not absorb uranium as much as bones consisting of hydroxyapatite, whose structures are changed with time by the replacement of atoms, as will be discussed in the next section.

It is known that α-rays produce defects of local high density, since the linear energy transfer (LET) is large. Alpha rays from the surrounding soils produce defects only at the surface of articles buried in the soil. The surface can be washed and removed by chemical etching with acids such as hydrogen chloride, as in TL dating. In ESR dating of shells, the effect of the surface damage caused by α-rays is small, as the defects produced by α-rays occur within only a few μm of the surface of a shell 1mm thick, as was shown in Figure 3.7. Thus, an annual radiation dose of about 0.05rad/yr would be due predominantly to the γ radiation from the environment. The sea shells are from Tohoku, the northern part of Japan, where the terrestrial γ radiation is about 30mrad/yr, very close to the present apparent annual dose of 50mrad/yr, as determined from the ESR of shells.

The absolute ESR ages obtained for the content of U^{238} and Th^{230}, using equation 3.12 and based on a finite medium in Figure 3.7, agree fairly well with the geological assessment or the ages determined by other methods. Figure 3.20 is a typical calculated result of a 'TD versus T' relation for shells with both internal and external total dose. The internal dose is only 10–20 per cent of the total dose for the α ray production efficiency (k) of 0.2 for shells with a low content of uranium. Hence, the external dose rate plays a major role for the ESR age of shells.

Figure 3.19: Relationship Between the Total Dose of Natural Radiation of Shells and the Geological Age Determined from Sediments and with C^{14} dating. The apparent annual radiation dose rate of 40mrad/yr is appropriate for the age estimation of shells at the northern part of Japan. The points • and × are the TDs of whale bone and shark tooth (Miocene), respectively

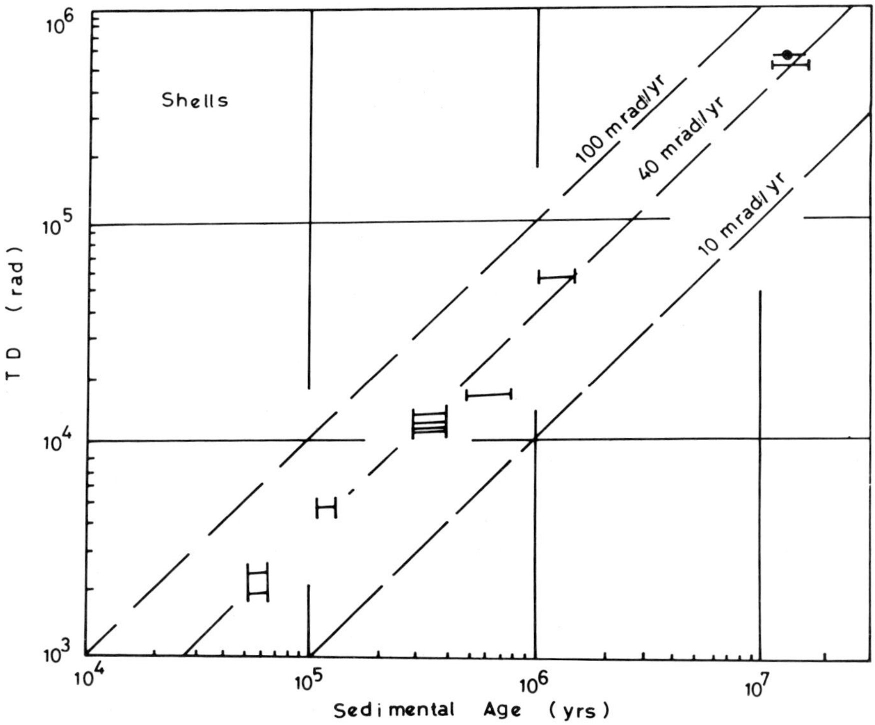

Modern shells contain less than 0.1 ppm of uranium in Tohoku. A model of uranium accumulation might even have to be used for open-system shells, since most of the uranium might have entered into the carbonate later. However, since the external dose rate plays a major part in the TD, the uranium content is assumed to be constant. Typical ESR ages, the content of uranium, and internal and external total dose are given in Table 3.20. These agree fairly well with those of the sediments, which were determined from Th^{230}/U^{234} dating of the closed system solitary corals.

Shells at Shell Mounds. Shells of archaeological remains, especially shells at shell mounds, have also been investigated. Generally, shells a few thousand years old have a small amount of radicals. The ESR signal of defects is masked by the presence of impurity signals associated with

Figure 3.20: The 'TD versus T' Curve for a Pleistocene Shell, Taking U^{238}-series Disintegration into Account, Using Equation 3.12. The internal dose (TD_{in}) by α and β rays and the external dose by the environmental γ rays are shown by dashed curves. The ages for equilibrium (T_e and disequilibrium (T_d) are deduced graphically. The small difference is due to the low content of U^{238} in the shell

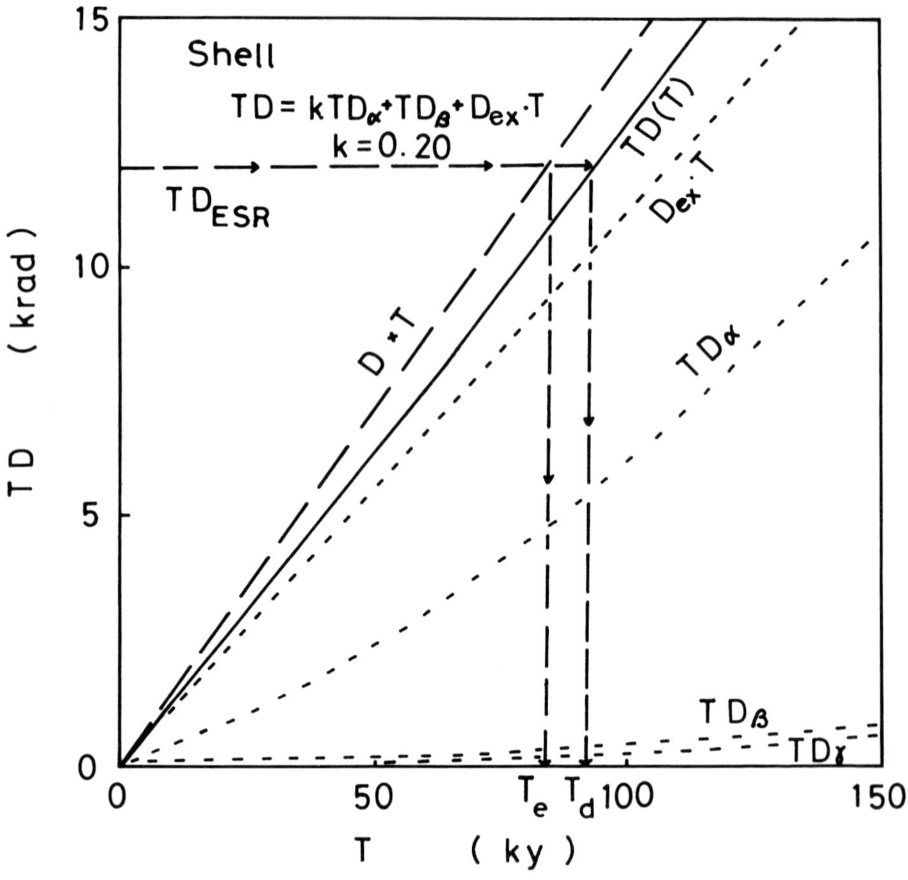

Mn^{++} or Fe^{+++}, as was described in the previous section. The signal-to-noise (S/N) ratio is low unless some signal averaging techniques are employed. Therefore, shell dating of ordinary archaeological remains is more difficult than that of geological shells.

Some shells, especially shells in fresh water, contain a small amount of Mn^{++} and can be dated with ESR with the present equipment if the shells are older than about 1,000 years. Preliminary results on ESR dating of shells at Jomon shell mounds in Japan, shells in pre-Inca days (about 2,000 years old) and Sambaqui shell mounds in Brazil (4,000 to 8,000 years old) indicate that ESR dating is not so simple. Organic radicals are pro-

Table 3.2: Comparison of ESR Ages and the Age of Sediments at Two Shell Beds. The age taking uranium series disequilibrium (T_d) is more reasonable than the age for secular equilibrium (T_e)

Sample	Radioactive elements U^{238}(ppm)	Radioactive elements Th^{232}(ppm)	TD (ESR) (krad)	D_{ex} [b] (mrad/yr)	TD_{in} [c] (krad)	TD_{ex} (krad)	ESR age (ky) T_e	ESR age (ky) T_d	Sediment
Uji	0.426 ± 0.01	0.02	11.0	112 ± 8	1.2	9.8	64	87	120 ± 9[a]
			12.0		1.3	10.7	70	95	
			15.0		1.8	13.2	87	118	
			17.0		2.0	15.0	98	133	
Hiradoko	0.85 ± 0.017	0.123	17.0	112 ± 8	4.4	12.6	98	112	115 ± 10
			17.5		4.5	13.0	100	115	119 ± 9
									125 ± 9

Notes:

a. Age of the shell bed obtained by Omura (1980) for corals using Th^{230}/U^{234} dating method;

b. External gamma ray dose rate measured by TLD ($CaSO_4$), converted from mR/y to mrad/y by multiplying by a factor of 0.86;

c. Internal total dose, $TD_{in} = KTD_\alpha + TD_\beta$, calculated from equation 3.12. External total dose $TD_{ex} = D_{ex}$. T was added to fit the experimental TD for the ESR age. Generally, the main contribution to TD for shells with a small uranium content is environmental gamma rays ($k = 0.2$ and $R = U^{234}/U^{238} = 1.15$ are used).

Source: Ikeya and Ohmura 1984.

duced in some cases by artificial irradiation, which makes the evaluation of TD difficult.

Egg Shell. Ostrich egg shells at Douara cave, Syria, have been dated with ESR. The TD of the egg shells was 4.2krad. Douara cave is being studied by anthropologists at Tokyo University (Akazawa, 1976). Bones, shells and gypsum have been measured with ESR.

Figure 3.21: ESR Spectra of Typical Corals. The signal position of CO_3^{--} type radicals is indicated by the letter C

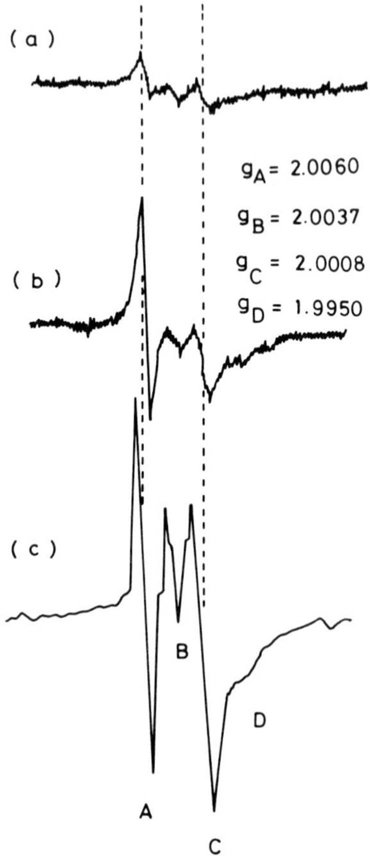

Notes:
a. 3,000-year-old coral;
b. 6,000-year-old coral;
c, 3,000-year-old coral irradiated for 1krad.

Corals at Marine Terraces. Corals are also important in marine geography or geology. Old sea levels and land movements are being investigated by dating corals at old terraces and beaches on coral reef islands at Ryukyu, Japan. Geologists using C^{14} dating have encountered the upper limit of this technique (Nakata, Takahashi and Koba, 1978) and applied ESR dating. Most of the corals, which contain calcite due to the recrystallisation of aragonite, showed large ESR signals associated with Mn^{++}. However, Ikeya and Omura (1983) found that the aragonite coral species (*Acropora* sp., *Portes* sp. etc) have relatively small concentrations of Mn^{++} or no Mn^{++} signals. Aragonitic coral samples can be dated with ESR most successfully, partly because corals are a closed-system to the environment. ESR ages based on U^{238} disequilibrium agree fairly well with those obtained by C^{14} and Th^{230}/U^{234} dating (Konoshi, Schlanger and Omura, 1971).

Figure 3.21 shows typical ESR spectra for a 3,000- and 6,000-year- old corals and for one irradiated with 1 krad. The magnetic field position of the signal is indicated by the dashed line and the letter C. The signal intensity is certainly increased as the coral becomes old and is irradiated by γ irradiation. The behaviour of the signal A is nearly the same as described above for the dating of shells.

Figure 3.22 shows the relationship of the relative ESR signal intensity of corals to the age determined by C^{14} dating. As the corals are from the same area, the relative efficiency of the defect production and the annual dose rate would be roughly the same, giving an ESR defect signal intensity proportional to the age. The result is remarkable because one can roughly estimate the relative age with ESR without evaluating the total dose of natural radiation (TD). If a few samples are dated using C^{14} dating or another method, the ESR signal intensity versus age calibration curve can be obtained. Corals in that area might be dated roughly with an ESR signal using this calibration curve by estimating the fluctuation of U^{238} content to be within 20 per cent. Generally, measurements of TD with ESR can be made only with an error of about 10 per cent, due to scattering of the data points, inaccurate exposure dose and the conditions of sample setting. It also takes a few hours to irradiate and obtain the TD. However, if one has a standard sample dated using C^{14} dating, then relative dating can be done within 10 minutes for one sample without irradiation.

The apparent annual dose rate of 0.07rad/y is very low, indicating that the effect of α rays is negligible ($k = 0$). This is due to insufficient build-up of Th^{230}. Theoretical TD versus T for 3 ppm of U^{238} and $k = 0.15$ and 0.20 are shown by the dashed curve in Figure 3.22. In this calculation the radiation dose from the initial incorporated Ra^{226} is also included, using the experimental value of the activity ratio $Ra^{226}/U^{238} = 0.027$ for modern corals (Komura, 1982). The TD by the initial Ra^{226} inclusion may be written as in equation 3.28:

$$TD_{226} = C_u\ (Ra^{226}/U^{238})\ (D_{226}/\lambda_{226})\ (1 - \exp(-\lambda_{226}T))$$

where D_{226} is the annual dose rate of Ra^{226} and its daughters. The effect is to shift the age about 100 years to the young side.

Figure 3.22: Relationship of the ESR Signal Intensity or the TD of Corals at an Area with an Uncorrected C^{14} Age. The theoretical curves of TD versus T using equation 3.12 for k = 0.15 and 0.20 are shown by dashed curves. A better agreement for k = 0.15 is obtained if the C^{14} age is corrected

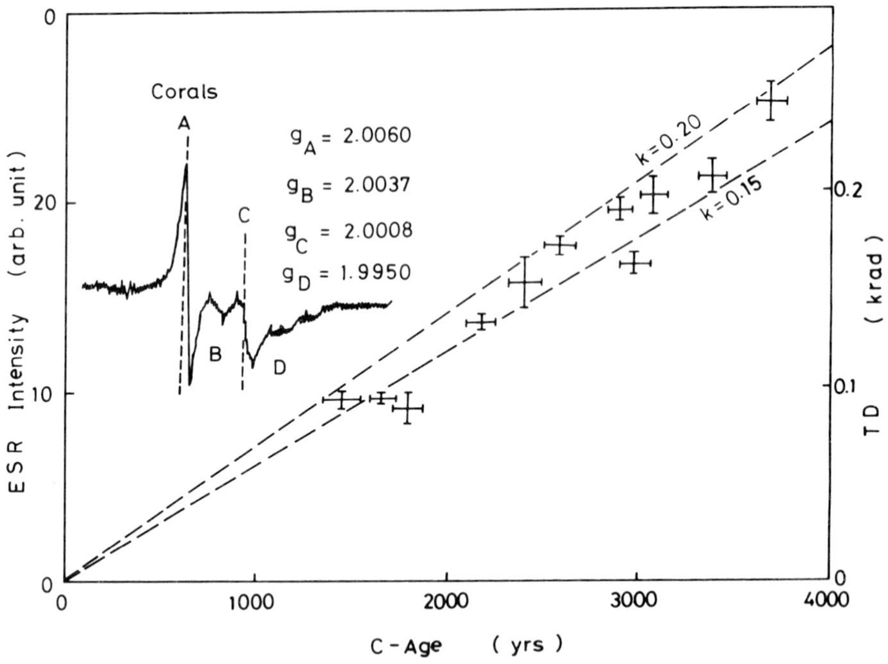

Figure 3.23 shows a calculated TD for the average content of U^{238} (3 ppm) in an infinite medium of the coral for k = 0 and k = 0.2, together with the dashed line assuming radiation equilibrium. The TD, considering the disequilibrium, coincides with the TD for k = 0, assuming an equilibrium for an age of thousands of years. The TDs for corals from high terraces are shown at the age determined by the Th^{230}/U^{238} dating method. Agreement seems reasonable considering the fluctuation of U^{238} and Th^{232} content within ± 20 per cent. The TD for corals beyond the detection limit with Th^{230}/U^{238} dating can come close to 500,000 or 800,000 years.

Figure 3.23: Relationship Between TD and Ages Determined by C^{14}, Th^{230}/U^{238}, and Pa^{231}/U^{235} Dating. The dashed line with the slope 1 is using the annual dose rate, assuming a radiation equilibrium for an average U^{238} content of 3 ppm and for k = 0.2 and k = 0. The solid curve is the theoretical one for k = 0.2 using equation 3.12, taking into account the disequilibrium of U^{238} disintegration

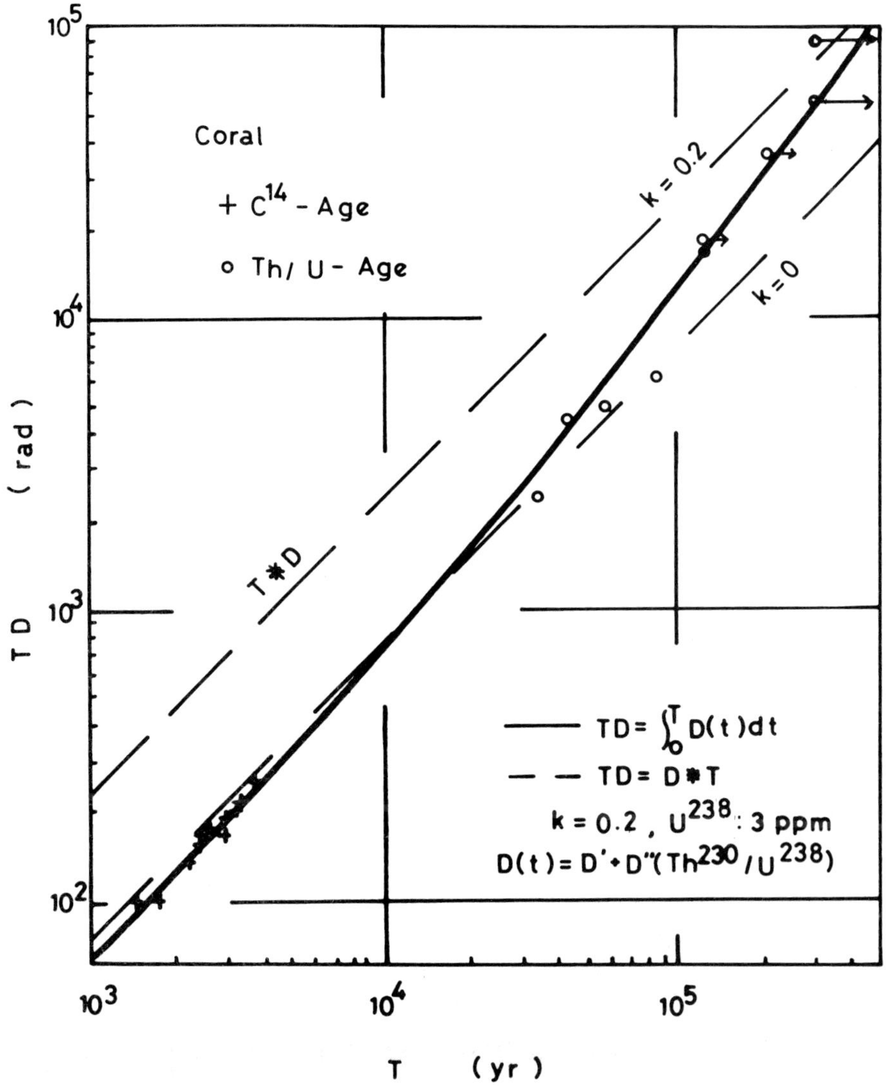

Table 3.3: Comparison of ESR Ages of Corals at Ryukyu Island with those by C^{14}, Th^{230}/U^{234} and Pa^{231}/U^{234} Methods, Taking into Account Disequilibrium of Th^{230} and Incorporation of Ra^{226}

Sample	Geology and locality (elevation)	U^{238} (ppm)	Th^{232} (ppm)	TD (krad)	(ESR age) k = 0.2	k = 0.15[c]	Radiometric ages (ky)	
76–4–1–4	Kikaijima			0.135 ± 0.015	1.92 ± 0.20	2.30 ± 0.20	C^{14} 2.17 ± 0.085	
75–3–27–12	Raised coral reef			0.155 ± 0.015	2.22 ± 0.20	2.61 ± 0.20	2.39 ± 0.085	
75–3–27–4	South eastern part	3.0[b]	0.05[b]	0.175 ± 0.017	2.49 ± 0.22	2.95 ± 0.20	2.57 ± 0.085	
75–3–27–7	(lower than 7m)			0.195 ± 0.020	2.80 ± 0.25	3.27 ± 0.29	3.06 ± 0.09	
76–3–31–2				0.250 ± 0.025	3.52 ± 0.30	4.17 ± 0.37	3.67 ± 0.09	
	Kikaijima						Th/U	Pa/U
CK–19	Araki Lm., western (25m)	3.0 ± 0.11	0.125	4.5 ± 0.3	43.2 ± 3	49.3 ± 4	41 ± 2	41 ± 2
CK–21	Younger Lm., southern (40m)	3.42 ± 0.06	0.02	4.9 ± 0.4	43.1 ± 3	49.5 ± 4	54 ± 2	64 ± 2
CK–24	Middle Lm., central (65m)	2.30 ± 0.10	0.044	6.2 ± 0.5	67.5 ± 6	76.7 ± 7	86 ± 4	85 ± 4
No. 30	Tokunoshima, southern (51m)	3.0[a]	0.05[a]	17.5 ± 1.0	125 ± 10	132 ± 11		
No. 25	Okierabu, Serikaku Ly., I(51m)	3.0[a]	0.05[a]	37.5 ± 4	202 ± 20	230 ± 25	>200	
Kz	Kikaijima, Bottom Lm., central (140m)	2.87	0.028	95 ± 10	460 ± 50	485 ± 50	∞	

Notes:

a. Samples supplied by Drs A. Omura, M. Koba and T. Nakata;

b. The average contents of U^{238} and Th^{230} for corals in this area calculated by Omura were used for some corals to obtain the tentative ages from the TDs. The error is within 20 per cent;

c. The initial incorporation of U^{224} and Ra^{226} are with activity ratios for U^{224}/U^{238} of 1.14 and Ra^{226}/U^{238} of 0.027. The α ray efficiency (k) is 0.2–0.15.

The ESR ages of corals for no radon loss (k = 0.15 and 0.20) are shown in Table 3.3. Recent γ ray spectroscopic study indicates that the loss of radon is negligible for corals. The very good concordance of ESR ages with those by other methods proves the excellence of ESR dating of these closed system corals. It also proves the applicability of the technique for the study of the plate tectonic movement of coral reef islands, which is well beyond the limit of Th^{230}/U^{234} dating.

Planktonic Foraminifera. The TD of a carbonate fossil of planktonic foraminifera has been determined as a function of the depth of a deep-sea core sediment (Sato, 1981, 1982). The apparent annual dose rates were obtained by tuning the TD to the expected ages determined from geomagnetic reversal (Matsuyama Boundary) and fauna analysis. The effect of the initial Th^{230} incorporation into the deep-sea sediment and the stability of the ESR signal have been commented upon recently (Wintle and Huntley, 1983; Sato, 1983). Optical bleaching effect on carbonate fossils is also discussed but would be negligibly small in general for ESR dating as ESR detects electrons in the depth of the material rather than the surface-trapped electrons of TL dating. Therefore, the main argument is over the saturation behaviour of the TD-versus-depth profile.

Fortunately, Sato (1982) referred to the equivalent content of U^{238} determined from γ rays of Ra^{226}. The top core value of 13.26 ppm would be due predominantly to the initial excess Th^{230} in the deep-sea core, since Th^{230} build-up from U^{238} is expected to be, in part, due to the half-life of 2.5×10^4 yr for U^{234}. The small content of U^{238} at a depth of 1.4m, 0.24 ppm, may be considered as the real content of U^{238}. Hence, the TD due to U^{238} (0.24 ppm), Th^{232} (1.2 ppm) and K (0.56 per cent) can be calculated as a function of age, considering an infinite medium for simplicity. The TD due to the initial Th^{230} incorporation is calculated similarly as the initial Ra^{226} incorporation in equation 3.28. The dose rate (D_{30}) from Th^{230} down to Pb^{206} has already been given in the calculation of TD in U^{238}-series disequilibrium.

Figure 3.24 shows the theoretical TD-versus-age relationship for rates of Rn^{222} loss (N) of 0, 50 and 100 per cent using the α ray efficiency (k- value) of 0.15. Sato's experimental data (1982) and the depth are also given. The agreement is fine for N=70 per cent, with no indication of the signal instability. The behaviour of the saturation is caused by the radiation due to the initial Th^{230}.

The rate of Rn^{222} loss may actually be much smaller. The theoretical model of an infinite medium is an overestimation at the site close to the surface. The radiation dose from sea water would be much smaller than from the sediment. It should be noted that TD was obtained experimentally by using the internal standard signal of Mn^{++} in the microfossil carbonate. The signal intensity of Mn^{++} is decreased 10–30

Figure 3.24: The TD-versus-age T (or the Depth) Profile for a Deep-sea Core Sediment, Considering the Initial Incorporation of Th^{230} and U^{238}-series Disequilibrium for Rates of Rn^{222} Loss (N) of 0, 50 and 100 per cent, Using the Contents of Radioactive Elements reported by Sato (1983). The experimental TD for planktonic foraminifera is also shown (Sato, 1982). The dashed and dotted curves are due to the TD by Th^{230} and U^{238} in disequilibrium

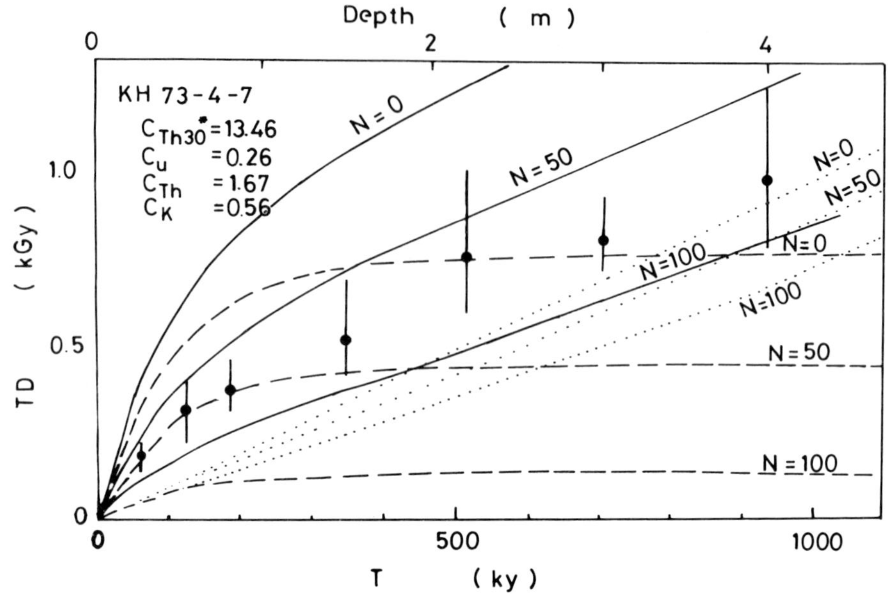

per cent in some cases. If this is the case, then TD is underestimated by the same percentage. The rate of Rn^{222} loss would then be actually less than 50 per cent. This rate would be reasonable if one considers the Pb^{210} activity in the soil, as discussed in studies of coral dating (Ikeya and Ohmura, 1983).

Summary of Carbonate Fossils

Shells and corals can be dated with ESR if one can make a proper assessment of radiation dose. For shells in sand or clay sediments, the external dose rate (D_e) can be measured with TLD, or equated to the γ ray dose rate calculated from the contents of U^{238}, Th^{232} and k in the sediment, assuming equilibrium. The internal total dose (TD_i) is (kTD_α + TD_β), calculated from the content of U^{238}, Th^{232} and K in the shell, assuming the radiation disequilibrium and using equation 3.12. A k value for cave deposits of about 0.15–0.20 is used for shells and corals, though such studies are required for carbonate fossils containing organic materials. The age is deduced from the experimental TD and the contents of radioactive elements, with a computer iteration.

An infinite medium of corals is considered for corals at terraces and deep-sea core sediments. The partial loss of radon is neglected in corals but has to be considered for planktonic foraminifera. The initial incorporation of Th^{230} must be considered for deep-sea cores. The ESR ages are thus quite satisfactory if the proper assessment of radiation is made. It may be concluded that ESR dating of carbonate fossils has now been established. A vast area of application lies ahead.

Fossil Bones and Teeth

Radiation produces defects and radicals with unpaired electrons in both the inorganic apatite part and the organic part of bones. Among the radiation-induced radicals, a prominent one in the apatite part is stable up to about 400°C or to a few million years at room temperature. The TD of standard fossil bones associated with anthropologically important sites are shown below, together with the age based on a theoretical model of a continuous uptake of uranium, taking into account the disequilibrium of U^{238} disintegration.

Structure of Bones

The inorganic structure of bones consists basically of microcrystalline hydroxyapatite, $Ca_{10}(PO_4)_6(OH)_2$. Each constituent element can be replaced chemically by other ions: Ca^{++} can be replaced by (a) divalent cations like Mg^{++}, Sr^{++}, Ba^{++}, Pb^{++}, Zn^{++}, Cd^{++}, Fe^{++} and Ra^{++}; (b) monovalent cations like Na^+, K^+ and H^+; and (c) trivalent cations like Al^{+++}, Y^{+++}, Ce^{+++}, La^{+++} and Dy^{+++}. The phosphate site PO_4^{---} can be replaced by anions such as SO_4^{--}, CO_3^{--}, HPO_4^{--}, SiO_4^{----}, and $H_4P_4^{----}$, etc. The hydroxyl site can be replaced by F^-, Cl^-, Br^-, as well as O^{--}, CO_3^{--} and H_2O. Therefore, the structure of apatite is extremely complicated (Brown and Chow, 1976). The organic part of the bone, collagen, is a protein that is extracted from fossil bones for age determination using C^{14} dating. Racemisation of the organic amino acids has also been used for dating fossil bones (Matsu'ura and Ueta, 1980; see also Masters, 1986, Chapter 2 of this volume).

When bones with collagen and apatite are left in the environment, acids may dissolve the apatite structure and fluorination proceeds. Bones also accumulate uranium from the environment (Charalambous and Papastefanous, 1977; Seitz and Taylor, 1974; Molleson and Oakley, 1966). The content of uranium and degree of fluorination have been studied in order to date bones. Results indicate that in nature there is a relatively rapid accumulation of uranium in the early stages from about 0 to 1 ppm, followed by a slow linear accumulation. The process of accumulation would depend on the environment as well as on the structure of the bones and enamels; soft spongy bones would accumulate more than solid rigid bones.

Atomic Model of Defects in Bone

Extensive ESR studies of radiation-induced defects in bones and tooth enamels, or in hydroxyapatite, have been carried out in the last decade in the field of biophysics, correlated with clinical and dental radiation measurements (Cevc and Schara, 1972; Doi *et al.*, 1979; Ostrowski, Goctawska and Stachowics, 1980). A recent review article on radicals in bones indicates that an atomic model of defects stable at room temperature is still an open question (Ostrowski, Goctawska and Stachowics, 1980). Table 3.4 shows some models of radiation-induced defects and their ESR parameters.

Table 3.4: ESR Parameters for Defects in Bones

Defects	g	Values	Comment	Reference
Hydrogen atom H^o	g A	= 2.002 = 50.2mT	Unstable at room temperature	Cole and Silver, 1963
Oxygen anion O^-	$g_{\parallel}$ $g_{\perp}$	= 2.002 = 2.07	Complementary defect to hydrogen atom	
F centre (?)	g	= 2.0019	Trapped electron at OH^- site	
h_t or CO_3^{---}	$g_{\parallel}$ $g_{\perp}$	= 1.9978 = 2.0036	Trapped electron at PO_4^{----} site (?)	Ostrowski *et al.*, 1980
CO_3^{---}	$g_{\parallel}$ $g_{\perp}$ $A(C^{13})$	= 2.0013 = 2.0052 = 17mT	In synthetic hydroxyapatite intentionally added with CO_2^{---}	Doi *et al.*, 1979
CO_3^-	$g_{\parallel}$ $g_{\perp}$	= 2.0168 = 2.01	Trapped hole only at 77k	
Alanine radicals	g A	= 2.0032 = 2.2 mT	Only in fossil teeth at Arago, Choukoutien, Petralona, Steinheim	Ikeya and Miki, 1981a, b

Ionising radiation produces hydrogen atoms (H^o) from OH^- ions in the apatite, $Ca_{10}(PO_4)_6(OH)_2$ (Cole and Silver, 1963). However, H^o is stable for only a short time at room temperature. No hydrogen atoms are observed in fossil bones exposed to continuous natural radiation. The paramagnetic O^--type ion is more stable than H^o. Carbonate ions,

CO_3^{--}, exist in hydroxyapatite as impurity ions. These impurity molecular ions trap electrons and holes. The resultant CO_3^{---} and CO_3^{-} ions, either isolated or associated with some impurity in the hydroxyapatite, give ESR signals (Doi *et al.*, 1979; Panepucci and Farach, 1977; Koberle *et al.*, 1977).

One plausible model for the ESR signal in fossil bones is molecular radical ions stabilised in the vicinity of impurities, analogous to a Y^{+++}–CO_3^{---} centre in a calcite. Another model for a stable defect in bones is an h_t centre in which an electron is trapped at the PO_4^{---} ion site (Ostrowski *et al.*, 1980). The detailed structures of defects in fossil bones are not clear at present.

Thermoluminescence (TL) studies of bone indicate that the oxidation of organic materials gives chemiluminescence (CL) and that grinding gives tribothermoluminescence (TTL) (Jasinska and Neiwiadomski, 1970). The peak temperature of the TL was around 400°C. The organic part is chemically removed from fossil bones; the TL peak for fossil bones was 460°C (Christodoulides and Fremlin, 1971). The TL peak temperature depends strongly on the rate of the temperature increase and direct comparison is difficult. However, it might be taken that the temperature T_p, where τ is one second, is more than 400°C. This corresponds to a limitation in dating of a few tens of millions of years, as has been mentioned above.

Sample of Fossil Bones

Fossil bones excavated from several anthropologically and palaeontologically important sites have been supplied to the author's laboratory for age determination by ESR.

A small piece of bone is broken off and the surface soil is removed before ESR measurement. A quartz flake is used, rather than a shaving blade made of iron or stainless steel, which sometimes gives a fictitious ESR signal if the intensity of the signal associated with the defects is small. Typical ESR spectra and the TD are shown in the following section.

ESR Spectra of Bones

Figure 3.25 shows a typical ESR derivative line of defects in the mineral part of bones. The bone specimen is from an excavation site at Arago, Tautavel in southern France. The spectrum from randomly oriented microcrystalline crystals of hydroxyapatite consists of signals with an angular dependent g value. The g value is expressed as in equation 3.29:

$$g(\theta) = (g_{\parallel}{}^2\sin^2\theta + g_{\perp}\cos^2\theta)^{1/2}$$

where $g_{\parallel}$ and $g_{\perp}$ are g values of electron spins for the direction of the magnetic field parallel and perpendicular to the principal axis of the crystalline field. The ESR spectrum of solid bones sometimes results in a different spectrum due to some preferential orientation of microcrystalline apatite along a certain direction; this spectrum is also observed in fresh bones irradiated by X-rays or gamma rays (Panepucci and Farach, 1977).

Figure 3.25: ESR Derivative Absorption of a Bone from Arago, Tautavel, France. The typical ESR signals of the mineral part of the bone are indicated by $g_{\parallel}$ and $g_{\perp}$. Signals associated with Mn^{++} are indicated

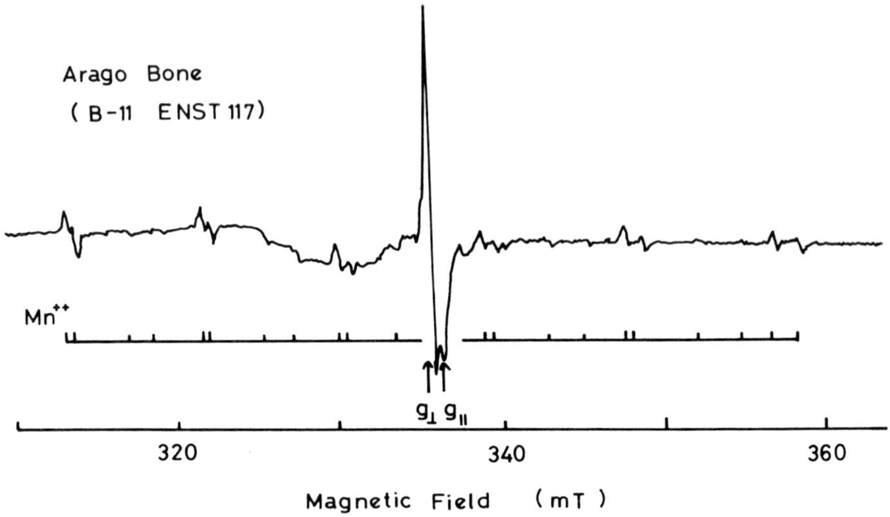

The characteristic spectrum of Mn^{++} observed in shells or in carbonate stalagmite materials can be also seen. The weak lines in both sides of the six hyperfine lines due to the nuclear spin of Mn^{++} are called the forbidden transition lines of Mn^{++}.

Figure 3.26 shows the ESR spectrum of Java bone. The spectrum consists of a signal around 160mT (1,600G) due to Fe^{+++}, Mn^{++} and a broad signal in addition to the signal due to defects in the mineral part of the bone. The broad signal might be partly associated with adsorbed clay-like particles. Old bones at Heidelberg and Java show nearly the same ESR spectrum.

Figure 3.26: ESR Spectrum of an Animal Bone from Java (Pucangan Layer). The signals associated with Fe^{+++} and Mn^{++} are indicated with the signal of the defects. A typical spectrum of an old bone at an open site

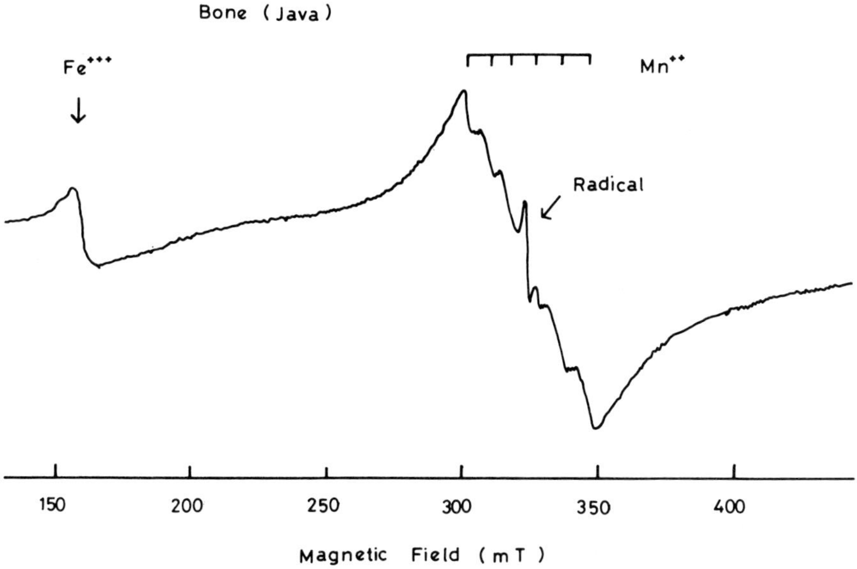

Figure 3.27 shows an ESR spectrum of a relatively young bone at Moroyo in Hokkaido associated with the Ohotsuku culture (1,000 years). The signal associated with the defect in the bone is extremely weak. The contribution from the broad signal must be subtracted, as is shown in the inserted figure. The apparent TD obtained by observing the enhancement of signal intensity is larger than TD $\simeq$ 1 krad, unless the contribution from the initial broad line is subtracted. An apparent annual dose rate of 0.1rad/y, which is much higher than expected from the U^{238} content, seems to explain the age of bones thousands of years old. The signal intensity is weak in general for such a bone. The signals associated with impurities predominate over the defect signal.

It is much easier to measure Pleistocene or Pliocene bones than to date archaeological bones less than 1,000 years old. The ESR signal associated with the decomposition of the organic part in relatively young bones gives an erroneous TD; a large TD is obtained due to overlapping of the signal. Some fresh bones also show an ESR signal associated with radicals due to

Figure 3.27: ESR Spectrum of a Bone Approximately 1,000 Years Old from Moroyo, Hokkaido. The contribution from the broad line (B) must be subtracted from the signal height (S), as is shown in the inserted figure

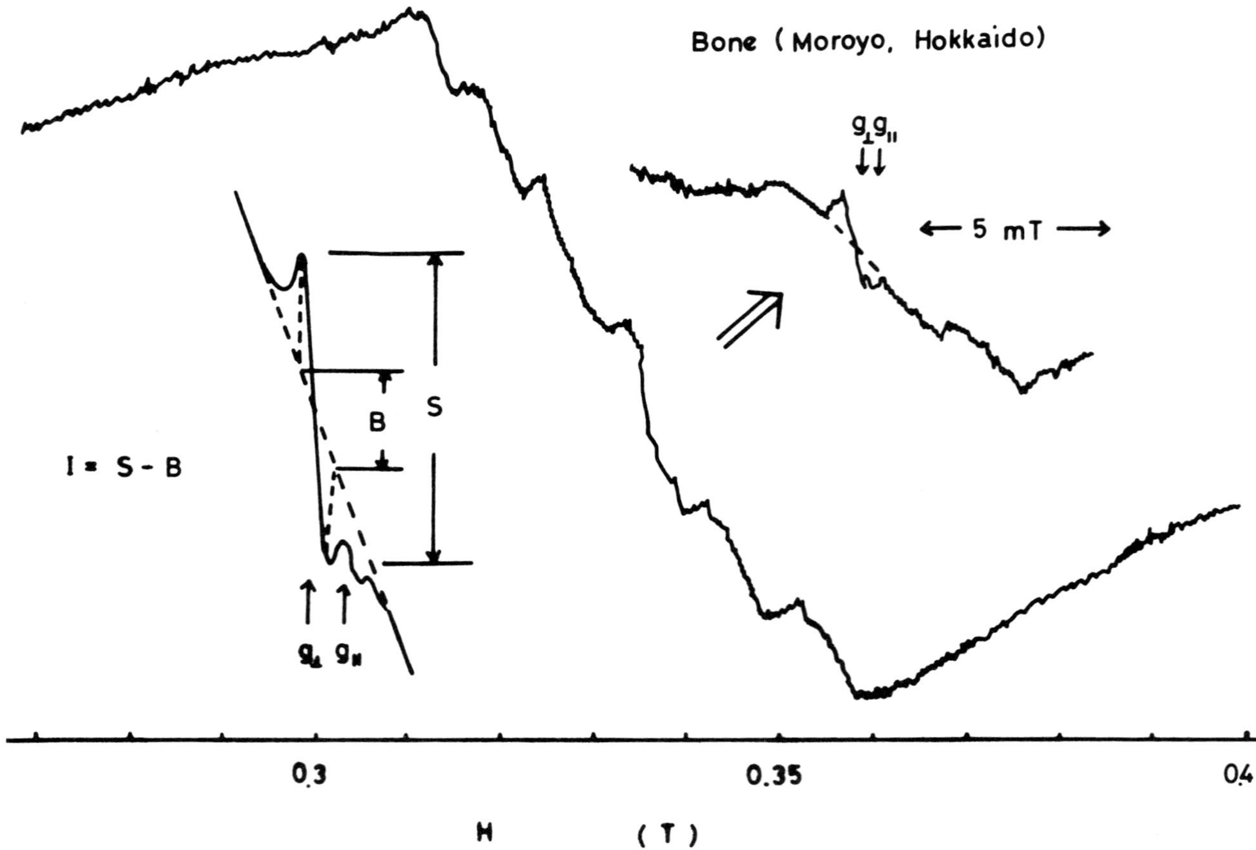

Figure 3.28: ESR Spectra of Fossil Bone at Petralona, Greece (a), and with Heating to 250°C (b). The intensity variation caused by heating the sample at several temperatures for 10 minutes is shown as a function of the temperature (c)

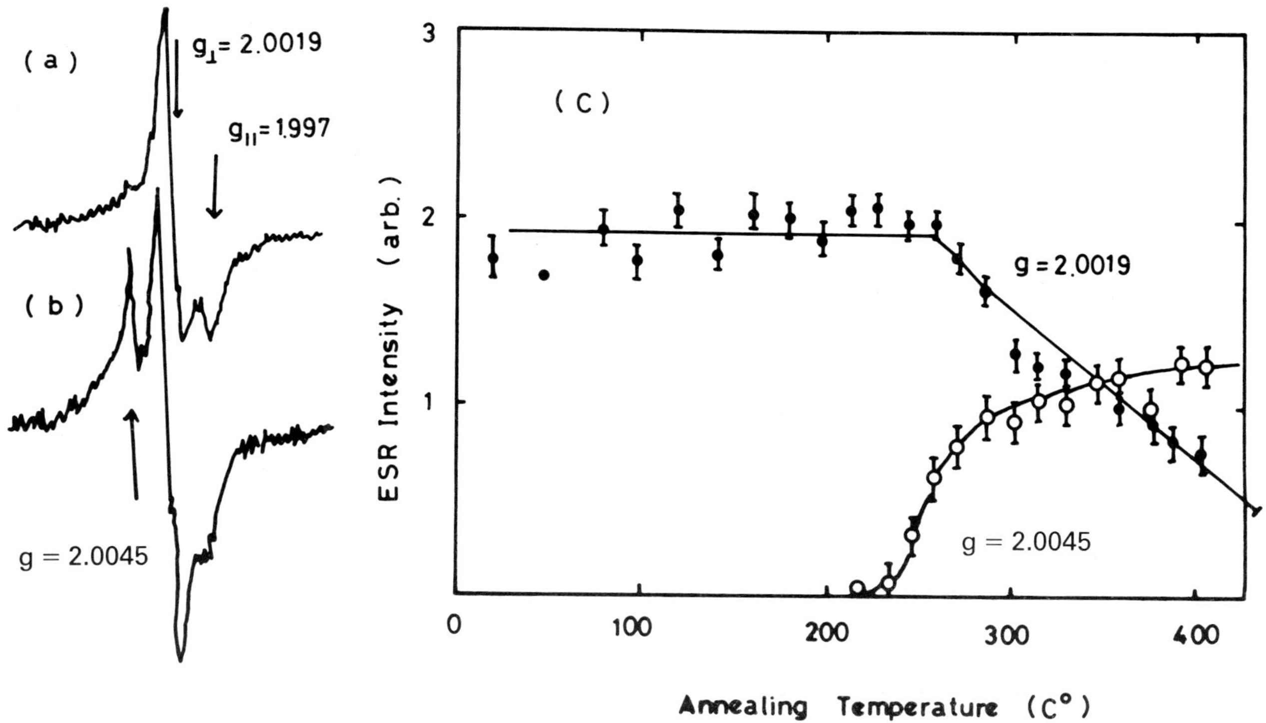

organic composition. The high TD explained by the high apparent annual dose of 1 rad/y is due to organic signals.

It should be noted that a burnt bone gives an ESR signal due to the presence of carbon colloids or organic radicals produced by insufficient oxidation of the organic part of the bone; this effect results in a falsely elevated TD. Figure 3.28a and 3.28b show an ESR signal of a fossil bone and one heated at 250°C. The new signal appearing at g.0045 is related to organic carbon; the position is very close to $g_{\parallel} = 1.997$ and $g_{\perp} = 2.0019$ associated with defects in bones. Figure 3.28c shows thermal annealing studies of ESR signals for a bone in an evacuated quartz cell. The results clearly indicate the appearance of a new signal by heating the bone. Chemiluminescence is also observed at this temperature.

The procedures required to obtain the palaeo dose (PD) or TD of fossil bones are described in the section on ESR dating (pp. 67ff). It must be noted that too high a dose rate of artificial irradiation reduces the defect production efficiency due to radiation heating and the high density effect. The interactions between defects during irradiation affect the defect production yield. This is demonstrated in Figure 3.29 for fossil bones from Dali, China. If the dose rate is 5×10^4rad/hr, then the TD obtained is larger than those for a dose rate below 10^4rad/hr. This is because the enhanced concentration is smaller at high dose rate than it should be, in agreement with the dose-rate dependence of TL response (Groom *et al.*, 1978). Thus, dose rates at much less than 10^4r/hr are being used. TDs of about 20krad are obtained by multiplying the roentgens by 0.9 to convert the absorption dose of a bone. This takes into account the mass absorption coefficient at an energy of 1.3MeV.

TD and Age Correlation

The TDs of several fossil bones of known origin have been obtained and compared with their anthropological ages. Figure 3.30 shows a logarithmic plot of the TD versus the known age of fossil bones for apparent annual radiation doses of 0, 0.2, and 1rad/yr, indicated by dashed lines with a slope of 1.0. The open circles indicate bones from limestone caves, where the environmental radiation is considered to be low. Nearly 0.1rad/y seems to hold for such samples. The black dots indicate bones from fertile soils; an annual dose rate of about 1.0rad/y seems appropriate here. Thus, one can get a rough idea about the apparent dose rate for different environments.

The radiation assessment is not straightforward if the concentration of uranium elements increases. Samples from the same excavation site give different TDs due to differences in the local environment, as observed at Arago or Petralona. The age information given by the supplier of the sample, based on studies of fauna or sediments, has been used for plotting

Figure 3.29: Enhancement of ESR Signals of Bones at Dali, China, for Different Dose Rates of γ Rays from the Source of Co^{60}. The production yield at a high dose rate is low, giving an erroneously large TD. A dose rate below 10^4R/hr is preferable, to avoid such high-dose rate effects

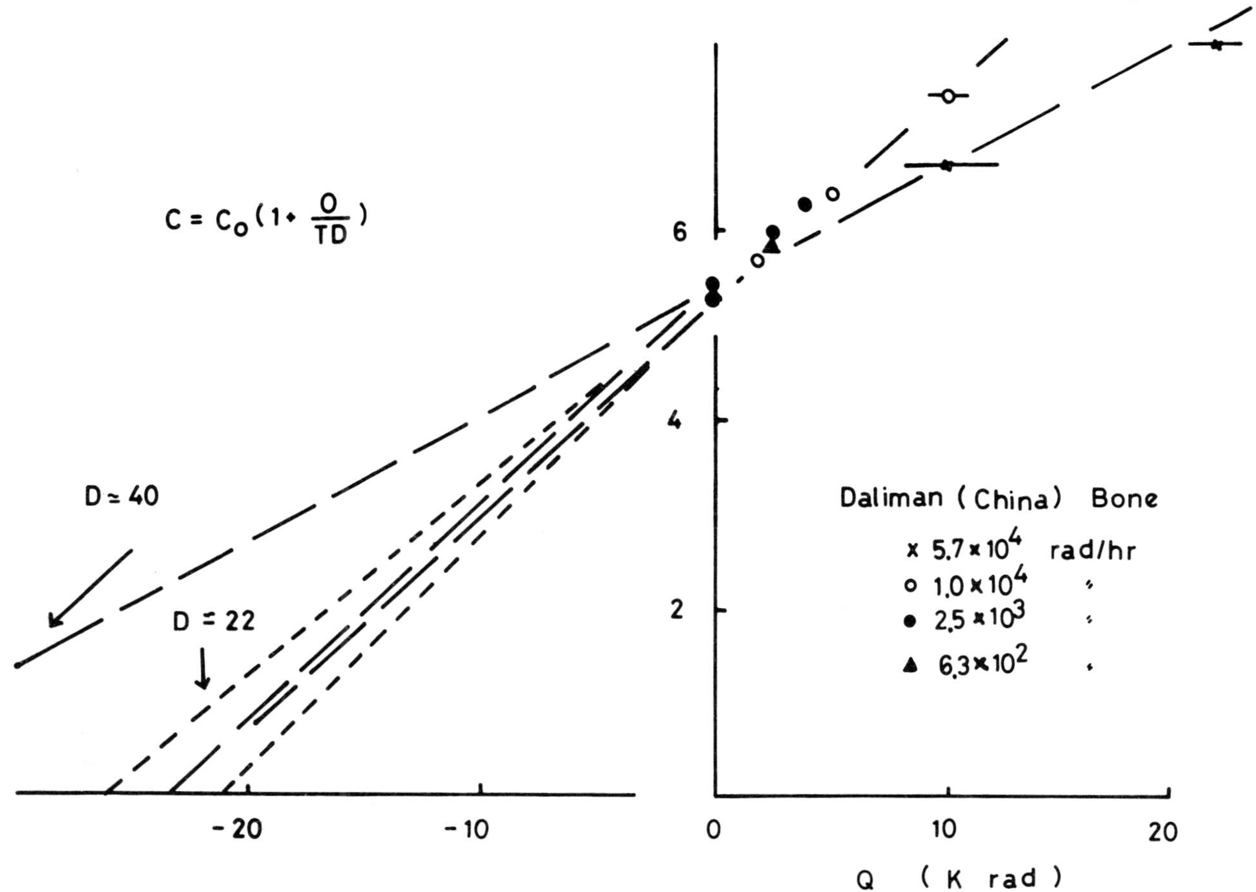

Figure 3.30: Logarithmic Plot of the TDs of Animal and Human Bones, as a Function of the Age (T), Known Historically or Determined by Other Methods. Apparent annual dose rates of 0.1, 0.2 and 1 rad per year are indicated by dashed lines

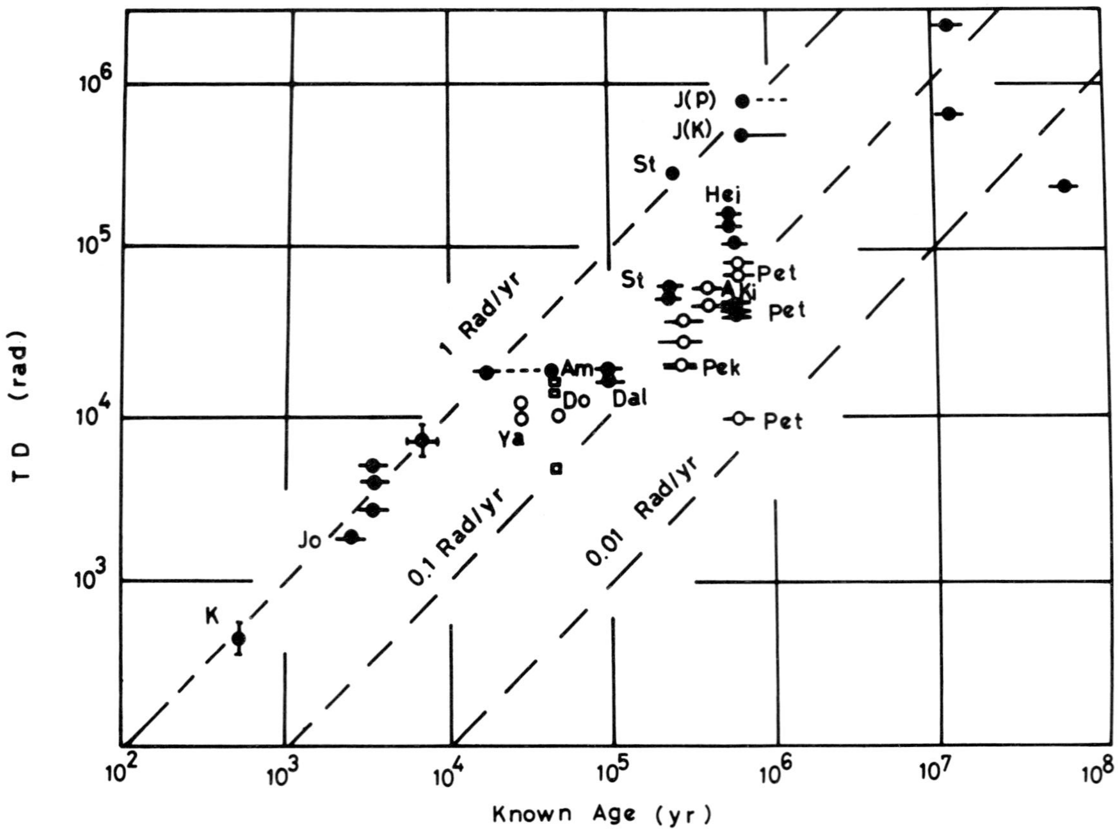

Figure 3.29: these suggested ages are not necessarily correct, as the ages seem to be biased toward older ages.

An apparent high dose rate of 1rad/y for young bones (Ikeya and Miki, 1980b; Mascarenhas, Filho and Ikeya, 1982) would be partly due to the incorporation of Ra^{226} from the environment. Since Th^{230} has not yet been built up, the dose rate from inherent U^{238} is small. The external γ ray dose rate is generally 0.1–0.4rad/y, at most. Ra^{226} and its daughters may create radicals up to the apparent dose of 1rad/y until about five times the half-life of Ra^{226} ($T_{1/2}$ = 1,860y). A high accumulation of $Ra^{226}/U^{238} \cong 2.0$ in fossil bones has been reported recently in γ ray spectroscopy studies (Komura, 1982). Thus, detailed study of radioactive elements becomes necessary, including the behaviour of leaching-out and incorporation. Precaution is required not to include the slightly broad ESR signal of organic radicals at g = 2.0045 ($\triangle H$ = 2mT) in the bone. This effect might be included for Holocene bones with the high apparent dose rate.

Models of Uranium Accumulation in Bone

Accumulation of radioactive elements in fossil bones makes evaluation of a past radiation dose difficult. It has been suggested that uranium accumulates from the environment rapidly in the first stage and then increases slowly (Seitz and Taylor, 1974). It might be considered that uranium has accumulated linearly at a constant rate for fossil bones older than 10,000 years, as the rough ages are estimated from the content of uranium. Then the time-dependent annual dose rate can be calculated considering the disequilibrium of Th^{230}. The concentration of uranium may be written as in equation 3.30:

$$C_U(t) = (C_U/T)t$$

where C_U is the present concentration of uranium in ppm and T is the age of the bone.

The increment of the total dose for the linear uptake of uranium, $\triangle TD_A$, by the increment of uranium concentration $\triangle C_U$ at time t is:

$$\triangle TD_A = \triangle C_U(t) \cdot TD(T\text{-}t)$$

where TD is the total dose as a function of time, taking Th^{230} build-up as is shown in equation 3.12. Thus, equation 3.31 is derived:

$$TD_A = \int_o^T TD\,(T - t)\,\frac{dC\,(t)}{dt} \cdot dt$$

$$= \frac{C_U}{T} \int_o^T TD\,(T - t)\,dt$$

$$= \frac{(C_U)}{T} \cdot \int_o^T TD\,(t)\,dt$$

Using equation 3.12, the TD for a bone with a linear accumulation of uranium can be calculated as shown in equation 3.32:

$$TD_{ac} = \frac{1}{2} DT + D_{34} \frac{r_o - 1}{\lambda_{34}} \left[1 - \frac{1 - e^{-\lambda_{34} T}}{\lambda_{34} T}\right]$$
$$-D_{30}'' \left\{ \frac{1}{\lambda_{30}} \left[1 - \frac{1 - e^{-\lambda_{30} T}}{\lambda_{30} T}\right] \right.$$
$$\left. - \frac{r_o - 1)}{\lambda_{34}} \left[1 - \frac{\lambda_{30}^2 \,(1 - e^{-\lambda_{34} T}) - \lambda_{34}^2 \,(1 - e^{-\lambda_{30} T})}{\lambda_{30}\lambda_{34}\,(\lambda_{30} - \lambda_{34})\,T}\right]\right\}$$

where the symbols have the same meaning as equation 3.12. Radiation equilibrium for Th^{232} series elements is assumed. TD for α and β rays are calculated with equation 3.32 as TD_{in} caused by the internal radiation dose. The external γ ray dose rate (D_{ex}) is measured by a thermoluminescent dosimeter (TLD).

A computer program based on this model of the linear accumulation of uranium is developed to calculate the age, T_{ac}, from the experimental TD, as is shown for the bone excavated at Mauer, Heidelberg in Figure 3.31. The ages obtained are shown in Table 3.5, together with others, (T_d) based on the uranium content in the fossil bone being constant from the beginning, assuming the k value of 0.15 and uranium-series disequilibrium. Unfortunately, available data on the external dose rate (D_e) and the uranium content for the fossil bones are limited. Therefore, a tentative age estimate was made on the data from the literature. Efforts are now going on to get a reliable measure of the uranium content in the fossil bone and to get the external dose rate at the site. If the content of U^{238} is sufficiently high, the main contribution is from the internal radiation of α and β rays. The external dose rate (D_e) variation from 50mR to 300mR does not affect the age of bones with 10 ppm of U^{238}.

Figure 3.31: Relationship Between the Total Dose of Natural Radiation (TD) and the Age (T) for a Fossil Bone at Mauer, Heidelberg, Using an External Dose Rate of 50mrad/y. The age can be obtained graphically or by a computer fitting, for the case of uranium-series equilibrium (T_e) and disequilibrium (T_d) using equation 3.12, and for the linear accumulation of uranium (T_{ac}) using equation 3.32

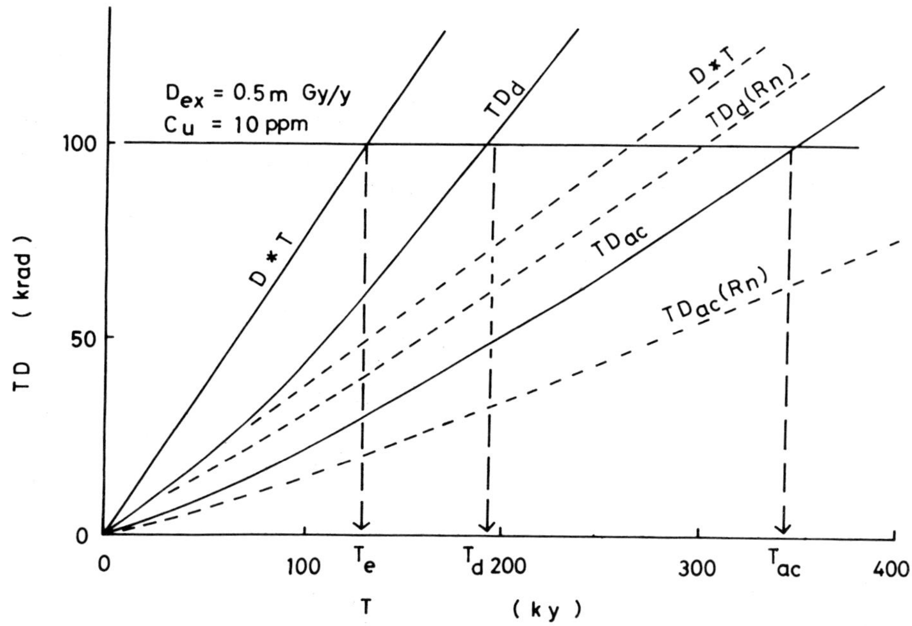

Standard Fossil Bones

The TDs of fossil bones from the following anthropologically or palaeontologically important sites have been determined by ESR in the author's laboratory as standard fossil samples.

Choukoutien. Choukoutien is well known as the site where *Homo erectus pekinensis* was excavated in 1929. The age of Peking man is between 200,000 and 500,000 years. The Science Academy of China donated to the author's laboratory 16 pieces of fossil bones from Choukoutien, including some excavated in 1934.

Table 3.5: Tentative ESR Ages for Anthropologically Important Bones: Ages for Disequilibrium of U^{238} Disintegration Series (T_d) and for a Linear Accumulation of U^{238} (T_{ac}) for No Radon Loss and 50 per cent Loss of Radon Without the Effect of Fluorination

Sample name	U^{238} (ppm)	Th^{232} (ppm)	K_2O (%)	D_{ex} (mrad/y)	TD (ESR) (krad)	T_d (ky)	T_{ac} (ky)	T_{ac}(50%) (ky)	Comment (personal communication and reference)
Choukoutien									D_{ex} estimated
Bone Ly 4					23				
Ly 6	14				40	88	155	179	
Ly 10	28.5	0.6	0.16	50?	42	54	97	110	U contents by
Horse molar	32.0	0.5	0.4	50?	80	81	148	189	rays (Komura, 1982)
Petralona									a. Hennig *et al.*,
Cranium	0.42(ft)[a]	0.3[a]	0.014[a]	170[a]?	27	142	169[a] 147	152	1981 too high D_{ex}
				53[b]		351	421	438	b. At undisturbed site
				72[c]		287	327	336	c. Recent D_{ex} by author's
Bone				200[d]	86[d]			430?	TLD at the site of cranium
				25[d]	10[d]			400?	d. Rough estimation by
Carbonate with trace of fire	D = 450 mrad/y (Liritz)				210	T = 430ky			Ikeya and Miki (1981 a, b) Ikeya and Poulianos (1979)
Arago, Tautavel									U content by Yokoyama,
III–G–134(ISG2)	36.4			60	176	137	215	299	1982, D_{ex} measured (TLD)
III–G–3308	20?			98	90	121	209	241	in CNRS report
I–K	9.9			98	94	191	327	397	
	22.6			80?	51	73	127	144	

Heidelberg, Mauer							Samples by Kraatz, 1980
Bone M_3	10 (L)	50	95	240	335	455	ft by Hashimoto (1982,
	26 (ft)			110	216	254	personal communication)
Bone 5m under	10 (L)	50	145	303	542	659	new method; accuracy?
Lettenbank	45 (ft)			102	187	221	D_{ex} by Aitken (TLD, personal communication) L by Oakley, Campbell and Moleson (1971)
Steinheim	13 (L)	50?	60	126	216	260	Samples by Adam (1976)
	37.5 (ft)		60	59	105	123	L by Oakley *et al.* (1971) ft by Hashimoto: see relative ages with Mauer bone
Java	65.9 (ft)			222	415	512	Samples by Matsu'ura
Kabuh	37.5 (L)	100?	600	341	635	795	(personal communication)
Pucangan	35 (L)	100?	900	506	952	1212	ft by Hashimoto L in Jacob (1975)
Amud	7.8 (L)	100?	20	65	98	106	by Sakanoue (1974, personal communication) 28 ky (Th/U) 40 ky (?)
Yamashitacho	15 (ft)	50?	10.5	28.5	49	52	ft by Matsu'ura
	17 (ft)	50?	12	29.1	51	54	(personal communication) C–14 age: 32ky

Notes:
Numerical values indicated by ? are only tentative, taken from the literature or the author's estimation. Revision of the ages may be done after accurate estimation of uranium content and fluorination effect. The α ray efficiency (k value) of 0.15 used tentatively may still be an overestimation for a locally high content of uranium in the fossil; the age is therefore underestimated. ESR ages are generally younger than those suggested by the supplier anthropologists, except for Amud and Yamashita-cho in which T_d seems better. Fluorination effect is as great as a model of uranium accumulation as ESR signal intensity is reduced. (ft): the content of U^{238} determined by fission tracks.

The TDs of deer bones and ivory at location 1 are 40 and 42krad, respectively. The lower jaw and tooth of a deer at locations 3, 4 and 18 show TD = 30krad. Horse molars covered with brown soil give a TD of about 80krad. Environmental radiation is being studied now. Secondary ion mass spectroscopy (SIMS) studies have indicated that the content of uranium is much higher and non-uniform in these horse molars than in solid bones. Considering the age of Choukoutien, the apparent annual dose rate of about 0.1rad/y in a limestone cave seems appropriate for the explanation of the age.

The Low Level Radiation Laboratory of Kanazawa University has determined U^{238}, Th^{232} and K contents in these samples with low background γ ray spectroscopy (Komura, 1982). The model of a continuous uptake of U^{238} and Th^{230} disequilibrium gives a still somewhat young age of around 190,000 years at location 1 assuming the rate of radon loss as 50%.

Mauer, Heidelberg. The lower jaw of *Homo erectus heidelbergensis* was discovered in 1907 at Mauer, Heidelberg. Figure 3.31 shows TDs for several bones and the approximate ages for D=0.2rad/y. The age is supposed to be around 700,000 to 500,000 years (Mindel-Günz Era). The Institute of Geology and Palaeontology in the University of Heidelberg has supplied Ikeya with animal bones excavated in 1930, 1935 and 1972.

The TDs obtained are approximately 100krad for bones at level M_3, and 140krad for bones 5m under Lettenbank. The environmental radiation has been measured with TLD by Aitken at Oxford as being around 50mrad/y (personal communication). The tentative age of the Mauer bones, assuming 10ppm of U^{238}, is given in Table 3.5. The apparent annual dose rate is about 0.15–0.2rad/y. Results based on U^{238} content by fission tracks on mica, though higher than expected, gave 240,000 years. The TD versus T relation is shown in Figure 3.30 (Ikeya, 1982b).

Steinheim on Murr. The Steinheim skull, supposed to be the same age as the Swancombe skull in England, belongs to the Riss-Mindel period, and is approximately 200,000 to 250,000 years old (Adam, 1976). The TDs of solid bones are about 50–60krad, nearly half that of the Mauer (Heidelberg) bones. One deviation of TD of 300krad has been obtained for an elephant skull which is porous, sponge-like, soft and chalky, and has remnant organic materials. The SIMS study indicates that the contents of U^{238} and UO_2 are an order of magnitude higher than for a solid piece of bone. A rough estimation of the age for the U^{238} content of 13ppm (Oakley, Campbell and Molleson, 1971) is shown in Table 3.5.

Petralona. ESR dating in palaeoanthropology started with material from Petralona cave, located 50km from Salonika, northern Greece (Ikeya, 1977, 1978a). The author has visited the excavation site three times and selected bone and stalagmite samples. The TDs of bones at several layers ranged from 10krad to 80krad. The site was first considered as Würm and then middle Pleistocene (Poulianos, 1971), and finally as Gunz-Mindel Interglacial (Poulianos, 1977). Aging of the Petralona skull from 70,000 to 700,000 years has caused a controversy over the real age of the Petralona cranium. As the author is neither an anthropologist nor a chronologist, comment upon the morphology of the cranium or stratigraphic or faunal dating cannot be made. However, the TD and ESR age of animal bones at the site can be presented.

The bone showing a high TD of 86krad was located in brown soil sediments. Since the dose rate was about 200mrad/y, the age of the bone would be approximately 430,000 years. A piece of bone covered with shiny and transparent crystalline calcite gave a TD of about 10krad. The latter dose rate would be much lower than 0.1rad/y. These results indicate that the Petralona bones are certainly much older than those of the Würm (10,000 to 70,000 years) (Ikeya and Miki, 1981a).

The age of the Petralona cranium was directly assigned to 200,000 years with ESR dating of the carbonate encrustation on the cranium (Hennig *et al.*, 1981). However, the high external radiation dose of 190mrad/y indicates that there may be some overestimation, possibly due to radiation from soils caused by the excavation. The external radiation is usually around 50–80mrad/y at Petralona cave (Ikeya, 1978b; Liritzis and Poulianos, 1980). Using this dose rate for the carbonate encrustation leads to an age of around 400,000 years, using the TD by Hennig *et al.* (1981) with ESR. Assessment of the annual dose rate for the trace of fire at Petralona is 450mrad/y (Liritzis and Poulianos, 1980). The TD of 210krad for the carbonate with the trace of fire (Ikeya and Poulianos, 1979) gives an age of 450,000 years. The Petralona cave dating controversy (Hennig *et al.*, 1982; Ikeya, 1982a; Liritzis, 1982; Poulianos, 1982) has not yet been resolved. Recent re-measurements of D_{ex} by TLD gives 72mrad/y at the excavated cranium site, and 53mrad/y at the ordinary site.

Arago. A systematic study of materials excavated from Arago, Tautavel, France, has been undertaken by the Institute of Human Palaeontology at the National Museum of Natural History in Paris, and also by several laboratories of chronology science in Europe, the United States and Japan (Lumley, 1979).

Preliminary measurements show that the TDs are approximately 40krad for bones, and 100krad for teeth. These should be compared with samples from a limestone cave rather than with those of Steinheim and Heidelberg. Although several investigators measured contents of radioactive elements in bones and cave travertines, refined discussion on the radiation exposure is difficult because of a complex radiation environment at Arago cave. ESR ages using results by Yokoyama (Yokoyama *et al.*, 1982) are shown in Table 3.5 and compared with those of Mauer bones (Ikeya, 1982b).

Java. Java is well known as the site of *Homo* (formerly *Pithecanthropus) erectus*. The upper formation of the Kabuh is 40m thick, while the Pucangan formation has an 80–100m thick sediment in the Sangiran area. K-Ar dates for tektite or other volcanic stones in Kabuh formation range from 710,000 BP to 830,000 years BP (Nishimura, 1981). The K-Ar age of pumice in the Pucangan formation was reported as 1.9 + 0.4 million years old (Jacob, 1975). Unfortunately, the TLD elements were lost at the excavation site.

The TDs are about 600krad for bones at the Kabuh layer and 900krad at the upper level of the Pucangan layer. The high dose would be due to a high concentration of U^{238}, which averages 32–35 ppm of U^{238} (Jacob, 1975). Deduced ages are 550,000 and 810,000 years, respectively, for the Kabuh and Pucangan bones. The internal dose was about seven times the external dose, tentatively assumed to be 200mR/y.

Amud, Douara and Yamashitacho. The TDs of bones from Amud and Douara in Syria and Yamachitacho in Japan are shown as examples of ESR dating of relatively young bones (Sakanoue and Yoshioka, 1974; Akazawa, 1976; Matusu'ura and Ueta, 1980). In all cases, the ages obtained following the model of a linear accumulation of U^{238} seem to be slightly older than those expected by anthropologists. The age, T_d, for the disequilibrium consideration seems appropriate in these cases, suggesting a possible rapid accumulation of U^{238} to a certain amount in a short time.

Summary of Bone Dating

The annual dose rate based on radiation equilibrium or disequilibrium gave ESR ages of standard fossil bones too young to be accepted by palaeoanthropologists. Bones accumulate uranium from the environment: the annual dose rate in the past is not the same as that estimated from the present concentration of radioactive elements. A model of a linear uptake of uranium for an open system is created for bones older than 10,000 years, taking into account the disequilibrium of U^{238}-series disintegration. The results give reasonable ages, though still younger than those expected. The loss of radon (Rn^{222}) for a powdered bone was 15–20 per cent. If the loss of Rn for a bone at a buried site is evaluated, the ages shift to an older value.

Change of bone structure may change the sensitivity of defect formation by radiation. It may even recrystallise partly in the environment, giving a young age, due to recovery from natural radiation damage. More systematic work on the assessment of radiation and on physical properties of bones are necessary to establish the method of ESR dating of fossil bones. The results shown in Table 3.5 are preliminary, based on a speculative radiation assessment. The effect of Ra^{226} incorporation at the initial stage and fluorination are being studied in the author's laboratory. Fluorination reduces the ESR signal intensity, thus giving the young ages in Table 3.5. A final revision of ESR ages for bones must be made in the near future.

Miscellaneous Materials

Fossil Woods and Amber

Silicified materials of petrified wood show a clear ESR signal of defects, which come from an electron at the oxygen vacancy (E′ centre) in SiO_2, as shown in Figure 3.31 (Ikeya, 1982a). Organic radicals of $CH_3\cdot$ and $CH_2\cdot$ show a triplet (three-line) and a quartet (four-line) spectrum with intensity ratios of 1:2:1 and 1:3:3:1, respectively. This is due to a magnetic

Figure 3.32: ESR Spectrum of Petrified Wood with Typical Organic Radicals. Three types of signals are observed: a signal at the central part; a triplet signal with the intensity ratio of 1:2:1; and a quartet with the ratio 1:3:3:1. The latter two signals are associated with $CH_2COOH\cdot$ and $CH_3\cdot$ radicals, respectively. Signals associated with defects in silicate are observed in (b)

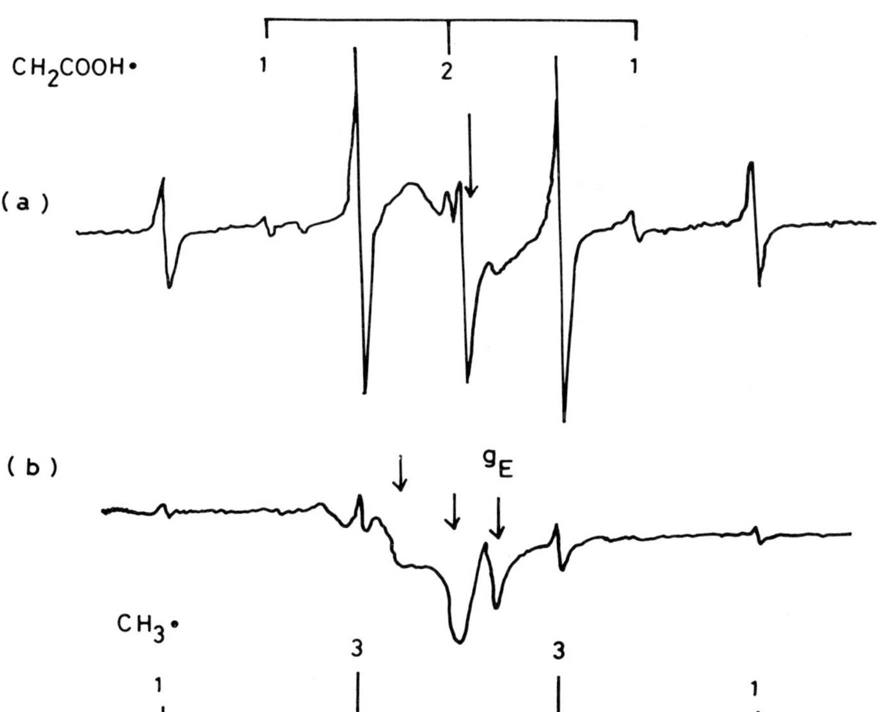

field caused by the nuclear spin states of protons. This is called hyperfine (hf) splitting by protons. Similar hf five-line (quintet) spectra, with an intensity ratio of 1:4:6:4:1, are associated with alanine radicals ($CH_3CHCOOH\cdot$) in tooth enamels and shells (Ikeya, 1981).

The apparent TD obtained by additive irradiation cannot be correlated with age unless the chemical process of silicification is known. The signal in a modern wood is, in some cases, so intense as to give a TD of a few krad (zero age): thus methods of reducing the zero age must be developed. The same can be said for ESR signals in amber and fossilised tree resin (Figure 3.32). Phytolioth (plant opal) is also an object of ESR studies.

Feathers, Hair, Wool and Mummy

Bird feathers, hair, wool and mummy tissues also give ESR signals due to lipid, pigment or keratin decomposition. Insects also give a huge ESR signal from radicals, which gives too large a 'zero age' from the beginning. Radicals in wool are studied in connection with yellowing of keratin and collagen by UV or high-energy irradiation (Smith, 1976). The TD equivalent to light illumination may be obtained by ESR. The dating method for potato chips (Figure 3.8) may be extended. It is still premature to conclude whether or not organic materials can be dated with ESR. Alanine radicals found in shells and tooth enamels (Ikeya, 1981) are sufficiently stable. These may be useful for assessing burial conditions or to trace geological or geographical origin.

Cave Deposits; Stalactites and Stalagmites

Secondary deposits in a limestone cave such as stalactites, stalagmites and cave floors (Ikeya, 1975, 1976; Apers *et al.*, 1980) are not biological materials. However, dating of these carbonates allows establishment of age limits for the occupation of the site by prehistoric man. Systematic studies on the dating of Arago and Petralona stalagmite materials by several methods are being carried out (Skinner, 1983).

Minerals

A vast area of ESR dating in geology is still open though it is outside the scope of this book. Zeller (1968) measured 20 minerals with ESR and TL and found that apatite and zircon show natural radiation damage. Mechanical destruction of natural radiation damage in quartz at fault gouge by slickensiding motion is utilised to date the age of geological fault formation (Ikeya, Miki and Tanaka, 1982). ESR dating using α recoil defects in ancient flints by calibrating with neutron irradiation may break the upper age limitation (Garrison *et al.*, 1981). Application of ESR dating to geothermal fields has recently started (Ikeya, 1983b).

Conclusion

Dating with electron spin resonance (ESR) has now been established as a new method in chronology science. Typical examples using some standard fossils of known origins are described to demonstrate the usefulness of this method, and will be helpful to those who start ESR dating. Further research on different objects from different viewpoints will bring new insights and also new problems to ESR dating. The accuracy of age determination will be increased with separate measurements of radiation.

Readers may have noted that ESR dating for a limited number of objects is not as accurate as C^{14} dating. This is because ESR dating has started recently and at present there is very little reported work available. Limited data from the laboratory in Yamaguchi University are presented to facilitate refinement of the method by other researchers with different backgrounds. Collaborative studies between physicists (or chemists) and archaeologists or geologists will make this dating technique complementary to C^{14} dating by covering the open time range between the old limit of C^{14} dating (50,000 years) and the lower limit of K-Ar dating.

This chapter, especially its latter part, must be completely rewritten in a few years to keep up with developments in the late 1980s. The author hopes that this chapter, the first description of ESR dating in a book of dating techniques, can stimulate the start of new projects in chronology science and be a source of information. A complete bibliography should be seen in the symposium Proceedings 'ESR Dating and Dosimetry'.

Acknowledgements

I wish to thank the anthropologists, archaeologists and geologists who supplied valuable samples for the development of this new technique of ESR dating. This work has been supported partly by a Grant-in-Aid for Specific Science 'Dating' from the Ministry of Education (1976–1978 and 1980–1982). Financial support by the Yamada Science Foundation, Asahi Newspaper Co. and the Yoshida Science Foundation is also appreciated. Part of this chapter is derived from lectures given to graduate students at Nagoya University (1979, 1982) and at the University of São Paulo (1980). The stay in Brazil was supported by the Research Foundation of the State of São Paulo. A Fellowship from the Alexander von Humbolt Foundation (1977–78, 1979) has enabled the collection of samples in Europe. The author wishes to thank his coworkers, T. Miki, R. Suehiro, T. Fujimoto and K. Ohmura for ESR measurements; Mr N. Kash, Professor W. Zimmerman and Dr M. Zimmerman for reading the manuscript; and Ms M. C. Kattoh for typing the manuscript.

References

Adam, K. D. (1976) 'Quartarforschung am Staatlichen Museum for Naturkunde in Stuttgart', *Stuttgarter Beitrage zur Naturkunde, 167*, pp. 1–14

Aiken, M. J. (1974) *Physics and Archaeology*, 2nd ed., Clarendon, Oxford

Akazawa, T. (1976) 'New Data on Middle Palaeolithic Assemblages from Douara Cave', *Bull. National Science Museum D (Anthropology), 2*, pp. 1–24

Apers, D. J., Debuyst, R., Dejehet, F. and Lombard, E. (1980) 'A Propos d'un Essai de Datation par RPE de Concretions Calcaires Originaires de Grottes Belges', *Radiochem. Radionanal. Lett., 45*, pp. 427–40

Bell, W. T. (1976) 'The Assessment of the Radiation Dose-rate for Thermoluminescence Dating', *Archaeometry, 18*, pp. 107–11

—— (1979) 'Thermoluminescence Dating: Radiation Dose-rate Data', *Archaeometry, 21*, pp. 243–5

Brown, W. E. and Chow, L. C. (1976) 'Chemical Properties of Bone Mineral', *Ann. Rev. Mater. Sci., 6*, pp. 213–35

Carrington, A. and McLachlan, A. D. (1967) *Introduction to Magnetic Resonance*, Harper and Row, New York

Cevc, P. and Schara, M. (1972) 'Electron Paramagnetic Resonance Study of Irradiated Tooth Enamel', *Radiat. Res., 51*, pp. 581–9

Charalambous, S. and Papastefanous, C. (1977) 'On the Radioactivity of Fossil Bones', *Nucl. Instr. Methods, 142*, pp. 581–8

Christodoulides, C. and Fremlin, J. H. (1971) 'Thermoluminescence of Biological Materials', *Nature, 232*, pp. 257–8

Cole, T. and Silver, A. N. (1963) 'Production of Hydrogen Atoms in Teeth by X-irradiation', *Nature, 200*, pp. 700–1

Daniels, F., Boyd, C. A. and Saunders, D. F. (1953) 'Thermoluminescence as a Research Tool', *Science, 117*, pp. 343–9

Debenham, N. C. (1983) 'Reliability of Thermoluminescence Dating of Stalagmite Calcite', *Nature, 304*, pp. 154–6

Doi, Y., Aoba, T., Okazaki, M., Takahashi, J. and Moriwaki, Y. (1979) 'Analysis of Paramagnetic Centers in X-ray-irradiated Enamel, Bone and Carbonate Containing Hydroxyapatite by Electron Spin Resonance Spectroscopy', *Calcif. Tissue Int., 28*, pp. 107– 12

Duchesne, J., Depireux, J. and van der Kaa, J. M. (1961) 'Origin of Free Radicals in Carbonaceous Rocks', *Geochimica et Cosmochimica Acta, 23*, pp. 209–18

Fleming, S. (1976) *Dating in Archaeology*, J. M. Dent and Sons, London

Fleisher, R. L., Price, P. B. and Walker, R. M. (1975) *Nuclear Tracks in Solids*, Berkeley, Los Angeles

Garrison, E. G., Rowlett, R. M., Cowan, D. L. and Holroyd, L. V. (1981) 'ESR Dating of Ancient Flints', *Nature, 290*, pp. 44–5

Groom, P. J., Durrani, S. A., Kazal, K. A. R. and McKeever, S. W. S. (1978) 'The Dose Rate Dependence of TL Response and Sensitivity in Quartz', *PACT., 2*, pp. 200–10

Hennig, G. J., Herr, W., Weber, E. and Xirotiris, N. I. (1981) 'ESR Dating of the Fossil Hominid Cranium from Petralona Cave, Greece', *Nature, 292*, pp. 533–6

——, Weber, E. and Xirotiris, N. I. (1982) 'Petralona Cave Dating Controversy', *Nature, 299*, pp. 282–4

Ikeya, M. (1975) 'Dating a Stalactite by Electron Paramagnetic Resonance', *Nature, 255*, pp. 48–50

—— (1976) 'Natural Radiation Dose in Akiyoshi Cavern and on Karst Plateau', *Health Physics, 31*, pp. 76–8

—— (1977) 'Electron Spin Resonance Dating and Fission Track Detection of Petralona Stalagmite', *Anthropos, 4*, pp. 152–68

—— (1978a) 'Electron Spin Resonance as a Method of Dating', *Archaeometry, 20*, pp. 147–58

—— (1978b) 'Spin Resonance Ages of Brown Rings in Cave Deposits', *Naturwissenschaften, 65*, p. 489

—— and Poulianos, A. (1979) 'ESR Age of the Trace of Fire at Petralona', *Anthropos, 6*, pp. 44–7

—— (1980) 'ESR Dating of Carbonates at Petralona Cave', *Anthropos, 7*, pp. 143–51
—— and Miki, T. (1980a) 'A New Dating Method with a Digital ESR', *Naturwissenschaften, 67*, pp. 191–2
—— and Miki, T. (1980b) 'Electron Spin Resonance Dating of Animal and Human Bones', *Science, 207*, pp. 977–9
—— and Miki, T. (1980c) 'Present Status of ESR Dating: Reply to Nambi's Comment', *Jap. J. Appl. Phys., 19*, pp. 1809–10
—— (1981) 'Paramagnetic Alanine Molecular Radicals in Fossil Shells and Bones', *Naturwissenschaften, 68*, pp. 474–5
—— and Ohmura, K. (1981) 'Dating of Fossil Shells with Electron Spin Resonance', *J. Geology, 89*, pp. 247–50
—— and Miki, T. (1981a) 'ESR Archaeological Dose of Petralona and Choukoutien Bones', *Anthropos, 8*, pp. 95–106
—— and Miki, T. (1981b) 'Archaeological Dose of Bones at Anthropologically Important Sites with Electron Spin Resonance', *Union Intern. Ciencias Prehist. Protohist., Xth Congreso (Comission II) Mexico.*, pp. 39–53 and pp. 21–38
—— (1982a) 'Electron Spin Resonance of Petrified Woods for Geological Age Assessment', *Jap. J. Appl. Phys., 21*, L28–30
—— (1982b) 'Petralona Cave Dating Controversy', *Nature, 299*, pp. 281–2
—— (1982c) 'A Model of a Linear Uranium Accumulation for ESR Age of Heidelberg (Mauer) and Tautavel Bones', *Jap. J. Appl. Phys. Letters, 21*, L690–L692
—— (1982d) 'Progress in ESR Dating of Fossils'., *PACT J., 9*, pp. 421–31
——, Miki, T. and Tanaka, K. (1982) 'Dating of a Fault by ESR on Intrafault Materials', *Science, 215*, pp. 1392–3
—— (1983a) 'A Trip to Interdisciplinary World: Electron Spin Resonance (ESR) Dating in Archaeology and Geology,' *JEOL News, 19A*, pp. 26–30
—— (1983b) 'ESR Studies of Geothermal Boring Cores at Hachobara Power Station', *Jap. J. Appl. Phys., 22*, 763–765
—— and Ohmura, K. (1983) 'Comparison of ESR Ages of Corals at Marine Terraces with C^{14} and Th^{230}/U^{234} Ages', *Earth Planet. Sci. Letters, 65*, pp. 34–8
—— (1984) 'ESR Age of Pleistocene Shells by Radiation Assessment', *Geochem. J. 18*, 11–17.
Ivanovich, M. and Harmon, R. S. (eds.) (1982) *Uranium Series Disequilibrium: Application to Environmental Problems*, Clarendon Press, Oxford
Jacob, T. (1975) *Catalogue of Fossil Hominids: Indonesia, Part III*, K. P. Oakely (ed.) British Museum, London, pp. 103–6
Jasinska, M. and Niewiadomski, T. (1970) 'Thermoluminescence of Biological Materials', *Nature, 227*, pp. 1159–60
Koberle, G., Terrile, C., Panepucci, H. C. and Mascarenhas, S. (1973) 'On the Paramagnetism of Bone Irradiated in Vivo', *An. Acad. Brasil Cienc., 45*, pp. 157–60
Komura, K. (1982) 'Dating by Non-Destructive Gamma Ray Spectroscopy', *Archaeology and Natural Science, 14*, pp. 1–24
Konishi, K., Schlanger, S. O and Omura, A. (1971) 'Neotectonic Rates in the Central Ryukyu Islands Derived from Th^{230} Coral Ages', *Marine Geology, 9*, pp. 225–40
Liritzis, Y. and Poulianos, A. N. (1980) 'A Radiation Survey of Petralona Cave' *Anthropos, 7*, pp. 252–9
Liritzis, Y. (1982) 'Petralona Cave Dating Controversy', *Nature, 299*, pp. 281–2
Lumley, H. de (1979) 'L'Origine de l'Homme Premiers Habitants de L'Europe', *L'Archaeologie, 36*, pp. 8–9
Mascarenhas, S., Filho, O. B. and Ikeya, M. (1982) 'ESR Dating of Human Bones from Brazilian Shell Mounds Sambaquis', *Am. J. Anthropology, 59*, pp. 413–17
Matsu'ura, S. and Ueta, M. (1980) 'Fraction Dependent Variation of Aspartic Acid Racemization Age of Fossil Bone', *Nature, 286*, pp. 883–4
McMorris, D. W. (1969) 'Trapped-electron Dating: ESR Studies', *Nature, 222*, pp. 870–1
Michels, J. W. (1973) *Dating Methods in Archaeology*, Seminar Press, Inc., New York
Miki, T. and Ikeya, M. (1978) 'Thermoluminescence and ESR Dating of Akiyoshi Stalactite', *Jap. J. Appl. Phys., 17*, pp. 1703–4
Miki, T. and Ikeya, M. (1980) 'Accumulation of Atmospheric Radon in Calcite Caves', *Health Phys., 39*, pp. 351–4

Molleson, T. I. and Oakley, K. P. (1966) 'Relative Antiquity of the Ubeidiya Hominid', *Nature, 209*, p. 1267
Morency, M., Emond, P. L. and Bitter, P. H. (1969) 'Dating Condonts Using Electron Spin Resonance', *Kansas Geol. Survey Bull., 199*, pp. 17–19
Nakata, T., Takahashi, T. and Koba, M. (1978) 'Holocene Emerged Coral Reefs and Sea Level Changes in Ryukyu Island', *Geogr. Rev. Japan, 51*, pp. 87–103
Nambi, K. S. V. (1979) 'On ESR Dating of Minerals', *Jap. J. Appl. Phys.*, 18, pp. 2319–20
Nishimura, S. (1981) 'On the Nuclear Track Dating of Tuffs and Volcanic Ash', *Nucl. Tracks, 5*, pp. 157–67
Oakley, K. P., Campbell, B. G. and Molleson, T. I. (1971) *Catalogue of Fossil Hominids: Part II Europe*, British Museum, London
O'Brien, M. C. (1955) 'The Structure of Color Centers in Smoky Quartz', *Proc. Roy. Soc.* (London), *231*, pp. 400–4
Omura, A. (1980) 'Uranium-series Age of the Hiradoko and Uji Shells Beds, Noto Peninsula, Central Japan', *Paleont. Society Japan Trans. Proc., 117*, pp. 247–53
Ostrowski, K., Goctawska, A. D. and Stachowics, W. (1980) 'Stable Radiation-induced Paramagnetic Entities in Tissue Minerals', *Free Radicals in Biology, 4*, pp. 321–44
Pake, G. E. (1962) *Paramagnetic Resonance*, Benjamin Inc., New York
Panepucci, H. and Farach, H. A. (1977) 'ESR Spectra of Quasirandomly Oriented Centers: Application to Radiation Damage Centers in Bone', *Medical Phys., 4*, pp. 46–8
Poole, C. O. and Farach, H. A. (1972) *The Theory of Magnetic Resonance*, Wiley and Sons, New York
Poulianos, A. N. (1971) 'Petralona — A Middle Pleistocene Cave in Greece', *Archaeology, 24*, pp. 6–11
—— (1977) 'Stratigraphy and Age of the Petralonian Archanthropus', *Anthropos, 4*, pp. 41–6
—— (1982) 'Petralona Cave Dating Controversy', *Nature, 299*, p. 281
Robins, G. V., Seeley, N. J., McNeil, D. A. C. and Symons, M. R. C. (1978) 'Identification of Ancient Heat Treatment in Flint Artefacts by ESR Spectroscopy', *Nature, 276*, pp. 703–4
Sakanoue, M. and Yoshioka, M. (1974) 'Uranium Series Dating of Bone Samples from Amud Cave Site', *Quat. Res. (Japan), 13*, pp. 220–4
Sato, T. (1981) 'Electron Spin Resonance Dating of Calcareous Microfossils in Deep-sea Sediment', *Rock Magnetism and Palaeogeophysics, 8*, pp. 85–8
—— (1982) 'ESR Dating of Planktonic Foraminifera', *Nature, 300*, pp. 518–21
—— (1983) 'ESR Studies of Planktonic Foraminifera', *Nature, 305*, p. 162
Seitz, M. G. and Taylor, R. E. (1974) 'Uranium Variations in a Dated Fossil Bone Series from Olduvai Gorge, Tanzania', *Archaeometry, 16*, pp. 129–35
Serway, R. A. and Marshall, S. A. (1967) 'Electron Spin Resonance Absorption Spectra of CO_3^- and CO_3^{---} Molecule Ions in Irradiated Single Crystal Calcite', *J. Chem. Phys., 46*, pp. 1949–52
Silk, E. C. H. and Barnes, R. S. (1959) 'Examination of Fission Fragment Tracks with an Electron Microscope', *Phil. Mag., 4*, pp. 970–1
Slichter, C. P. (1963) *Principle of Magnetic Resonance*, Harper and Row, New York
Skinner, A. F. (1983) 'Overestimation of Stalagmite Calcite ESR Dates due to Laboratory Heating', *Nature, 304*, pp. 152–4
Smith, G. J. (1976) 'Effect of Light at Different Wavelengths on ESR in Wool', *Textile Res. J., 46*, pp. 510–2
Uebersfeld, J., Etienne, A. and Combrisson, J. (1954) 'Paramagnetic Resonance: A New Property of Coal-like Materials', *Nature, 174*, p. 614
Wintle, A. G. (1978) 'A Thermoluminescence Dating Study of Some Quaternary Calcite', *Can. J. Earth Sci., 15*, pp. 1977–86
—— and Huntley, D. J. (1983) 'ESR Studies of Planktonic Foraminifera', *Nature, 305*, pp. 161–2
Yokoyama, Y., Nguyen, H. V., Quaegebeur, J. P. and Poupeau, G. (1980) 'Some Problems Encountered in the Estimation of Annual Dose-Rate in the Electron Spin Resonance Dating of Fossil Bones', *Proc. 1980, Specialist Seminar on TL Dating, PACT J., 6*, pp. 103–15
——, Quaegebeur, J. P., Bibron, R., Leger, C., Nguyen, H. V. and Poupeau, G. (1982)

'Electron Spin Resonance (ESR) Dating of Tautavel Man's Site', *C. R. Acad. Sci. Paris, 294*, pp. 759–64

Zeller, E. J., Levy, P. W. and Mattern, P. L. (1967) 'Geological Dating by Electron Spin Resonance', *Proc. Symposium, Radioactive Dating and Method of Low Level Counting*, pp. 531–40

—— (1968) 'Use of Electron Spin Resonance for Measurement of Natural Radiation Damage' in D. J. MacDougall (ed.), *Thermoluminescence of Geological Materials*, Academic Press, London, pp. 271–9

Further references up to early 1986 are in the bibliography of the 'Proceedings of the first International Symposium on ESR Dating' held at Yamaguchi University in 1985. Available as *ESR Dating and Dosimetry*.

4 ANCILLARY TECHNIQUES

a. DENDROCHRONOLOGY

Jeffrey S. Dean

Many ingenious techniques have been devised for the temporal placement of non-calendrically dated phenomena, but only one of these is capable of providing absolute dates in terms of actual calendar years (Smiley, 1955). That method of absolute time placement is dendrochronology, the study of annual growth layers in trees. The purpose of this chapter is to discuss dendrochronology as a tool for the dating of past events. Dendroclimatology, the subdiscipline of dendrochronology concerned with the relationships between the environment and tree growth and with the reconstruction of past environmental conditions from tree-ring data (Fritts, 1976), is not discussed in detail.

Historical Background

Dendrochronology, as a distinct scientific discipline, is a relatively recent development, although observation of the ring structure and related phenomena in trees has a long history (Studhalter, 1955). The first reference to the growth rings of trees occurs in the writings of Theophrastus of Eresus, who lived around 373 to 287 BC. The annual nature of tree rings in temperate regions was first remarked by Leonardo da Vinci, forgotten, and rediscovered in the seventeenth century by several individuals. The basic principle of dendrochronology, that of crossdating (the identification of contemporaneous rings in two or more trees), was first recognised in 1737 by Duhamel and Buffon, who discovered a frost-damaged ring 29 rings in from the bark on each of several freshly cut trees. Subsequent investigation by a number of scholars confirmed the existence of frost injury to the ring for 1709 in several species from widely separated localities in northern Europe. In 1827, Alexander C. Twining first recognised between-tree correlation of ring-width variation in a group of hemlock sections from Connecticut, and hypothesised that tree-ring widths might be used to reconstruct past climatic conditions (Studhalter, 1955, pp. 56–57). Twining's observations went unnoticed, and crossdating was rediscovered several times, most notably by Charles Babbage in England in 1838 (Heizer, 1956), by Jacob Kuechler in Texas in 1859 (Stallings, 1937), by J. C. Kapteyn in Germany and the Netherlands in the 1880s (Schulman, 1937), and finally by Andrew Ellicott Douglass in 1904, who employed it as the basis for the development of dendrochronology as it is known today.

In about 1900, A. E. Douglass, an astronomer at the Lowell Observatory in Flagstaff, Arizona, was studying cyclic fluctuations in the intensity of sunspot activity and the possible effects of such fluctuations on terrestrial climate. His efforts in this regard were impeded by the fact that weather records did not extend back far enough into the past to enable him to confirm the existence of periodic climatic fluctuations that might be correlated with the eleven-year cycle of sunspot activity. In 1901, as a result of his observations on a buckboard trip through the vast pine forests of northern Arizona, he conceived the idea that size variations in the annual growth rings of the trees in this arid region might be correlated with climate and might therefore be used to isolate long-term cyclic fluctuations in the weather. In 1904 he examined transverse sections from six ponderosa pine trees that had grown near Flagstaff. He observed that conspicuous characteristics of ring-width variation could be matched from tree to tree, and thus he rediscovered crossdating (Douglass, 1937, pp. 7).

Although Douglass first recognised crossdating in 1904, he did not realise its full significance until 1911 when he observed ring-width variations identical to those of the Flagstaff trees in a group of pine sections from Prescott, Arizona, some 100 kilometres from Flagstaff. This realisation stimulated Douglass to begin a systematic tree-to-tree comparison of ring-width variation in the Flagstaff area, and by 1914 he had developed a 500-year crossdated ponderosa pine chronology for the area. In an effort to secure longer ring series he turned to the long-lived sequoias of California, and by 1919 developed a sequoia chronology that extended back to 1300 BC.

In 1914 Douglass's dendrochronological research took a new turn when the anthropologist Clark Wissler of the American Museum of Natural History recognised the potential of dendrochronology for dating archaeological materials. In 1919, the Museum sent Douglass beam sections from the large prehistoric pueblo ruin at Aztec, New Mexico. Douglass recognised crossdating among these sections, and he soon constructed a short composite ring chronology that was 'floating' in time, for it did not crossdate with the dated Flagstaff or sequoia sequences. A year later, Douglass crossdated samples from the ruin of Pueblo Bonito, located 80 kilometres south of the Aztec ruin, with the Aztec sequence and established that parts of Pueblo Bonito were built 40 to 50 years before construction began at Aztec. The relative dating, in terms of an actual number of years, of the two ruins showed that the floating Aztec–Pueblo Bonito chronology could be extended toward the present by crossdating in tree-ring sequences from progressively younger archaeological sites until a recognisable overlap with the Flagstaff series was achieved.

This development intrigued Neil M. Judd, who was excavating Pueblo Bonito for the National Geographic Society. Judd's interest in obtaining absolute dates for this ruin prompted the Society to support an attempt to close the gap that separated the Aztec–Pueblo Bonito chronology from the Flagstaff chronology. Two 'Beam Expeditions' (in 1923 and 1928) collected a large number of archaeological tree-ring specimens, which permitted the extension of the Flagstaff chronology back to AD 1260 and the construction of a 585-year prehistoric floating chronology that could not be crossdated with the dated sequence. A third Beam Expedition was organised to sample sites that possessed pottery types known to fall in the time period represented by the gap between the two ring series.

On 11 June 1929 in Show Low, Arizona, a fragment of charred and rotted wood was unearthed from a large prehistoric pueblo. This sample's inner rings overlapped the outer end of the floating chronology, and its outer rings crossdated with the inner rings of the dated sequence (Douglass, 1929, 1935; Haury, 1962). The bridging of the gap between the dated and the floating sequences permitted the immediate assignment of calendar dates to scores of archaeological sites, and dendrochronology came of age as a dating tool.

Having built a continuous tree-ring sequence extending from AD 700 to 1930 and having demonstrated the utility of the tree-ring method for archaeological dating, Douglass turned his attention to other dendrochronological matters and left the bulk of archaeological dating in the hands of a succession of colleagues and students. In the 1930s, W. S. Stallings of the Laboratory of Anthropology in Santa Fé, New Mexico, expanded Southwestern dendrochronology into the Rio Grande valley of New Mexico and provided a foundation for the dating of archaeological sites in this important centre of Pueblo Indian development (Smiley, Stubbs and Bannister, 1953; Stallings, 1933). In the 1960s, Southwestern dendrochronology was extended into northern Mexico with the dating of archaeological samples from the site of Casas Grandes in northern Chihuahua (Scott, 1966). Since 1930, the Southwestern ring chronology has been extended back to 322 BC and serves as the standard for the dating of the thousands of new samples recovered annually from archaeological sites in the region. The current totals of nearly 50,000 dates from approximately 5,000 sites provide a degree of chronological control that has helped establish the American Southwest as one of the primary archaeological laboratories in the world. Beyond the Southwest, archaeological tree-ring dating has been successfully applied in the Arctic, the North American woodlands and Great Plains, the British Isles, Scandinavia, northern Europe, Russia and Turkey.

Dendrochronological studies of living trees have kept pace with archaeological developments, and nearly every region of the world has been sampled. Such chronology building efforts were initially concentrated in western North America, and at present hundreds of local tree-ring chronologies from throughout this vast region provide comprehensive data bases for studies of past environmental variability on continental and subcontinental scales of analysis. An important result of this activity was the discovery that gnarled bristlecone pine trees growing at high elevations in the basin and range province of the western US far exceed the giant sequoias and redwoods in longevity (Schulman, 1956, 1958), the oldest known living bristlecone pine being 4,900 years old (Currey, 1965).

Such ancient living trees combined with the weathered remnants of long-dead bristlecone pines provide the material for a continuous ring chronology of nearly 8,800 years (Ferguson, 1969, 1970c, 1979). Since about 1965, the collection of samples from living trees in Mexico, eastern North America, the British Isles, northern and central Europe, and Asia Minor, has greatly expanded the dendrochronological database for much of the northern hemisphere (Hecht, 1985). More recent collecting in South America, southern Africa, Australia, New Zealand, and Tasmania is beginning to produce tree-ring chronologies for the southern hemisphere (Hecht, 1985).

Though diverted into chronology building and archaeological dating for many years, Douglass maintained his interest in the effects of climate on the growth of trees. He demonstrated a positive correlation between winter precipitation and ring width (Douglass, 1914, pp. 112–115; 1919, pp. 65–73), and his colleague Edmund Schulman expanded the scope of dendroclimatic research to encompass much of North America (Schulman, 1956, pp. 39–50). Recent research into the inter-relationships between tree growth and environmental variability (Berger *et al.*, 1979; Blasing and Fritts, 1975, 1976; Brett, 1978; Eckstein, 1978a; Fritts, 1966, 1970, 1971, 1974, 1976, 1978, 1981; Fritts *et al.*, 1971; Hughes *et al.*, 1978; Kairiukstis, 1978) has produced refined conceptual models of tree growth-environmental relationships and sophisticated mathematical techniques for producing quantitative estimates of environmental variables from tree-ring data.

Considerable success has been achieved in using ring width data to reconstruct annual and seasonal variability in precipitation and temperature (Fritts, 1974, 1981; Rose, Dean and Robinson, 1981), atmospheric pressure pattern anomalies (Blasing and Fritts, 1975, 1976; Blasing and Lofgren, 1980; Fritts, 1971; Fritts *et al.*, 1971), the Palmer Drought Severity Index (Cook and Jacoby, 1979; Meko, Stockton and Boggess, 1980; Mitchell, Stockton and Meko, 1979), and surface runoff (Smith and Stockton, 1981; Stockton, 1975; Stockton and Fritts, 1971). Ring density variation also has been related to environmental fluctuations (Conkey, 1979; Parker and Henoch, 1971; Polge, 1970, 1978; Schweingruber, Bräker, and Schär, 1978; Schweingruber *et al.*, 1978) and the prospects for future densitometric reconstructions of past environmental variability are good. Variations in ring width and density have been related to atmospheric pollution caused by industrial waste emissions (Nash, Fritts and Stokes, 1975; Polge, 1970; Vins, 1970). Studies that involve relationships between various aspects of tree growth and the trees' environment are the province of dendroclimatology, a subject discussed fully by Fritts (1976).

In order to perpetuate Douglass's dendrochronological research, the University of Arizona in 1937 established the Laboratory of Tree-Ring Research, which for several decades was the only facility in the world dedicated solely to tree-ring studies, and which remains one of only a handful of institutions engaged in all facets of dendrochronology: dating, chronology building, dendroclimatology, dendrohydrology, and tree physiology. Douglass's Southwestern achievements stimulated individual efforts in the 1930s to apply dendrochronology in other regions, notably the American Arctic and Great Plains and southern Germany. World War II interrupted these activities, and world dendrochronology languished for a time afterwards. In the 1960s, a few individual scholars began to develop local dendrochronological programmes. The advent of quantitative dendroclimatology in the late 1960s spurred a worldwide interest in dendrochronology that has yet to abate. At present, the discipline is experiencing the most rapid growth in its history, and participation in tree-ring research is at an all-time high. Currently, formal dendrochronological research is pursued in at least 45 facilities distributed among Australia, Belgium, the British Isles, Canada, China, Czechoslovakia, Denmark, East Germany, France, Italy, Japan, New Zealand, Norway, Poland, Sweden, Switzerland, the USA, the USSR and West Germany.

Concepts And Principles

Dendrochronology has been defined as 'the method of employing tree-rings as a measurement of time . . . and the process of inferring past climatic conditions that existed when the rings were being formed' (Bannister, 1969, pp. 191). This chapter is focused on the first aspect of dendrochronology: the study of tree rings as chronological phenomena and their use in dating past events. The basic principles and techniques of tree-ring dating are considered in detail elsewhere (Baillie, 1982; Bannister, 1969; Dean, 1978a; Ferguson, 1970a; Stokes and Smiley, 1968), and they are only summarised here.

It is well known that a transverse section of a tree trunk exhibits a series of 'rings' concentric around the pith at the centre. Perhaps not so well known is the fact that morphological attributes of these rings often vary systematically, and this variation, when it correlates from tree to tree, is the foundation of dendrochronology. In the trees with which dendrochronology has been concerned, primarily temperate zone and arctic coniferous and deciduous species, a single ring is added each year, a fact that permits the assigning of exact calendar dates to all increments of growth when the date of only one of the rings is known.

Figure 4.1: Transverse Section of a Coniferous Tree-Ring Series

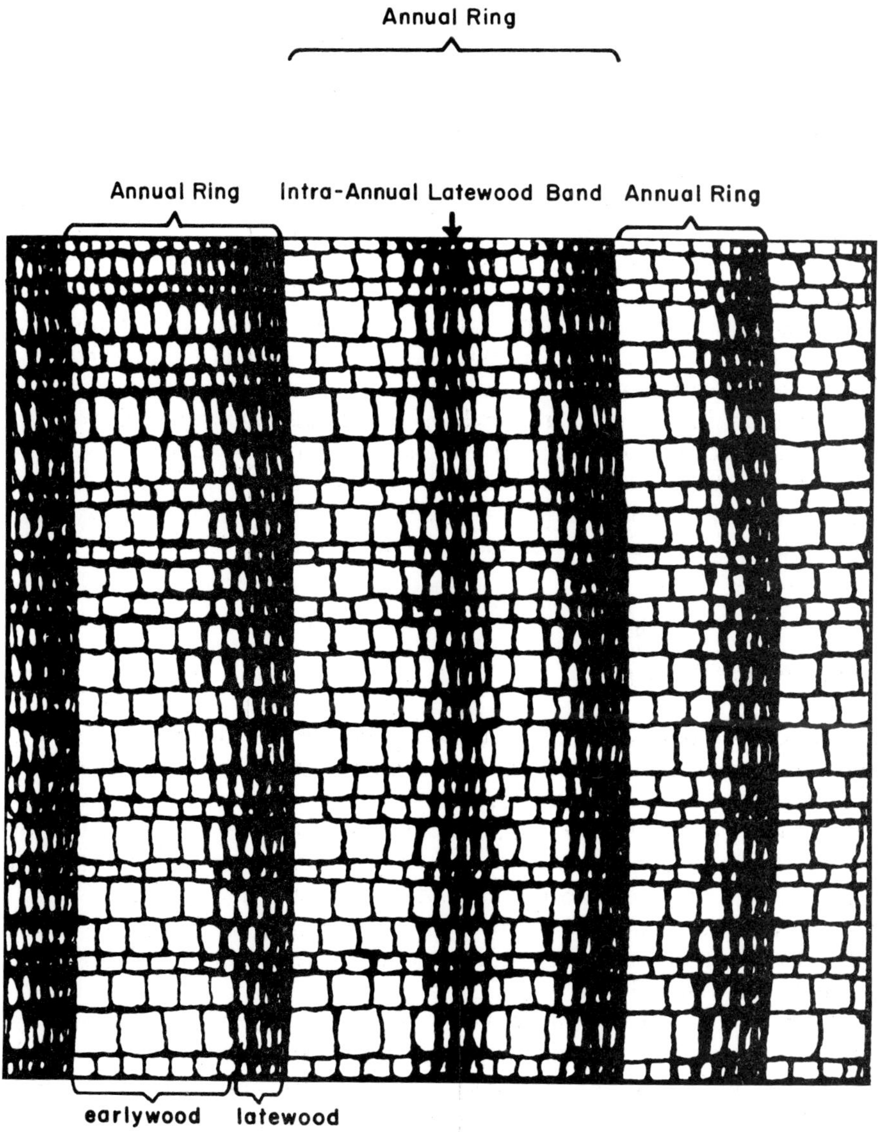

In the conifers, an annual ring is made up of two parts (Figure 4.1): (1) an inner band composed of large thin-walled tracheid cells called the 'earlywood', and (2) a darker outer band composed of smaller thick-walled tracheids known as the 'latewood'. Cessation of radial growth for the year produces a sharp line at the outer edge of the latewood, which marks the boundary of the annual ring. Occasionally, more than one band each of earlywood and latewood are produced in a single year (Glock, Studhalter and Agerter, 1960). Usually, bands of latewood cells within a single ring can be distinguished easily under magnification, for they grade into the following earlywood cells and do not exhibit a sharp terminal line. On occasion, however, these bands are extremely difficult to recognise visually, and crossdating is the principal clue to their presence. Intra-annular latewood bands of this second type are often called false or multiple rings to distinguish them from the more easily detected type. Densitometric studies of tree-ring structure show promise of providing quantitative techniques of identifying false rings even in the absence of crossdating (Polge, 1970).

Angiosperm ring structure is generally more complex and varied than that of conifers (Baillie, 1982; Figure 2.1, Plate 1; Fritts, 1976, pp. 58–63). In the 'ring porous' angiosperms, the concentration of large water-conducting vessels (pores) in the earlywood demarks the boundary of the annual rings. When very small rings are involved, this boundary may be obscured by the crowding of pores consequent on the lack of latewood development (Baillie, 1982, Plates 1f, 2a). In 'diffuse porous' species, vessels of relatively uniform size are distributed evenly throughout the earlywood and latewood, and the annual ring boundary is denoted by other morphological features. Rays that trend radially from pith to bark occur in some angiosperms. Occasionally, such rays disrupt the vesicular structure sufficiently to obscure annual ring boundaries. The problems inherent in annual ring recognition in angiosperms have stimulated an emphasis on conifers in dendrochronological research. This comparative neglect of deciduous species has been remedied only relatively recently by research in eastern North America, Europe, and the southern hemisphere.

Two different types of ring-width series, representing end points on a continuum of variation rather than mutually exclusive categories, are recognised (Figure 4.2). Complacent series are characterised by rather uniform ring width and little ring-to-ring variation. Sensitive series exhibit considerable variation in width from one ring to the next and are much more suitable for dendrochronological analysis. Sensitivity in tree-ring series increases from the forest interior, where optimum growing conditions prevail, to the forest borders, where external factors are more limiting to growth (Fritts *et al.*, 1965). Both complacent and sensitive ring series exhibit a characteristic growth trend; that is, the absolute size of the rings gradually diminishes from near the centre to the outside of the cross-section, independently of the between-ring variation.

Figure 4.2: Types of Tree-Ring Series and Crossdating: A, Complacent Series; B and C Sensitive Series that Crossdate with One Another

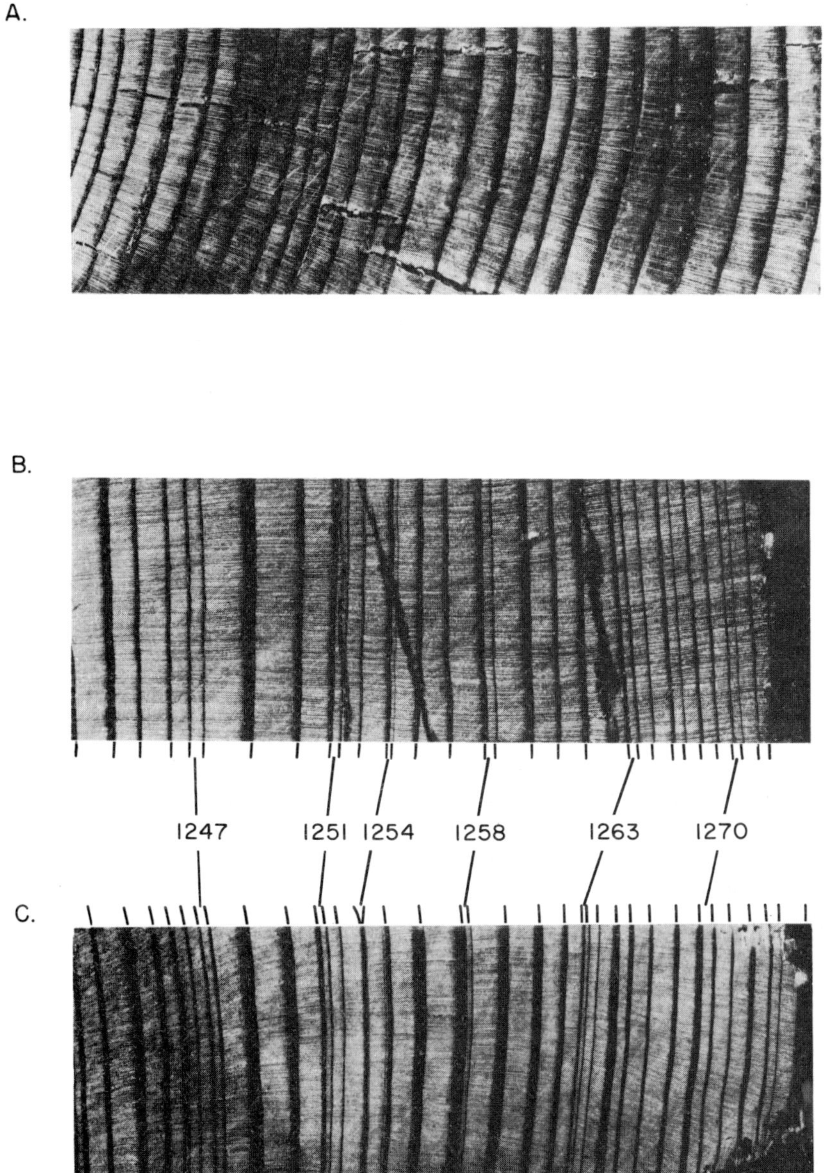

The fundamental principle of dendrochronology is crossdating, which may be generally defined as the existence of characteristics in tree-ring structure that permit the identification in many different trees of rings that

were formed contemporaneously with one another. Crossdating may be recognised on the basis of several different types of phenomenon. Gross features of ring morphology such as frost damage and intra-annular latewood bands may crossdate from tree to tree, although such crossdating is limited to restricted stands of trees. Measurements of earlywood and latewood growth and the ratio between the two have been used to match different ring series. Ring density covaries among trees and can be used to establish crossdating (Fletcher and Hughes, 1970; Parker, 1970; Parker and Henoch, 1971; Polge 1970, 1978; Schweingruber, Bräker and Schär, 1978; Schweingruber *et al.*, 1978). Most commonly, however, crossdating has been recognised on the basis of the covariation of ring widths among sensitive ring series from different trees within a given area (Figure 4.2). Mathematically, this type of crossdating is expressed as the correlation (Pearson's r) between standardised ring-width indices from two or more discrete ring sequences.

Crossdatable patterns of ring-width variation are most apparent in the narrow rather than the wide rings. This is so primarily because the limiting environmental conditions that produce narrow rings are more likely to influence tree growth over broad areas than are 'normal' or optimum conditions, which often merely permit local site factors or the inherited capabilities of individual trees to control the amount of growth. In some regions conspicuously large rings are utilised in the crossdating process. European dendrochronologists working with trees that exhibit relatively slight ring-width variability have developed statistical crossdating methods that utilise concordances in the direction of size changes in adjacent rings (Bauch and Eckstein, 1970; Huber and Giertz, 1970) or the correlations between paired ring series (Baillie and Pilcher, 1973). Despite the varying emphasis on small or large rings and on total variability or directional concordances as indicators of crossdating in different areas, the covariation of all the ring widths in different ring sequences remains the primary criterion for crossdating.

Rigorous application of the principle of crossdating is the basis of dendrochronology and is necessary for two fundamental reasons: (1) it permits the assignment of a calendar date to every ring in a series when the date of only one ring is known, and (2) it is the means whereby ring series of unknown age can be dated by matching them with sequences of known date.

Were it not for the existence of locally absent and multiple ('false') rings, a simple ring count from one ring of known date would suffice to date all the other rings in a given sequence, and careful crossdating would not be necessary. However, locally absent or unrecognised multiple rings render counts either too low or too high respectively, and crossdating is the only means of recognising these potential errors.

Locally absent rings occur in years when unfavourable conditions limit radial growth to such an extent that the rings for those years are present in only a small part of the tree stem. For example, the dissection of a ponderosa pine tree from the Flagstaff area revealed the ring for the year 1902, which typically is narrow in the conifers of the region, to occur only as a lens 0.07mm wide on a 2.0mm arc of a section 10.1m above the ground (Schulman and Baldwin, 1939, p. 13). In cases such as this, a core or even a full section from the tree is likely to represent a portion of the stem in which no growth for the critical year occurred, and that ring will be absent from the study sample.

Crossdating with other trees provides a means of detecting the total or partial absence of rings from a particular specimen. The sequence of narrow rings in the study sample will match those of the control specimens (chronology) with which it is being crossdated until, at a point at which the control specimens exhibit a narrow ring, the expected corresponding small ring will not be evident in the study sample. Beyond this point the ring sequence of the study sample will fall one ring short of matching the sequence of narrow rings in the control specimens. Thus crossdating reveals that a ring is missing from the study sample and indicates the locus of the absent ring. In many cases a search of the indicated area on the study sample or an examination of a different sample from the same tree will reveal the 'missing' ring to be present locally.

Multiple rings are indicated when too many rings appear in a sample as compared to other trees with which it crossdates. Thus, crossdating assures the dating of individual rings because it permits the identification and localisation of multiple and locally absent rings, both of which render ring counts inaccurate. Locally absent and false rings are the principal reasons that apparent crossdating must be affirmed by the analysis of many trees from different types of growth site.

Crossdating also permits the sequence of narrow rings in a sample of unknown age to be matched to a control sequence in which every ring has been assigned a calendar date. Matching the two ring series permits the assignment of an absolute date to each ring in the previously undated sample. If one were to deal only with counts of complacent ring series, there would be no way to correlate an undated series with a dated control sequence. Only samples with at least one ring of known year could be 'dated' in this manner, and the method would be useless for the dating of anything but living trees, the years of whose outermost rings were already known.

An outgrowth of the application of the principle of crossdating is the building of long-range ring chronologies from overlapping samples of increasingly greater age (Figure 4.3). In constructing such a chronology one begins with trees of known date — living trees sampled for this purpose — and works out the crossdating as far back in time as a sufficient number of overlapping samples can carry it. The chronology is then extended beyond the range of living trees by crossdating in material from successively older historic and prehistoric contexts. In this way the Southwestern and Central European chronologies were extended back to 322 BC and 717 BC, respectively, by the use of archaeological material (Dean, 1975; Becker and Delorme, 1978), and the bristlecone pine chronology of almost 8,800 years was built by incorporating ring series from long-dead trees and eroded remnants of such trees (Ferguson, 1969, 1970c, 1979c).

Figure 4.3: Building a Tree-Ring Chronology from Progressively Older Crossdated Ring Sequences

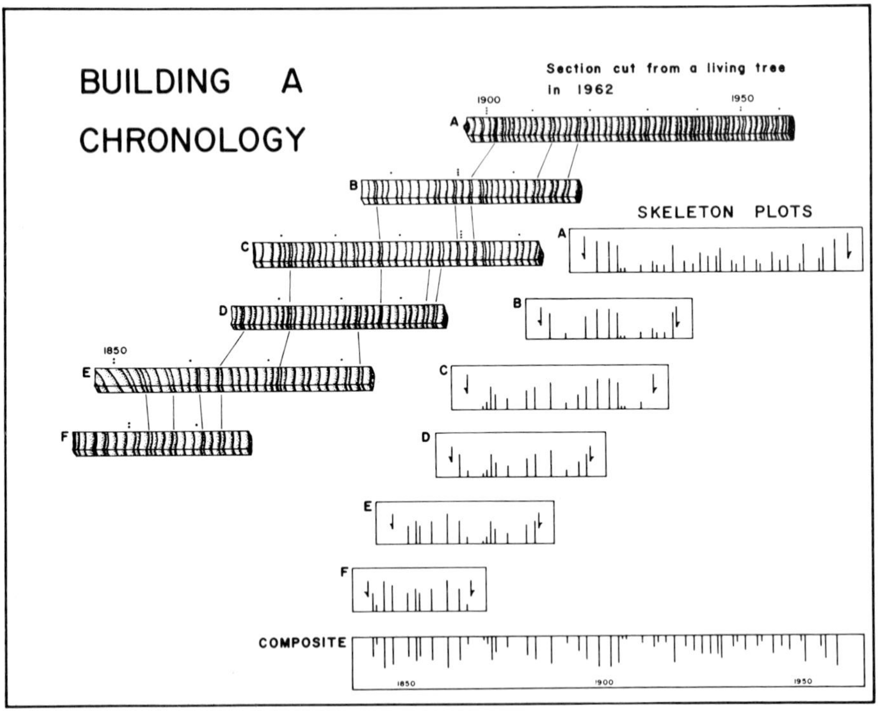

Such composite or master chronologies provide standards of reference against which ring sequences of unknown age can be compared, matched and precisely dated. Composite chronologies also form the databases for dendroclimatic reconstructions of past environmental variability. Although chronology building superficially appears to be a fairly simple procedure, it actually is a painstaking effort that requires the careful comparison of large numbers of individual samples and the use of meticulous checking procedures to ensure accuracy. Many years of work may be expended before the effort pays off in the form of a usable chronology.

The maximum area covered by a composite tree-ring chronology varies in response to local factors and generally must be determined empirically. Correlations among regional chronologies in western North America decrease fairly gradually with increasing distance in an east-west direction, with significant correlations at distances up to 1,100 kilometres. Correlations fall off much more abruptly in a north-south direction with increasing distance (Fritts, 1963, pp. 4–5). The basic Southwestern master chronology is generally valid for Arizona, southern Utah, southwestern Colorado, and western New Mexico. In contrast, chronologies in the Great Plains of the United States often cover only a few hundred square kilometres (Weakley, 1950). The central European chronological pattern originally recognised from the Carpathian Mountains to east of the Rhine (Becker, 1978), has recently been extended to include Northern Ireland and northern Germany (Pilcher *et al.*, 1984). More localised crossdating prevails in other areas of northern Germany, the Netherlands, southern Scandinavia, and the British Isles (Bauch, 1978a; Eckstein, 1972, 1978a; Fletcher, 1987a; Huber and Giertz, 1970; Pilcher, 1973; Pilcher and Baillie, 1980). Intraregional chronological variability sometimes hinders dating on regional master sequences, and the construction of geographically restricted chronologies is necessary for archaeological dating and for localised dendroclimatic reconstructions.

In any given area, the dendrochronological dating of past human or natural events can be accomplished only if certain, rather specific, conditions prevail (Bannister, 1969, pp. 193–4; Dean, 1978a, p. 142). First, wood or charcoal must occur in association with the phenomena to be dated. Second, the cellular and ring structure of the wood must be preserved. Wood itself may survive for long periods in dry caves, in massive open structures, buried in alluvium, and in wet environments such as bogs and lakes. Nevertheless, wood is perishable, and were it not that charring preserves the ring structure, much prehistoric tree-ring dating would not have been accomplished for want of usable material. Charcoal, which is practically indestructable unless crushed, occurs in burned archaeological sites and in deposits produced by natural conflagrations such as forest and range fires. Occasionally, mineralisation preserves the cellular structure well enough that the rings of petrified wood can be observed.

The third requirement for tree-ring dating is that crossdating must be apparent both in the living trees in the area and in the wood or charcoal samples to be dated. It is important in this regard that the samples of unknown age and the living trees contain enough rings to substantiate the crossdating. The final requirement is the presence of living trees old enough to extend the dated chronology back into the past far enough to overlap with at least a portion of the ring sequence represented in the undated material. In the absence of this condition, calendar dates cannot be assigned to the rings of the ancient sequences, although the relative dating of past events is possible. Thus, one who plans to employ dendrochronology in an area where developed chronologies are not available must construct a living tree chronology, even when his primary interest is the dating of past events.

Sampling

The nature of dendrochronological analysis imposes rather specific conditions on the collection of study samples from living trees, archaeological timbers or geological wood. The techniques and equipment used in the collection of such tree-ring samples are discussed in detail by Stokes and Smiley (1968, pp. 21–36), and only general comments are presented here.

The main objective in collecting living tree samples for chronology building is to select trees with sensitive crossdatable ring series and to avoid trees with complacent or erratically variable ring records. Suitable trees tend to be found on 'stress sites' where external constraints on growth are strongest and most uniform over wide areas (Fritts *et al.*, 1965). In semi-arid and arid regions, stress sites are most abundant on the edges of a tree species' distribution, on the lower elevational border of a forest, where available moisture is the primary limiting variable, and near the upper timberline, where growth is limited primarily by temperature. In such regions, the favourable conditions of the forest interior, where general external constraints have a minimal effect on tree growth, produce complacent ring series or ring series with idiosyncratic variability caused by localised factors such as competition, disease, and insect infestation.

In the Arctic, ring series sensitive to temperature are found in trees from a wide range of habitats, especially those along the forest margins. In temperate regions, such as eastern North America and Europe, where more favourable growing conditions prevail, stress sites occur in restricted locales where topography, drainage, substrate, and other local factors combine to enhance the effects of climatic variability on tree growth (Becker, 1978). Thus, the selection of growth sites suitable for acquiring sensitive, crossdatable tree-ring series is a major concern in dendrochronological research (Fritts, 1976, pp. 17–19). Regional, local and site factors must be carefully assessed in choosing study loci so that time and effort are not wasted on the indiscriminate sampling of trees with ring characteristics unsuitable for dendrochronological analysis.

The most common method of sampling living trees is to core them with a

Swedish increment borer, an instrument that removes from the trunk a 4mm diameter core representing one radius of the cross-section. Such coring does not harm the trees, and they can be periodically resampled to keep the chronologies up to date and to provide data for continued growth studies. A set of two opposed cores is taken from each tree, and trees are usually sampled in groups of 20 to 30 for each growth site. In areas where dendrochronology is being newly applied or where crossdating is difficult, it may be necessary to procure full sections in order to establish cross-dating, a procedure that is otherwise undesirable because it kills the trees. X-ray densitometric analysis requires that samples be taken as nearly perpendicular as possible both to the ring boundaries and to the vertical grain of the wood. As a result, elaborate coring techniques involving extra sharp manual and power-driven borers and special devices for holding the borer at right angles to the bole have been developed to achieve the necessary precision (Parker, 1970; Parker and Meleskie, 1970; Polge, 1970). Collections from living trees should be accompanied by data on the location of the tree, on the nature of the tree (height, diameter, crown density, damage, disease, etc.), on the site on which it is growing (slope, soil composition, drainage, exposure, etc.), and on the tree's relationship to nearby trees and other vegetation. These data facilitate subsequent relocation and identification of the trees for further study and are helpful in assessing non-climatic conditions that might produce growth anomalies.

The collection of tree-ring samples from archaeological or geological contexts presents a different set of problems. Here the sample is pre-selected by the agencies, human or natural, that caused the timbers to be emplaced in their ancient setting, and the collector has no control over the sites on which the trees grew. Because of this pre-selection and because it is virtually impossible to determine in the field which of the available logs possess datable ring series, at least one sample should be collected from each timber. Selection should be performed in the laboratory after analysis has shown which specimens date and which do not.

The collection of prehistoric tree-ring samples should be oriented toward four basic objectives. First, the cellular structure must be conserved so that each ring can be observed, plotted, and measured. Second, the innermost and outermost rings on each sample must be preserved; the germination and death dates of a tree cannot be ascertained from a sample whose interior and exterior rings have been destroyed by careless handling. Third, collection should be made with regard for the techniques of specimen preparation and analysis used in dendrochronology. Fourth, supplementary data that might be of value in assessing the dates should be recorded. Exact provenience, use–category, and the samples' relationship to other timbers and features with which they are associated, are among the types of data that are important in this regard.

The methods used in the collection of prehistoric specimens depend largely on the state of preservation of the wood. Cross-sections or partial sections can be sawed from solid wood. Sawed samples are difficult to transport and inconvenient to store; consequently, in cases in which dating rather than chronology-building is the primary objective, coring is more efficient. The coring of ancient timbers is accomplished with a tubular borer, which consists of a steel cylinder, 12mm or 25mm in diameter and 20mm to 45mm long, with a set of saw teeth at one end and an adapter to be fitted to the chuck of a carpenter's brace or electric drill at the other end. The tool is powered by the operator or by an electric drill driven by house current, a generator, or batteries, as the situation dictates. Each sample should be given a number that is keyed to notes containing detailed provenience and supplementary information.

Charcoal and rotted wood samples should be carefully removed from the matrix in which they are found, so that the rings and other attributes are not damaged. Once exposed and thoroughly dried, the specimens should be tightly bound with string, to which is affixed a tag bearing the sample number. These samples must be treated with a preservative, but any preparation that hardens or destroys the cellular structure must be avoided or it will be impossible to achieve a usable study surface. Best results have been obtained with a saturated solution of paraffin in gasoline. The solution penetrates into the cells and the gasoline evaporates, leaving the paraffin to strengthen and preserve but not harden the cell walls. After soaking in the paraffin–gasoline solution, the samples should be thoroughly dried, wrapped in cotton batting, and packed for transfer to the laboratory. Water-logged archaeological or geological samples are particularly difficult to handle because they tend to disintegrate when dried. Many techniques, including freezing and continued saturation, have been used to preserve such samples long enough to measure the rings.

Detailed supplementary data are critical to the analysis and evaluation of the dates derived from any collection of tree-ring samples. Necessary data include the name or designation of the recovery site, its precise location, its place in the local temporal sequence, and a description of the local environment. These data help the dendrochronologist to choose appropriate master chronologies for dating, limit the range of time in which dating might be expected, and facilitate the selection of comparative materials from nearby sites. In addition, detailed information on each timber should be recorded, including its spatial and functional relationships to the context in which it occurs and to other timbers with which it is associated. Finally, data relative to the condition of each timber should be noted: degree of weathering, presence of bark, beetle galleries, tool marks, transportation scars, and any other features that may possibly be of interpretive value. Such data greatly enhance the interpretive significance of the dates derived from the samples.

Dating

The analysis and dating of tree-ring specimens involves several specialised techniques that are only briefly discussed here, for they are adequately covered in the literature (Bannister, 1969; Dean, 1978a; Ferguson, 1970a; Stokes and Smiley, 1968). Samples are prepared for study by a series of operations designed to expose their cellular and ring structure. Wood is usually power-sanded with a series of grits graded from coarse to fine. A razor-cut surface is sometimes applied to bring out minute details. Charcoal and rotted wood are prepared either by producing a sharp fractured surface or by slicing with a razor blade. The surfaced samples are studied with a hand lens or a microscope.

Figure 4.4: Graphical Representations of Ring-width Variability

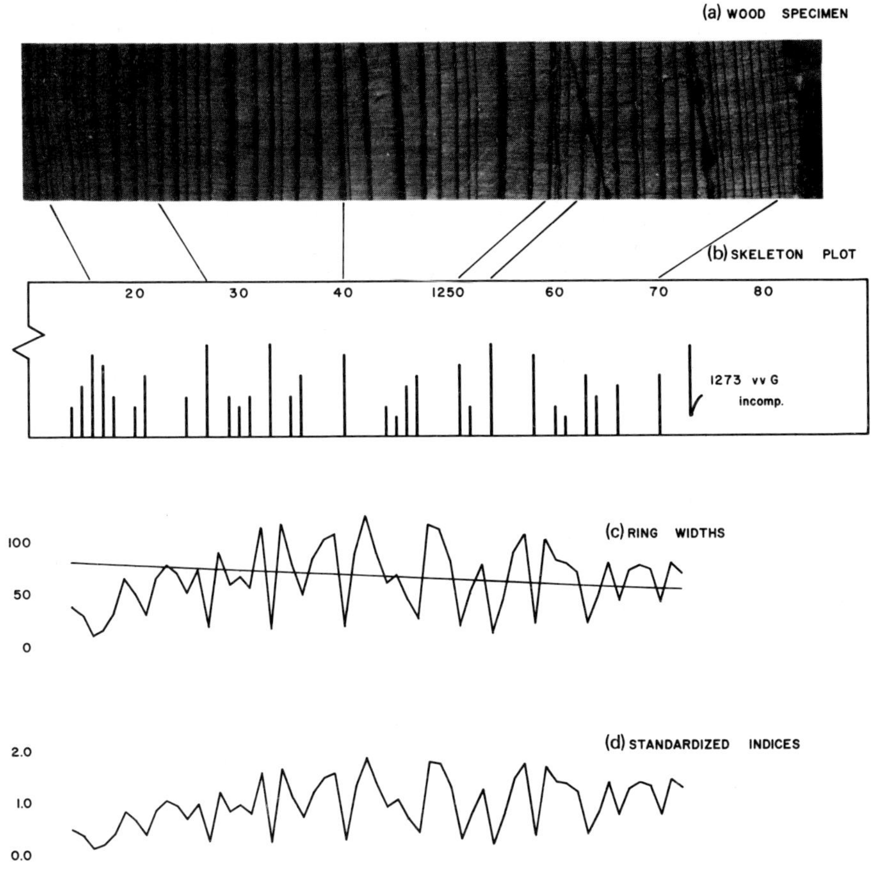

Figure 4.5: Skeleton Plot Dating of Samples from Betatakin Ruin, Northeastern Arizona

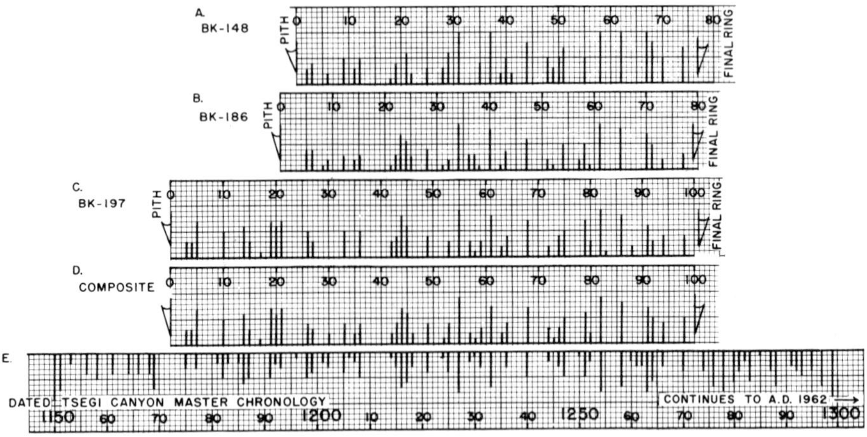

Most prehistoric dating in North America is done by the skeleton plot method (Dean, 1978a; Stokes and Smiley, 1968), a technique that provides a quick, subjective representation on paper of a specimen's ring pattern at a standard scale so that samples can be conveniently compared with one another to establish crossdating (Figures 4.4 and 4.5). The skeleton plot has the added advantage of not reflecting long-term growth trends that are irrelevant to the crossdating.

All the samples from a site are skeleton plotted and the plots are compared with one another to establish internal crossdating. Once this has been done a composite plot that represents a visual average of the individual crossdated plots is made. The composite plot is then compared, year by year, to the appropriate regional master chronology until the point is found where the two match. When this has been accomplished, a calendar date can be assigned to each of the previously undated samples (Figure 4.5).

The skeleton plot technique is not always used in dating. Samples from living trees are often compared directly with one another, since the problem is not dating an unknown but working out the chronological sequence of narrow rings through crossdating. Visual comparison of plotted ring width measurements (Figure 4.4c) are commonly used for dating bristlecone pines and other long-lived species that lack a pronounced growth trend. In some cases, the mathematical correlations between ring-width indices of the undated and dated sequences are calculated for a series of successive matches until the proper match is indicated by a significantly high coefficient of correlation (Fritts, 1963; Scott, 1966, pp. 71–2). In parts of Europe where crossdating is limited by the relative complacency of the ring series, dendrochronologists have devised statistical dating methods that rely on the directional or absolute covariation of measured ring series to establish crossdating (Baillie and Pilcher, 1973; Bauch and Eckstein, 1970; Huber and Giertz, 1970). Densitometric crossdating, although it has as yet only rarely been used for dating samples of unknown age (Jozsa, Parker and Bramhall, 1980; Jozsa *et al.*, undated), holds great promise for dating ring series that cannot be matched on the basis of ring-width variations.

Whatever the dating technique employed, ring-width and density data are presented in quantitative form. Ring widths are measured to the nearest 0.01mm (Robinson and Evans, 1980), and these raw data are usually converted to standardised growth indices (Fritts, 1963, 1976, pp. 24–5; Fritts, Mosimann and Bottorff, 1969; Graybill, 1979). Standardisation eliminates the growth trend and between-sample differences in average ring size, both of which represent individual variability that obscures the variability common to all samples. By enhancing the common variability, standardisation facilitates crossdating and maximises the environmentally induced component of annual tree growth that is the subject of dendroclimatic analysis. Indices are produced by fitting a trend line to the ring-width data (Figure 4.4c) and calculating the percentage departure of each ring-width value from the trend line.

Originally, trend lines were fitted by eye or with the aid of running means (Schulman, 1956). Computer-fit trend lines usually involve negative exponential curves, straight lines of various slopes, or polynomial functions (Graybill, 1979). Recently, other mathematical techniques for deriving trend lines, such as the compound increment function (Warren, 1980; Warren and MacWilliam, 1981) and the cubic smoothing spline function (Cook and Peters, 1981) have been employed. Once individual sample indices have been calculated, they are averaged, year by year, to form composite chronologies that represent particular species, sites, locales and regions. Index chronologies are presented in tables or as graphs of indices (Figure 4.4d), formats that reflect the true nature of ring-width variation and facilitate comparison and the sophisticated statistical operations involved in dendroclimatic analysis.

Ring density data are acquired through more complicated processes that vary from laboratory to laboratory (Jones and Parker, 1970; Parker, 1970; Parker and Meleskie, 1970; Polge, 1970; Schweingruber, Bräker and Schär, 1978; Schweingruber *et al.*, 1978). Samples are reduced to a standard thickness with precision-cutting or abrading tools. Each prepared sample is passed at a uniform rate beneath a precisely controlled X-ray beam oriented parallel to the vertical plane of the wood grain in the sample. The X-ray beam passes through the wood and exposes a photographic film situated beneath and parallel to the sample to produce a negative image of the sample. The X-ray negative is then analysed by a scanning microdensitometer, which quantifies and records the densities of a series of points within each ring along the radius represented by the sample. The plotted trace of the densitometric data reveals the density structure of each ring in the sample. Similarities in the density structures of individual rings in different samples provides a means of crossdating different trees (Conkey, 1979; Schweingruber *et al.*, 1978).

Densitometric analysis is complicated by the fact that different aspects of the density spectrum can be used. For example, the minimum, maximum, and mean densities of the total ring or of components of the ring (such as the earlywood and latewood) have proved useful for different purposes. Density values from different trees can be combined into crossdated composite chronologies analogous to composite sequences based on ring widths. The time, specialised equipment and expense constraints of densitometric analysis have slowed its development as a tool for dating samples of unknown age. In most areas where densitometric dendrochronology has been attempted, ring-width dating is also possible. As a result, densitometry has been focused on dendroclimatic applications, especially since ring density appears to respond to climatic factors different from those that control radial growth. Because density varies from ring to ring even when size does not, the greatest potential of densitometry for dating lies in regions where ring-width dating is impossible or difficult. It is safe to predict that densitometric analysis will make its greatest contribution to the dating of past events in temperate and perhaps subtropical regions where radial growth is not severely limited by external environmental factors.

Chronological Analysis of Tree-ring Collections

The dating of a tree-ring sample is just the initial step in the process of dating the past event(s) with which the specimen is associated. The application of a date or a series of dates to an event of unknown age involves the careful assessment of many factors such as the date itself, the species of the sample, the provenience of the sampled log, the associational context in which it was found and its physical attributes.

The long association of dendrochronology with archaeology has produced considerable sophistication in the evaluation of tree-ring dates. The availability of many dates from a variety of well-controlled archaeological contexts has served to illuminate the critical problems that arise in applying the dates. Therefore, the following presentation draws heavily on archaeological tree-ring dating as a reservoir of examples pertinent to the points that are raised. The standards and techniques developed for the evaluation of archaeological tree-ring dates are relevant to the application of such dates to other types of phenomenon and, indeed, have important implications for the evaluation of dates derived by different chronometric methods.

One of the primary considerations in any effort to determine the temporal placement of an event through the use of tree-ring dates is the nature of the date itself. A tree-ring date is not expressed in terms of statistical probability; it is an absolute calendar date assigned to the outermost ring on a sample by the process of matching each ring on the sample to a corresponding ring on a dated master chronology. Thus, every ring is correlated with the year in which it was produced. Under ideal circumstances, the outside ring on the sample represents the final annual increment of growth in the tree's life.

Despite the fact that all tree-ring dates are absolute, some are more relevant to the dating of associated materials than are others. This involves the distinction between 'cutting' and 'non-cutting' dates. A cutting date records the last year of growth before the tree died or was killed by human or other agency. Cutting dates are indicated by the presence on the samples of bark, of galleries produced by phloem-eating beetle larvae, of a characteristic patina that develops under the bark of trees, or by an overall lack of weathering denoted by the uniformity of the outer ring around the perimeter of a full or partial section.

A non-cutting date, although absolute, does not record the final year of a tree's life. Burning, weathering, decay, intentional shaping, and other processes that remove outside rings or the attributes that identify cutting dates produce this type of situation. A non-cutting date is indicated by the absence of those characteristics that signify cutting dates, or by the presence of evidence for weathering or shaping of the timber. Obviously, non-cutting dates are less useful than cutting dates for prehistoric dating; however, they do permit the establishment of a lower temporal limit for the associated material. Furthermore, the uncertainty of non-cutting dates may be offset by the clustering of many such dates from a single provenience unit (Bannister, 1962; Dean, 1978a, b).

A second factor of importance in the evaluation of prehistoric tree-ring dates is the context in which the timbers that produce the dates are found. Crucial in this regard are the spatial, functional and associational relationships of the dated log to the unit in which it occurred and to other features, including other timbers that were found in the same unit. To give an archaeological example, it is not sufficient to know merely that a dated sample comes from a particular site or even from a specific room within a site. For adequate analysis it must also be known whether the specimen comes from a firepit, from a wall, or from a roof; and if it is from a roof, whether it is from a primary, secondary or lesser beam. Only this kind of provenience control permits the maximum use of tree-ring data.

Another critical factor that must be controlled is sampling. The adequacy or inadequacy of a particular sample from a site can be gauged only on the basis of the nature of the collection itself and the context within which it is found. Since the sampling possibilities of a particular site cannot usually be determined in advance, random sampling procedures are inappropriate, and it is necessary to sample all the timbers that are available.

Sampling problems remain, even with a 100 per cent collection from a particular site. A few small fragments of charcoal from an archaeological site obviously do not constitute a reliable sample of the total number of wooden artifacts that were used during the occupation of the site. On the other hand, a large collection of specimens from a burned structure or well-preserved dry site may approach 100 per cent of the structural wood used in the site. The latter sample, of course, is much more reliable than the former. Nevertheless, even in cases where an adequate sample of structural wood is available, sampling error may be an important factor. If only one date is derived, it makes little difference if there are one or 100 specimens in the collection. Thus the adequacy of the sample can, in the last resort, be determined only after laboratory analysis is completed.

Given adequate control of the above factors, the assessment of archaeological tree-ring dates involves the recognition and elimination of potential dating anomalies, which are defined as temporal disjunctions between the date and the event to which the date is to be applied (Dean, 1978b). Anomalies arise from a wide variety of sources. The use of dead wood, of stockpiled materials, or of timbers salvaged from older contexts produce dates that are earlier than the event to be dated. The repair of an existing feature produces dates that are later than the initial construction of the feature. Anomalies such as these are detected on the basis of discordant relationships between the dates and the other archaeological data from the site. Stratigraphic inconsistencies, or the presence of archaeological materials known to fall in time intervals different from that of the date, indicate anomalies. Physical attributes of the timbers, such as whether one log is more or less weathered than the other timbers in the same feature, often specify potentially anomalous dates.

The most powerful indicator of anomalies is the relationships of the dates to one another. Most dates from a particular provenience unit tend to cluster at the year in which they were cut for use in the feature. Dates that fall outside such a cluster can be identified as potentially anomalous, and evidence for their relevance to the feature in question can be examined more closely. Once dating anomalies have been detected and resolved, precise dates can be applied to various components of the site. This accomplished, the dated features can be arranged to form a chronological synthesis of the site as a whole.

An instructive example of the site-dating process is provided by the analysis of 159 tree-ring dates from Betatakin, a thirteenth century 'cliff dwelling' in northeastern Arizona (Dean, 1969, pp. 40–84). Nearly perfect preservation in the dry rock shelter assured exact provenience control on each sample and afforded a wealth of supplementary data relevant to the evaluation of the dates. As indicated in Figure 4.6, Betatakin was not constructed as a unit. Rather, it grew by the accretion of 'room clusters', each consisting of at least one habitation room and several storage chambers, opening onto an unroofed courtyard.

Figure 4.6: Floor Plan of Betatakin Ruin, Showing Chronology of Village Growth

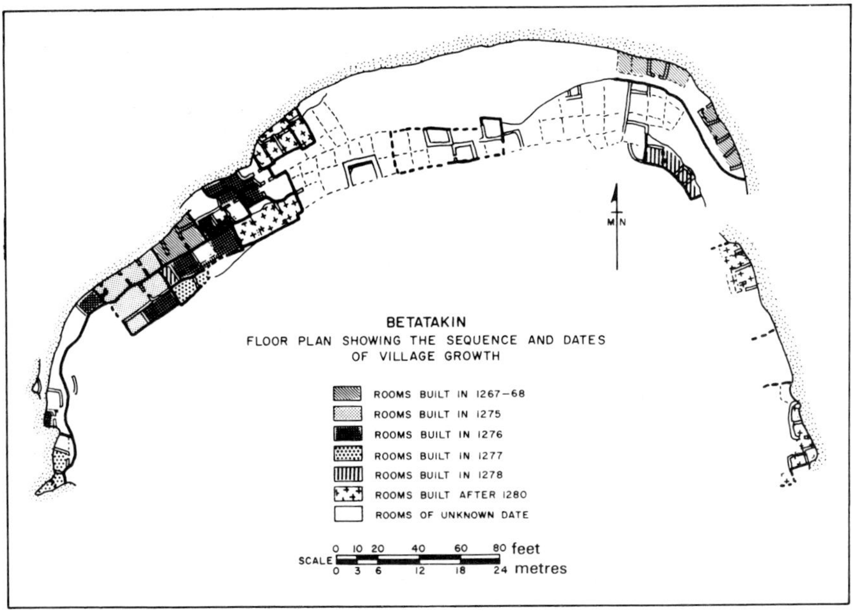

Betatakin was founded in 1267 when at least two room clusters were erected at widely separate loci on the bedrock floor of the rock shelter. After the addition of another cluster in 1268, a hiatus ensued in construction activity that lasted until 1275. Twice during this interval, in 1269 and 1272, trees were felled and reduced to standard-size primary and secondary roof beams that were stockpiled in anticipation of future building activity. None of the stockpiled beams were used until 1275, the first year of a four-year building boom that saw the addition of at least nine room clusters to the village. Betatakin grew less rapidly after 1278. A minimum of five room clusters were built after 1280 to bring the pueblo to its present configuration in 1286. Construction ceased after 1286, and ceramic evidence indicates that Betatakin was abandoned by 1300. Thus, well-controlled tree-ring dates accompanied by comphrehensive

Figure 4.7: Sequence of Dated Archaeological Features from Northern Arizona. Horizontal lines indicate range of dates; vertical bars indicate clusters of dates

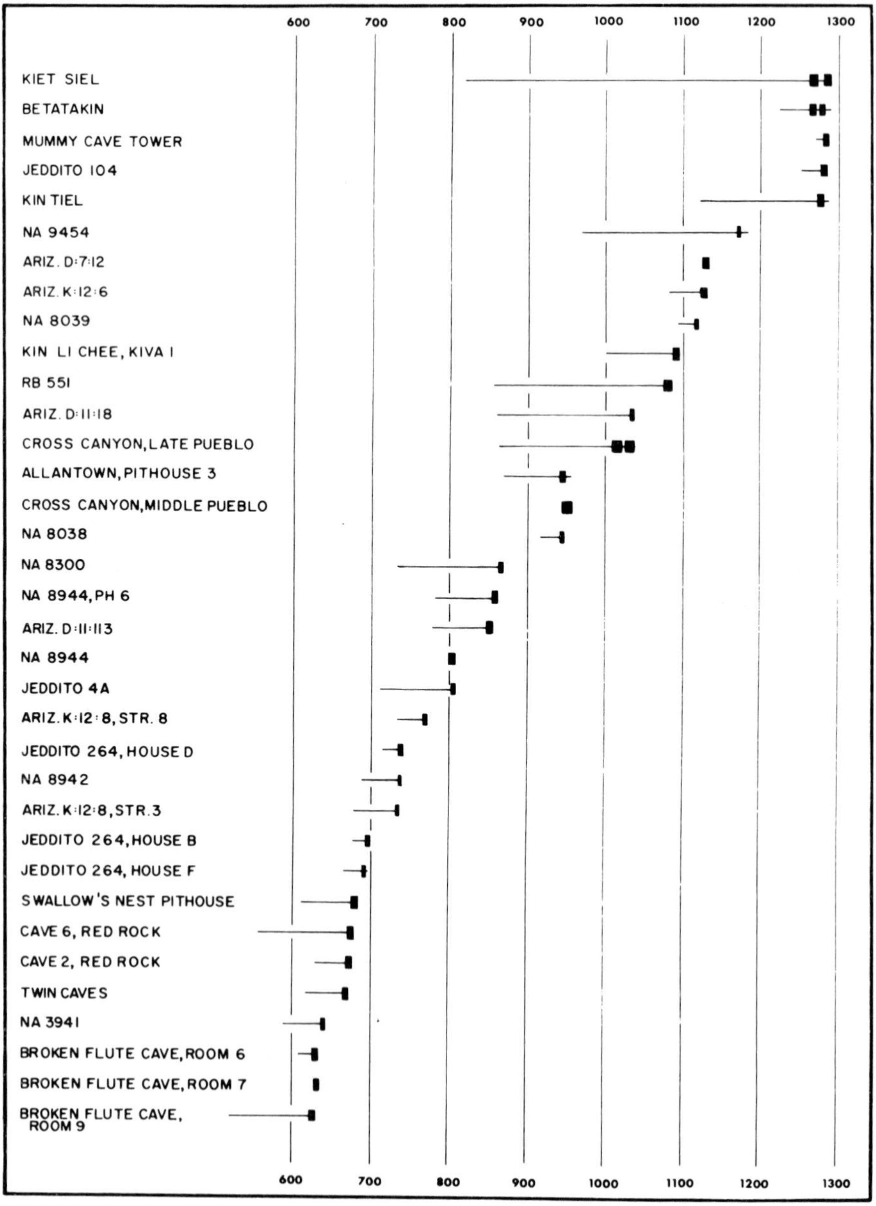

supplementary data allow the sequence of events that characterised the entire occupation span of this fairly large pueblo to be placed within a span of no more than 33 years.

Archaeological tree-ring dating is carried even further to construct areal and regional chronological syntheses, a process that involves the comparative analysis of absolute temporal inter-relationships among sites and groups of sites. High resolution site and feature sequences, such as those illustrated in Figure 4.7, provide unequalled opportunities for the investigation of the processes of sociocultural stability and change on local scales (Dean, 1970, pp. 147–51; Harrill and Breternitz, 1976) or region scales (Cordell and Plog, 1979, pp. 418–24). Conversely, holding absolute time constant permits the study of contemporaneous spatial variability in past human behaviour through the comparison of absolutely contemporaneous sites and features (Dean, 1969, pp. 189–99).

The use of tree-ring dates in archaeology has several implications for the evaluation of dates derived by other chronometric techniques. If it is assumed that the date, however derived, is correct, it is evident that the most critical factor in assessing the date is the nature of the relationship between the object from which it is derived and the event to which it is to be applied. Thus, it is vitally necessary to know whether the dated object is directly or indirectly associated with the feature to be dated (Bannister, 1962). Information such as this can result only from the utilisation of precise provenience controls.

The second problem of prehistoric dating that has been made explicit by archaeological tree-ring dating is that materials of extreme age may be incorporated in much younger contexts and will yield dates far too early for the event being studied. The use in construction of timbers salvaged from older structures, of stockpiled timbers, of dead wood, or of driftwood will produce erroneously early dates whether derived by dendrochronological or radiocarbon analysis. The potential magnitude of errors of this type is indicated by the discovery of eroded bristlecone pine remnants that have been shown by radiocarbon dates to be up to 9,000 years old (Ferguson, 1969, p. 14).

A related problem is that the removal of exterior rings by weathering, burning or shaping causes the date to fall too early for the event to which it is to be related. A classic example of this situation occurred in the dating of Tumulus MM near Gordion, Turkey (Young, 1958), where a radiocarbon date from the exterior of a log in the wall of the burial chamber was much earlier than the historically known date of the tomb. Subsequently, a section removed from the same timber revealed that the radiocarbon sample had been taken from a portion of the log where aproximately 100 rings had been removed by weathering or shaping (Kohler and Ralph, 1961, p. 361). The species of the wood sample is an important variable in this regard, for the weathering of timbers from long-lived trees that produce hundreds or even thousands of rings has a much greater potential impact than does the weathering of timbers from short-lived species.

The most important implication of archaeological dating is that there definitely is strength in numbers. A single date of any kind is of limited value, for any of a number of factors can render the date inapplicable to the event under consideration. The foremost indication of such anomalies is the distribution of multiple dates from a single feature. If several dates from different but associated samples fall in a narrow range of time, the dates can be applied to the event with a high degree of confidence. By the same token, one or two dates that are out of line with the majority probably are not truly applicable to the feature. Thus, the twin keys to the adequate evaluation of any dates are the precise control of context and associations, and the existence of many independent dates from the feature to be dated.

Non-chronological Analysis Of Tree-ring Collections

Among the kinds of non-chronological information that can be extracted from tree-ring collections are reconstructions of past environmental conditions that existed when the sites that yielded the wood samples were produced. One approach to palaeoenvironmental reconstruction involves the application of dendroclimatic techniques of analysis and reconstruction to the ring records of samples recovered from archaeological, geological, or other contexts. This approach has been attempted only recently because of technical problems in ensuring statistical comparability between the living-tree and prehistoric segments of composite chronologies. Dean and Robinson (1977) used a method developed by Fritts (1965) to produce a series of 129 decade maps that represent the temporal and spatial patterning of relative dendroclimatic variability from AD 680 to 1970 in the northern part of the American southwest. More recently, Rose, Dean and Robinson (1981) performed a quantitative reconstruction of palaeoclimate in the area of Santa Fé, New Mexico, that resulted in estimates of annual and spring precipitation for each year from AD 985 through 1970. Similar quantitative retrodictions for other areas of the Southwest are in process.

Another type of environmental information can be derived from the assemblage of tree species represented in a collection of tree-ring samples. The species present give some clues to the environment at the time that the feature being investigated was produced, although in archaeology selection by the inhabitants of the site is an important variable in this type of analysis. Comparisons of archaeological species assemblages with those of the present day can isolate environmental changes that have occurred since the sites were occupied. For example, in Tsegi Canyon, Arizona, there are significant differences between the prehistoric and modern species distributions at the thirteenth century site of Kiet Siel, while no such differences exist at the contemporaneous Betatakin Ruin only 8 km away (Dean, 1969). In this case, human selection and differential preservation can be eliminated, and the obvious inference is that there was an environmental shift near Kiet Siel that did not occur at Betatakin. In a non-cultural context, the presence in the Clark Mountains, California, of bristlecone pine needles in fossil pack rat nests radiocarbon-dated at 20,000 to 30,000 years ago, indicates a significant environmental change, for bristlecone pines are presently limited to elevations much higher than that of these pack rat middens and, in fact, are now altogether absent from the Clark Mountains (Mehringer and Ferguson, 1969).

A second kind of non-chronological information that can be derived from archaeological tree-ring collections concerns human behaviour (Bannister, 1962, 1965; Robinson, 1967). Behavioural inference rests on the fact that objects of wood associated with archaeological sites are artifacts manufactured and used by the sites' inhabitants. By analysing archaeological tree-ring collections as populations of artifacts, a wide range of information can be obtained on human behaviour toward trees as a natural resource and toward wood as a raw material. Among the many behavioural practices that can be recognised through analyses of properly sampled tree-ring collections are: the selection of certain species for particular uses (Dean, 1969, pp. 80–1), the stockpiling of timbers for future use (Bannister, 1965, p. 151), the reuse of beams salvaged from older structures (Ahlstrom, Dean and Robinson, 1978, pp. 37–9), the use of repair timbers (Bannister, 1962, p. 509), the use of dead wood for constructional purposes (Giddings, 1942, pp. 3–8), the peeling and shaping of beams (Robinson, 1967, pp. 47, 49; Scott, 1966, pp. 35–6) and the periodic harvesting of limbs from 'cultivated' living trees (Nichols and Smith, 1965, pp. 57–64).

The interspecies distribution of incomplete and complete terminal rings in a collection of tree-ring samples often indicates seasonality in the felling of trees (Dean, 1969, pp. 77–80; Robinson, 1967, pp. 74–9) and sometimes can be used to differentiate between communal and individual tree-cutting activities (Dean, 1969, p. 80). Marks made by tools on structural beams yield valuable information on the implements and techniques used in felling trees and working wood. Such tool marks on dated beams can be used to date the introduction of different types of implements, such as stone or metal axes, into a particular area (Robinson, 1967, pp. 27–42).

The possibilities of drawing non-chronological inferences from non-archaeological tree-ring collections are limited by the general lack of the variety of associational relationships that form the basis for this type of analysis. Nevertheless, it is possible to infer conditions of deposition and distance of transport from geological tree-ring collections and to specify events such as forest fires (Stokes and Dieterich, 1980) and insect infestations, in the histories of forests.

Applications of Tree-ring Dating

The success of archaeological tree-ring dating in the southwest of North America inspired numerous attempts to extend the method into other regions and to apply it to different classes of phenomena. Predictably, archaeological dating was the focus of most initial attempts to apply dendrochronology outside the southwest of North America. Generally, the dating of other kinds of events and dendroclimatic reconstruction came later. Since 1960, such efforts have proliferated to the extent that it truly can be said that dendrochronology has experienced its maximum expansion during the last 20 years. Today, dendrochronological research encompasses the globe. The following brief survey of tree-ring dating is selective rather than exhaustive and is intended to give only a general indication of the range of accomplishments in the discipline.

In North America, apart from the southwest, systematic archaeological tree-ring dating has been accomplished in the Great Plains, the Mississippi Valley and the Arctic. In the Great Plains, H. E. Weakly (1946, 1949, 1950), W. F. Weakly (1971), and Will (1946) constructed tree-ring chronologies extending back as far as the twelfth century, which have been used for dating prehistoric sites in Nebraska and South Dakota. Since 1970, archaeological tree-ring dating activity in the Great Plains has ceased; nevertheless, the region retains great potential for the dendrochronological dating of past human and natural events.

Early tree-ring research in the Mississippi River basin by Hawley (1941) and Bell (1953) produced some living-tree chronologies and dates for a nineteenth century log cabin, meagre results that caused a long hiatus in eastern woodland tree-ring dating. Research in the 1970s demonstrated the capability of dendrochronology in the region through the dating of 24 nineteenth century wooden structures in Arkansas (Stahle, 1979). Recently documented crossdating in bald cypress, a long-lived species that occurs in historic and prehistoric sites, augurs well for the future of archaeological tree-ring dating in the lower Mississippi Valley. Considerable tree-ring dating has been accomplished in Alaska and the American Arctic, initially by Giddings (1941, 1942, 1948) and later by Oswalt (1949, 1952) and Van Stone (1953). Several local chronologies were built, some extending back into the tenth century. Many Eskimo sites along the Kobuk and Kuskokwim Rivers and on the shores of Kotzebue Sound, Hooper Bay and St Lawrence Island were dated, some as early as the fourteenth century. Archaeological tree-ring dating in this region has lapsed since the middle 1950s, but the Arctic has great potential for further dating research. Modest success with the densitometric dating of seventeenth and eighteenth century samples from sites in the northwestern USA (Jozsa *et al.*, 1980) and southwestern Canada (Jozsa *et al.*, undated) may foreshadow the systematic densitometric dating of ancient wood in the temperate regions of North America.

Nowhere else in the world has dendrochronology expanded so rapidly as in Europe during the last 25 years. It is impossible to detail here the many dating accomplishments of European dendrochronologists. A few examples represent the totality of these burgeoning developments. Several general summary papers and volumes convey an accurate picture of the growth and current scope of tree-ring dating in the region. Baillie (1982, pp. 37–43) provides a succinct history of the origin and development of dendrochronology in Europe. A convenient baseline for assessing the recent growth of the discipline there is provided by Volume 21 of the *Tree-Ring Bulletin*, which summarises European dendrochronology as of 1956. Eckstein (1972) describes the wide variety of tree-ring research that was being pursued by a large number of individuals and institutions only 16 years later, and Eckstein and Wrobel (1983) summarise subsequent developments. The many papers assembled in the *British Archaeological Reports* volume entitled 'Dendrochronology in Europe' (Fletcher, 1978b) and the dendrochronological handbook for archaeologists (Eckstein, Baillie and Egger, 1984) exemplify the current robust state of this rapidly developing field.

European dendrochronology was initiated in Munich in the 1930s when Huber, stimulated by Douglass's work, began constructing long-range tree-ring chronologies in southern Germany (Huber and Giertz, 1970; Huber and von Jazewitsch, 1956; Liese, 1978). In a classic application of chronology building, Huber and his students and colleagues crossdated progressively older archaeological tree-ring series into their living-tree sequences to extend the 'central European' composite chronology back into the sixth century (Becker and Delorme, 1978; Hollstein, 1979). In the process, dates were derived for hundreds of postmedieval and medieval structures and for earlier Roman-Iron Age sites. Hollstein's (1979, p. 162) elegant dating of construction events at Trier Cathedral exemplifies the high degree of temporal resolution that dendrochronology brings to archaeological analysis. Recently, the apparent linkage of the dated chronology with long-floating composite ring sequences from Bronze Age and Neolithic sites in Germany and Switzerland, from subfossil oaks in geological contexts such as the gravels of the Danube River and from bog oaks in northern Germany, produced a provisional chronology of 6,000 years (Pilcher *et al.*, 1984, p. 151).

The high degree of intraregional chronological variability that prevails in northern Germany and the Netherlands (Bauch, 1978a; Eckstein, 1978a) has prohibited the construction of a single regional master tree-ring chronology equivalent to that of central Europe. Despite this limitation, tree-ring dates have been derived from medieval and more recent structures and from oak backing panels on paintings by artists such as Rembrandt and Rubens (Bauch, 1978a,b; Eckstein 1978a; Eckstein, Brongers and Bauch, 1975). The most notable archaeological dating so far is that accomplished at Haithabu, an important Viking commercial centre on the Baltic Sea coast of Schleswig. Recently, a floating chronology of 554 years, the product of the analysis of more than 4,000 samples from Haithabu, was linked with the dated regional sequence. This achievement carried the Schleswig-Holstein tree-ring chronology back to AD 436 and provided firm dates ranging from AD 800 to 1020 for a large number of features at Haithabu (Eckstein, 1978b).

Developments in the British Isles have kept pace with those on the Continent. Tree-ring research by a number of individuals and institutions has produced several chronologies, of which some have only local significance, some demonstrate crossdating throughout the British Isles (Hillam, 1980), and some correlate with European sequences (Fletcher, 1978a; Hillam, 1980). Chronologies from England, the oldest of which extends back into the fifth century AD (Baillie, 1982, pp. 193–5), have permitted the dating of many structures, painting panels and other wooden artifacts (Fletcher, 1978a, b; Hillam, 1980; Morgan and Schofield, 1978; Siebenlist-Kerner, Schove and Fletcher, 1978). Tree-ring samples from wooden trackways across marshlands in Somerset produced a 314-year floating chronology that permits the relative dating of the walkways (Morgan, Coles and Orme, 1978). Radiocarbon placement of these trackways in the first three millennia BC establishes the potential for the eventual development of extremely long, continuous tree-ring sequences in England.

In Northern Ireland, a continuous chronology that reaches back into the first century BC (Baillie, 1982, p. 193) provides a foundation for the temporal placement of numerous medieval structures including several early medieval horizontal mills that date from the seventh to the tenth centuries (Baillie, 1973, 1977a, 1982, pp. 175–95). Similarly, medieval structural elements and boat timbers recovered from excavations in Dublin in the Irish Republic have yielded dates (Baillie, 1977b, 1978, 1982, pp. 164–9). Like other areas of Europe, Northern Ireland possesses an ancient floating chronology that ultimately may permit the derivation of an extremely long continuous regional tree-ring record (Pilcher, 1973; Pilcher *et al.*, 1977). This 2,990-year floating sequence, derived from subfossil oaks preserved in peat deposits, is known on the basis of radiocarbon determinations to date in the 3,900 to 900 BC interval. In southern Scotland, the construction of a 1,030-year oak chronology (AD 946–1976) produced dates for several medieval structures, including Glasgow Cathedral, and helped establish the existence of crossdating throughout the British Isles (Baillie, 1977c, 1982, pp. 159–63).

Recently, cooperative research by European dendrochronologists has achieved a chronology building breakthrough of major proportions (Pilcher *et al.*, 1984). Cross-correlations among the Northern Ireland sequences, the northern Germany bog oak series, a sequence from western

Germany, a corrected version of the south central Germany sequence and a number of shorter ring series from Northern Ireland and Germany established a continuous 7,272-year tree-ring chronology for western Europe that extends back to 5289 BC. This sequence is comparable in length to the 8,800-year bristlecone pine record and provides for the first time an independent check of the bristlecone pine-based radiocarbon calibration. Furthermore, this chronology building accomplishment provides a foundation for the absolute dating of Bronze Age and Neolithic sites from Germany and Switzerland (Egger, 1983).

Although recent dendrochronological research in the USSR has been focused on climatic relationships and reconstructions (Kairiukstis, 1978), in the 1960s Russia was the locus of one of the most successful archaeological applications of dendrochronology. In this study, Kolchin analysed wood recovered from the medieval levels of Novgorod, located 160 kilometres southeast of Leningrad (Thompson, 1967). A floating chronology of 579 years was developed from pine logs recovered from 28 superimposed levels of now-buried corduroy roads that formed the streets of the medieval city. Crossdating between this floating chronology and ring series derived from wood in five historically dated churches established absolute dates for the floating sequence. As a result, the lowest street level was dated at AD 953 and the topmost at 1462. In addition, a number of well-preserved wood houses and workshops associated with the various street levels were dated.

Archaeological tree-ring research in Anatolia began in the early 1960s when the Laboratory of Tree-Ring Research developed an 806-year floating chronology from juniper logs in Tumulus MM near Gordion (Bannister, 1970). Historical dating of the tomb to about 725 BC established that the floating chronology extended back into the sixteenth century BC and raised intriguing possibilities for future relative and perhaps absolute tree-ring dating in Anatolia. Building on this potential, Kuniholm has collected thousands of samples from living trees and archaeological sites on both sides of the Aegean Sea. His research has produced dated living tree chronologies as well as several floating archaeological sequences from sites and structures of Byzantine, Phrygian, Hittite, Bronze Age, and Neolithic affiliation (Kuniholm and Striker, 1983). Unification of these chronology segments would produce not only absolute dates for the archaeological features that contributed samples to the floating sequences, but would create yet another extremely long, continuous tree-ring chronology extending far back into the BC period.

Geological events can be dendrochronologically dated, provided wood or charcoal is associated with the relevant deposits. Tree-ring dated stumps and archaeological sites buried in the Holocene alluvium along streamcourses on the Colorado Plateau provide an absolute chronology for alternating episodes of aggradation and erosion during the last 2,000 years

(Euler *et al.*, 1979). Similar studies have been undertaken in Alaska (Giddings, 1938) and the Netherlands (Vogel, Casparie and Munaut, 1969) where floating ring chronologies from buried trees provide relative dates for the deposition of silt and peat deposits. The degree of exposure of tree-ring dated roots of living bristlecone pine trees has been used to measure rates of slope degradation in the White Mountains of California (LaMarche, 1968).

A classic instance of geological dendrochronology is the dating of the initial eruptions of Sunset Crater, a cinder cone near Flagstaff, Arizona (Breternitz, 1967; Smiley, 1958). Distinctive red ash ejected from the volcano occurs in the fills of prehistoric pithouses that were occupied before the eruption and is absent from structures built after the eruption. Tree-ring dates from pre-eruptive and post-eruptive pithouses combined with a marked discontinuity in the growth patterns of contemporaneous trees identified two early eruptions: one between the tree growing seasons of AD 1064 and 1065, and one between the 1066 and 1067 growing seasons.

Dendrochronology has been frequently utilised to assign absolute or approximate dates to events in the histories of plant communities. Occurrences such as logging, fires, frost injury, insect infestation, defoliation and porcupine damage can be recognised and dated as a result of the distinctive markers they leave in the ring series of the affected trees (Baillie, 1982, Plates 1f, 2c; Erdman, 1970; Estes, 1970; LaMarche, 1970; Spencer, 1964; Stokes and Dieterich, 1980). Elevational changes in the upper and lower boundaries of forests can be measured and dated by analysing tree species recovered from archaeological (Smiley, 1961) or geological contexts, by dating dead tree remnants that lie beyond the present limits of a species' distribution (LaMarche and Mooney, 1967), and by dating pioneer trees that are in the process of colonising areas beyond the borders of mature forests (Dunwiddie, 1977). Successional events in the histories of forests can be reconstructed by dating dead trees within the forest boundaries and by analysing the age distributions of the different species that make up the forest (Estes, 1970).

An instructive example of the contribution of dendrochronology to the study of seemingly unrelated phenomena is Giddings' (1952) attempt to trace ocean currents by the analysis of driftwood from the Arctic shores of Alaska and western Canada. The existence of three distinct, well-documented Arctic ring chronologies enabled him to identify the source areas of many of the drift logs by matching their ring patterns with one or another of the regional chronologies. He found that driftwood west of Point Barrow was predominantly Alaskan with an admixture of Siberian larch. East of Point Barrow, no Alaskan or Siberian wood was recognised, and all crossdating was with the MacKenzie River delta chronology. He was thus able to delimit two currents: one flowing northward along the west coast of Alaska and another flowing westward along the northern coast of Canada, the two currents apparently meeting near Point Barrow and swinging northward.

Because of its accuracy, precision and high resolution, dendrochronology is used as a standard for the evaluation and calibration of other chronometric methods. The most notable of these efforts is the dendrochronological calibration of the radiocarbon dating system (Damon, Lerman and Long, 1978; Damon, Long and Grey, 1966; Ferguson, 1969, 1970b,c; Klein *et al.*, 1982; Olsson, 1970; Ralph and Michael, 1967; Ralph, Michael and Han, 1973; Stuiver and Suess, 1966; Suess, 1965, 1967, 1970). Radiocarbon age determinations run on tree-ring dated wood samples reveal the relationships between the C^{14} time scale and absolute time. Because the 8,800-year bristlecone pine sequence is the longest continuous tree-ring chronology, radiocarbon-tree-ring calibration relies heavily on that species; however, samples from other contexts — such as sequoia and redwood trees, European oaks, and historically dated archaeological materials — have contributed substantially to this effort.

The absolute time scale of the tree-ring chronologies serves as a control for evaluating the accuracy of and variations in radiocarbon determinations run on the absolutely dated wood samples. These studies reveal systematic differences between known sample ages and calcuated C^{14} ages. With increasing age beyond 2,000 years the radiocarbon dates become progressively too young so that C^{14} dates of samples in the 5000 BC range are about 750 years too young (Michael and Ralph, 1970, p. 111; Ralph and Michael, 1967, p. 10). Superimposed on this long-term trend are higher frequency oscillations (Suess, 1970) that are the basis for studies of variability in the production of C^{14} in the upper atmosphere and of fluctuations in the atmospheric and terrestrial reservoirs of radioactive carbon.

This dendrochronological-radiocarbon research has produced several methods for estimating the true age of C^{14}-dated material. One technique involves the use of correction factors, which are calculated from the deviations of the radiocarbon time scale from absolute time, in order to bring C^{14} dates closer to the true ages of the dated samples (Damon *et al.*, 1974; Klein *et al.*, 1982; Ralph *et al.*, 1973). Another technique (Ferguson, Huber and Suess, 1966; Suess and Strahm, 1970) has been applied when floating, crossdated tree-ring chronologies of at least a few hundred years can be constructed from wood recovered from the prehistoric sites involved. Wood samples that represent a succession of known points on the floating tree-ring sequence are radiocarbon-dated. The sinuous curve of the high frequency departures of the C^{14} dates from the true annual intervals separating the sample points is compared with the curve of the tree-ring calibrated radiocarbon time scale.The interval in which the 'wiggles' of the floating curve match the high frequency sinuosities of the tree-ring calibrated curve denotes the true period of time spanned by the former.

Dendrochronological calibration of the radiocarbon time scale has important implications for archaeological chronologies in regions in which dating depends on C^{14} determinations. For example, the revelation that radiocarbon dates become progressively too young with increasing age has sparked a heated debate over the chronology of European prehistory. Some scholars have taken the revisions of the radiocarbon time scale as an opportunity to redefine the relationships between the prehistoric societies of northern Europe and the high civilisations of the Mediterranean basin (Renfrew, 1973). Other prehistorians have been moved to either reject the utility of C^{14} dates or to argue that, for obscure reasons, the bristlecone pine calibration does not apply in lower elevation areas such as Europe and the Near East (McKerrell, 1975). Much of the impetus behind the current wave of European interest in dendrochronology stems from a desire to construct millennia-long, low elevation tree-ring chronologies that can be used to test independently the bristlecone pine calibration. A few such tests have been made using wood from floating chronologies in the BC period (Pearson *et al.*, 1977) and dated wood representing the early portion of the continuous central European oak sequence (Suess, 1978, pp. 1–2).

Although some of these tests appeared to refute the validity of the high frequency 'wiggles' (Pearson *et al.*, 1977), all supported the general long-term departure of radiocarbon dates from absolute time. Subsequent analysis tends to support the existence of high frequency fluctuations around the general trend (Baillie, 1982, pp. 225–8). Much more research based on tree-ring dated samples is necessary to resolve these important issues. The recent construction in western Europe of a 7,272-year chronology should provide sufficient independent data from a wide enough variety of habitats to advance materially resolution of the relationships between radiocarbon years and the scale of absolute time.

The foregoing selective, and incomplete, survey of some of the ways in which dendrochronology has been employed is intended primarily to illustrate the broad range of problems, situations, and geographical areas to which tree-ring dating can be applied. The studies mentioned have only scratched the surface of the potential contributions of dendrochronology. New applications of the method are continually being devised and tested. It is safe to predict that as more intensive and sophisticated tree-ring studies are made, the value of dendrochronology to the fields enumerated above and to problems not yet envisioned will continue to grow.

References

Alstrom, R. V. N., Dean, J. S. and Robinson, W. J. (1978) *Tree-Ring Studies of Walpi Pueblo*, Laboratory of Tree-Ring Research', The University of Arizona, Tucson

Baillie, M. G. L. (1973) 'A Recently Developed Irish Tree-Ring Chronology', *Tree-Ring Bulletin*, *33*, pp. 15–28

—— (1977a) 'The Belfast Oak Chronology to AD 1001', *Tree-Ring Bulletin*, *37*, pp. 1–12.

—— (1977b) 'Dublin Medieval Dendrochronology', *Tree-Ring Bulletin*, *37*, pp. 13–20

—— (1977c) 'An Oak Chronology for South Central Scotland', *Tree-Ring Bulletin*, *37*, pp. 33–44

—— (1978) 'Dating of Some Ships' Timbers from Woodquay, Dublin', in J. Fletcher (ed.), 'Dendrochronology in Europe', *British Archaeological Reports, International Series 51*, pp. 259–62

—— (1982) *Tree-Ring Dating and Archaeology*, Croom Helm, London; The University of Chicago Press, Chicago

—— and Pilcher, J. R. (1973) 'A Simple Crossdating Program for Tree-Ring Research', *Tree-Ring Bulletin*, *33*, pp. 7–14

Bannister, B. (1962) 'The Interpretation of Tree-Ring Dates'. *American Antiquity*, *27* (4), pp. 508–14

—— (1965) 'Tree-Ring Dating of the Archaeological Sites in the Chaco Canyon Region, New Mexico', *Southwestern Monuments Association Technical Series*, *6*, part 2, pp. 117–201

—— (1969) 'Dendrochronology', in D. Brothwell and E. Higgs (eds.), *Science in Archaeology: A Survey of Progress and Research* (revised and enlarged), Thames and Hudson, London

—— (1970) 'Dendrochronology in the Near East: Current Research and Future Potentialities', *Proceedings of the Seventh International Congress of Anthropological and Ethnological Sciences*, *5*, pp. 336–40

Bauch, J. (1978a) 'Tree-Ring Chronologies for the Netherlands', in J. Fletcher (ed.) 'Dendrochronology in Europe,' *British Archaeological Reports, International Series 51*, pp. 133–7

—— (1978b) 'Dendrochronology Applied to the Dating of Dutch, Flemish, and German Paintings', in J. Fletcher (ed.), 'Dendrochronology in Europe', *British Archaeological Reports, International Series 51*, pp. 307–14

—— and Eckstein, D. (1970) 'Improvement of a Dendrochronological Method for Dating in Central Europe', in J. H. G. Smith and J. Worrall (eds.), 'Tree-Ring Analysis with Special Reference to Northwest America', *The University of British Columbia Faculty of Forestry Bulletin*, no. 7, pp. 37–40

Becker, B. (1978) 'Dendrochronological Zones of Central European Forest Communities', in J. Fletcher (ed.), 'Dendrochronology in Europe', *British Archaeological Reports, International Series 51*, pp. 101–14

—— and Delorme (1978) 'Oak Chronologies for Central Europe: Their Extension from Medieval to Prehistoric Times', in J. Fletcher (ed.), 'Dendrochronology in Europe', *British Archaeological Reports, International Series 51*, pp. 59–64.

Bell, R. E. (1953) 'Dendrochronology in the Mississippi Valley', in J. B. Griffin (ed.), *Archaeology in the Eastern United States*, University of Chicago Press, Chicago, pp. 345–51

Berger, A. L., Guiot, J., Mathieu, L. and Munaut, A. V., 'Tree-Rings and Climate in Morocco', *Tree-Ring Bulletin*, 39, pp. 61–75

Blasing, T. J., and Fritts, H. C., (1975) 'Past Climate of Alaska and North-western Canada as Reconstructed from Tree Rings', in G. Weller and S. A. Bowling (eds.), 'Climate of the Arctic', *Proceedings of the 24th Alaskan Science Conference, August 1973*, pp. 48–58

—— (1976) 'Reconstructing Past Climatic Anomalies in the Northern Pacific and Western North America from Tree-Ring Data', *Quaternary Research*, 6, pp. 563–79

Blasing, T. J., and Lofgren, G. R., (1980) 'Seasonal Climatic Anomaly Types for the North Pacific Sector and Western North America', *Monthly Weather Review*, 108, pp. 700–19

Breternitz, D. A. (1967) 'Eruptions of Sunset Crater: Dating and Effects', *Plateau, 40* (2), pp. 72–6

Brett, D. W. (1978) 'Dendrochronology of Elm in London', *Tree-Ring Bulletin*, *38*, pp. 35–44

Conkey, L. E. (1979) 'Response of Tree-Ring Density to Climate in Maine, U.S.A.', *Tree-Ring Bulletin, 39*, pp. 29–38
Cook, E. R. and Jacoby, G. C., Jr. (1979) 'Evidence for Quasi-Periodic July Drought in the Hudson Valley, New York', *Nature, 282*, pp. 390–2
Cook, E. R. and Peters, K. (1981) 'The Smoothing Spline: A New Approach to Standardizing Forest Interior Tree-Ring Width Series for Dendroclimatic Studies', *Tree-Ring Bulletin, 41*, pp. 45–53
Cordell, L. S. and Plog, F. (1979) 'Escaping the Confines of Normative Thought: A Reevaluation of Puebloan Prehistory', *American Antiquity, 44*, pp. 405–29
Currey, D. R. (1965) 'An Ancient Bristlecone Pine Stand in Eastern Nevada', *Ecology, 46*, pp. 564–6
Damon, P. E., Ferguson, C. W., Long, A. and Wallick, E. I. (1974) 'Dendrochronological Calibration of the Radiocarbon Time Scale', *American Antiquity, 39*. pp. 350–66
Damon, P. E., Lerman, J. C. and Long, A. (1978) 'Temporal Fluctuations of Atmospheric ^{14}C: Causal Factors and Implications', *Annual Review of Earth and Planetary Sciences, 6*, pp. 457–94
Damon, P. E., Long, A. and Grey, D. C. (1966) 'Fluctuation of Atmospheric C-14 During the Last Six Millennia', *Journal of Geophysical Research, 71*, pp. 1055–63
Dean, J. S. (1969) 'Chronological Analysis of Tsegi Phase Sites in North-eastern Arizona', *Papers of the Laboratory of Tree-Ring Research*, no. 3, The University of Arizona Press, Tucson
—— (1970) 'Aspects of Tsegi Phase Social Organization: A Trial Reconstruction' in W. A. Longacre (ed.), *Reconstructing Prehistoric Pueblo Societies*, University of New Mexico Press, Albuquerque, pp. 140–74
—— (1975) *Tree-Ring Dates from Colorado W: Durango Area*, Laboratory of Tree-Ring Research, The University of Arizona, Tucson
—— (1978a) 'Tree-Ring Dating in Archaeology' in J. D. Jennings (ed.), 'Miscellaneous Collected Papers 19–24,' *University of Utah Anthropological Papers*, no. 99, University of Utah Press, Salt Lake City, pp. 129–63
—— (1978b) 'Independent Dating in Archaeological Analysis' in M. B. Schiffer (ed.), *Advances in Archaeological Method and Theory, 1*, Academic Press, New York, pp. 223–55
—— and Robinson, W. J. (1977) *Dendroclimatic Variability in the American Southwest, A.D. 680 to 1970*, Laboratory of Tree-Ring Research, The University of Arizona, Tucson
Douglass, A. E. (1914) 'A Method of Estimating Rainfall by the Growth of Trees', in E. Huntington, 'The Climatic Factor as Illustrated in Arid America', *Carnegie Institution of Washington Publication 192*, p. 101–21
—— (1919) 'Climatic Cycles and Tree Growth I: A Study of the Annual Rings in Relation to Climate and Solar Acitivity', *Carnegie Institution of Washington Publication 289*, vol. I
—— (1929) 'The Secret of the Southwest Solved by Talkative Tree Rings', *National Geographic Magazine, 56* (6), pp. 736–70
—— (1935) 'Dating Pueblo Bonito and Other Ruins of the Southwest', *National Geographic Society Contributed Technical Papers, Pueblo Bonito Series*, no. 1, National Geographic Society, Washington
—— (1937) 'Tree Rings and Chronology', *University of Arizona Bulletin*, vol. 8, no. 4, *Physical Science Bulletin*, no. 1, The University of Arizona, Tucson
Dunwiddie, P. W. (1977) 'Recent Tree Invasion of Subalpine Meadows in the Wind River Mountains, Wyoming', *Arctic and Alpine Research, 9*, pp. 393–9
Eckstein, D. (1972) 'Tree-Ring Research in Europe', *Tree-Ring Bulletin, 32*, pp. 1–18
—— (1978a) 'Regional Tree-Ring Chronologies along Parts of the North Sea Coast' in J. Fletcher (ed.), 'Dendrochronology in Europe', *British Archaeological Reports, International Series 51*, pp. 117–24
—— (1978b) 'Dendrochronological Dating of the Medieval Settlement of Haithabu (Hedeby)' in J. Fletcher (ed.), 'Dendrochronology in Europe', *British Archaeological Reports, International Series 51*, pp. 267–74
——, Baillie, M. G. L. and Egger, H. (1984) *Handbook for Archaeologists No. 2: Dendrochronological Dating*, European Science Foundation, Strasbourg
——, Brongers, J. A. and Bauch, J. (1975) 'Tree-Ring Research in the Netherlands', *Tree-Ring Bulletin, 35*, pp. 1–13
—— and Wrobel, S. (1983) 'Dendrochronologie in Europa', *Dendrochronologia, 1*, pp. 9–20

Egger, H. (1983) 'Die Absolute Datierung der Saône-Rhöne-Kulture und der Bronzezeit der Westschweiz', *Dendrochronologia, 1*, pp. 37–44

Erdman, J. A. (1970) 'Pinyon-Juniper Succession after Natural Fires on Residual Soils of Mesa Verde, Colorado', *Brigham Young University Science Bulletin, Biological Series*, vol. 11, no. 2, Brigham Young University, Provo

Estes, E. T. (1970) 'The Dendrochronology of Black Oak (*Quercus velutina* Lam.), White Oak (*Quercus alba* L.), and Shortleaf Pine (*Pinus echinata* Mill.), in the Central Mississippi Valley', *Ecological Monographs, 40*, pp. 295–316

Euler, R. C., Gumerman, G. J., Karlstrom, T. N. V., Dean, J. S. and Hevly, R. H. (1979) 'The Colorado Plateaus: Cultural Dynamics and Paleoenvironment', *Science, 205*, pp. 1089–1101

Ferguson, C. W. (1969) 'A 7104-Year Annual Tree-Ring Chronology for Bristlecone Pine, *Pinus aristata*, from the White Mountains, California', *Tree-Ring Bulletin, 29* (3–4), pp. 3–29

—— (1970a) 'Concepts and Techniques of Dendrochronology', in R. Berger (ed.), *Scientific Methods in Medieval Archaeology*, University of California Press, Los Angeles, pp. 183–200

—— (1970b) 'Bristlecone Pine Chronology and Calibration of the Radiocarbon Time Scale' in J. H. G. Smith and J. Worrall (eds.), 'Tree-Ring Analysis with Special Reference to Northwest America', *The University of British Columbia Faculty of Forestry Bulletin*, no. 7, pp. 88–91

—— (1970c) 'Dendrochronology of Bristlecone Pine, *Pinus aristata*: Establishment of a 7484-Year Chronology in the White Mountains of Eastern-Central California, U.S.A.', in I. U. Olsson (ed.) *Radiocarbon Variations and Absolute Chronology: Proceedings of the Twelfth Nobel Symposium*, Almqvist and Wiksell, Stockholm, pp. 237–59

—— (1979) 'Dendrochronology of Bristlecone Pine, *Pinus longaeva*', *Environment International, 2*, pp. 209–14

——, Huber, B. and Suess, H.E. (1966) 'Determination of the Age of Swiss Lake Dwellings as an Example of Dendrochronologically-Calibrated Radiocarbon Dating', *Zeitschrift fur Naturforschung, 21a*, pp. 1173–7

Fletcher, J. (1978a) 'Oak Chronologies for Eastern and Southern England: Principles for Their Construction and Application: Their Comparison with Others in North-West Europe' in J. Fletcher (ed.), 'Dendrochronology in Europe', *British Archaeological Reports, International Series 51*, pp. 139–56

—— (1978b) 'Tree-Ring Analysis of Panel Paintings' in J. Fletcher (ed.), 'Dendrochronology in Europe', *British Archaeological Reports, International Series 51*, pp. 303–6

—— Ed. (1978) 'Dendrochronology in Europe', *British Archaeological Reports, International Series 51*

—— and Hughes, J. F. (1970) 'Uses of X-Rays for Density Determinations and Dendrochronology' in J. H. G. Smith and J. Worrall (eds.), 'Tree-Ring Analysis with Special Reference to Northwest America', *The University of British Columbia Faculty of Forestry Bulletin*, no. 7, pp. 41–54

Fritts, H. C. (1963) 'Computer Programs for Tree-Ring Research', *Tree-RingBulletin, 25* (3–4), pp. 2–7

—— (1965) 'Tree-Ring Evidence for Climatic Changes in Western North America', *Monthly Weather Review, 93*, pp. 421–43

—— (1966) 'Growth Rings of Trees: Their Correlation with Climate', *Science, 154*, pp. 973–9

—— (1970) 'Tree-Ring Analysis and Reconstruction of Past Environments' in J. H. G. Smith and J. Worrall (eds.), 'Tree-Ring Analysis with Special Reference to Northwest America', *The University of British Columbia Faculty of Forestry Bulletin*, No. 7, pp. 92–8

—— (1971) 'Dendroclimatology and Dendroecology', *Quaternary Research, 1*, pp. 419–49

—— (1974) 'Relationships of Ring Widths in Arid-Site Conifers to Variations in Monthly Temperature and Precipitation', *Ecological Monographs, 44*, pp. 411–40

—— (1976) *Tree Rings and Climate*, Academic Press, New York

—— (1978) 'Tree Rings, A Record of Seasonal Variations in Past Climate', *Naturwissenschaften, 65*, pp. 48–56

—— (1981) 'Statistical Climatic Reconstructions from Tree-Ring Widths', in A. Berger (ed.), *Climatic Variations and Variability: Facts and Theories*, D. Reidel Publishing Company, Dordrecht, pp. 135–53

—— and Blasing, T. J., Hayden, B. P. and Kutzbach, J. E. (1971) 'Multivariate Techniques for Specifying Tree-Growth and Climate Relations and for Reconstructing Anomalies in Paleoclimate', *Journal of Applied Meteorology, 10*, pp. 845–64

——, Mosimann, J. E. and Bottorff, C. P. (1969) 'A Revised Computer Program for Standardizing Tree-Ring Series', *Tree-Ring Bulletin, 29* (1–2), pp. 15–20

——, Smith, D. G., Cardis, J. W. and Budelsky, C. A. (1965) 'Tree-Ring Characteristics Along a Vegetation Gradiant in Northern Arizona', *Ecology, 46*, pp. 393–401

Giddings, J. L., Jr (1938) 'Buried Wood from Fairbanks Alaska', *Tree-Ring Bulletin, 4* (4), pp. 3–5

—— (1941) 'Dendrochronology in Northern Alaska', *University of Alaska Publications*, vol. 4, *University of Arizona Bulletin*, vol. 12, no. 4, *Laboratory of Tree-Ring Research Bulletin*, no. 1, University of Alaska and University of Arizona, Tucson

—— (1942) 'Dated Sites on the Kobuk River, Alaska', *Tree-Ring Bulletin, 9* (1), pp. 2–8

—— (1948) 'Chronology of the Kobuk-Kotzebue Sites', *Tree-Ring Bulletin, 14* (4), pp. 20–32

—— (1952) 'Driftwood and Problems of Arctic Sea Currents', *Proceedings of the American Philosophical Society*, vol. 96, no. 2, pp. 129–42

Glock, W. S., Studhalter, R. A. and Agerter, S. R. (1960) 'Classification and Multiplicity of Growth Layers in the Branches of Trees at the Extreme Lower Forest Border', *Smithsonian Miscellaneous Collections*, vol. 140, no. 1

Graybill, D. A. (1979) 'Revised Computer Programs for Tree-Ring Research', *Tree-Ring Bulletin, 39*, pp. 77–82

Harrill, B. G. and Breternitz, C. D. (1976) 'Chronology and Cultural Activity in Johnson Canyon Cliff Dwellings, Interpretations from Tree-Ring Data', *Journal of Field Archaeology, 3*, pp. 375–90

Haury, E. W. (1962) 'HH-39: Recollections of a Dramatic Moment in South-western Archaeology', *Tree-Ring Bulletin, 24* (3–4), pp. 11–14

Hawley, H. (1941 'Tree Ring Analysis and Dating in the Mississippi Drainage', *University of Chicago Occasional Papers in Anthropology*, no. 2, University of Chicago Press, Chicago

Hecht, A. D. (ed.) (1985) *Paleoclimate Analysis and Modeling*, John Wiley and Sons, New York.

Heizer, R. F. (1956) 'The First Dendrochronologist', *American Antiquity, 22*, pp. 186–8

Hillam, J. (1980) 'A Medieval Oak Tree-Ring Chronology from Southwest England', *Tree-Ring Bulletin, 40*, pp. 13–20

Hollstein, E. (1979) *Mitteleuropaische Eichenchronologie*, Mainz am Rhein

Huber, B. and Giertz, V. (1970) 'Central Euorpean Dendrochronology for the Middle Ages' in R. Berger (ed.), *Scientific Methods in Medieval Archaeology*, University of California Press, Los Angeles, pp. 201–12

Huber, B. and von Jazewitsch, W. (1956) 'Tree Ring Studies of the Forestry-Botany Institutes of Tharandt and Munich', *Tree-Ring Bulletin, 21*, pp. 28–30

Hughes, M. K., Leggett, P., Milsom, S. J. and Hibbert, F. A. (1978) 'Dendrochronology of Oak in North Wales', *Tree-Ring Bulletin, 38*, pp. 15–23

Jones, F. W. and Parker, M. L. (1970) 'G.S.C. Tree-Ring Scanning Densitometer and Data Acquisition System', *Tree-Ring Bulletin, 30*, pp. 23–31

Jozsa, L. A., Parker, M. L., and Bramhall, P. A. (1980) *Feasibility of Tree-Ring Dating at Ozette*, Forintek Canada Corporation, Western Forest Products Laboratory, Vancouver

Jozsa, L. A., Parker, M. L., Bramhall, P. A., Kellogg, R. M. and Rowe, S. (n.d.) *Wood and Charcoal Sample Analysis for the Kitwanga National Historic Site*, Forintek Canada Corporation, Western Forest Products Laboratory, Vancouver

Kairiukstis, L. A. (1978) 'Dendroclimatic and Dendrochronological Investigations in the USSR' in J. Fletcher (ed.), 'Dendrochronology in Europe', *British Archaeological Reports, International Series 51*, pp. 55–8

Klein, J., Lerman, J. C., Damon, P. E. and Ralph, E. K. (1981) 'Calibration of Radiocarbon Dates: Tables Based on the Consensus Data of the Workshop on Calibrating the Radiocarbon Time Scale', *Radiocarbon 1982, 24* (2), pp. 103–50

Kohler, E. L. and Ralph, E. K. (1961) 'C-14 Dates for Sites in the Mediterranean Area', *American Journal of Archaeology, 65* (4), pp. 357–67

Kuniholm, P. I. and Striker, C. L. (1983) 'Dendrochronological Investigations in the Aegean and Neighboring Regions, 1972–1982', *Journal of Field Archaeology, 10*, pp. 411–20

LaMarche, V. C. Jr (1968) 'Rates of Slope Degradation as Determined from Botanical Evidence White Mountains California' in 'Erosion and Sedimentation in a Semiarid En-

vironment', *United States Geological Survey Professional Paper 352–I*, pp. 341–77
—— (1970) 'Frost-Damage Rings in Subalpine Conifers and Their Application to Tree-Ring Dating Problems', in J. H. G. Smith and J. Worrall (eds.), 'Tree-Ring Analysis with Special Reference to Northwest America', *The University of British Columbia Faculty of Forestry Bulletin*, no. 7, pp. 99–100
—— and Mooney, H. A. (1967) 'Altithermal Timberline Advance in Western United States', *Nature, 213*, pp. 980–2
Liese, W. (1978) 'Bruno Huber: The Pioneer of European Dendrochronology' in J. Fletcher (ed.), 'Dendrochronology in Europe', *British Archaeological Reports, International Series 51*, pp. 1–10
McKerrel, M. (1975) 'Correction Procedures for C-14 Dates' in T. Watkins (ed.), *Radiocarbon: Calibration and Prehistory*, Edinburgh, pp. 47–100
Meko, D. M., Stockton, C. W. and Boggess, W. R. (1980) 'A Tree-Ring Reconstruction of Drought in Southern California', *Water Resources Bulletin, 16*, pp. 594–600
Mehringer, P. J. and Ferguson, C. W. (1969) 'Pluvial Occurrence of Bristlecone Pine (*Pinus aristata*) in a Mohave Desert Mountain Range', *Journal of the Arizona Academy of Science, 5*, pp. 284–92
Michael, H. N. and Ralph, E. K. (1970) 'Correction Factors Applied to Egyptian Radiocarbon Dates from the Era Before Christ' in I. U. Olsson (ed.), *Novel Symposium 12: Radiocarbon Variations and Absolute Chronology*, Almqvist and Wiksell, Stockholm, pp. 109–19
Mitchell, J. M. Jr, Stockton, C. W. and Meko, D. M. (1979) 'Evidence of a 22-Year Rhythm of Drought in the Western United States Related to the Hale Solar Cycle since the 17th Century' in B. M. McCormac and T. A. Seliga (eds.), *Solar Terrestrial Influences on Weather and Climate*, D. Reidel Publishing Company, Dordrecht, pp. 125–43
Morgan, R. A., Coles, J. M. and Orme, B. J. (1978) 'Tree Ring Studies in the Somerset Levels' in J. Fletcher (ed.), 'Dendrochronology in Europe', *British Archaeological Reports, International Series 51*, pp. 211–22
Morgan, R. and Schofield, J. (1978) 'Tree-Rings and the Archaeology of the Thames Waterfront in the City of London' in J. Fletcher (ed.), 'Dendrochronology in Europe', *British Archaeological Reports, International Series 51*, pp. 223–38
Nash, T. H. III, Fritts, H. C. and Stokes, M. A. (1975) 'A Technique for Examining Non-Climatic Variation in Widths of Annual Tree Rings with Special Reference to Air Pollution', *Tree-Ring Bulletin, 35*, pp. 15–24
Nichols, R. F. and Smith, D. G. (1965) 'Evidence of Prehistoric Cultivation of Douglas-Fir Trees at Mesa Verde' in D. Osborne (ed.) 'Contributions of the Wetherill Mesa Archaeological Project', *Memoirs of the Society for American Archaeology*, no. 19, pp. 57–64
Olsson, I. U. (ed.) (1970) *Nobel Symposium 12: Radiocarbon Variations and Absolute Chronology*, Almqvist and Wiksell, Stockholm
Oswalt, W. (1949) 'Dated Houses at Squirrel River, Alaska', *Tree-Ring Bulletin, 16* (1), pp. 7–8
—— (1952) 'The Archaeology of Hooper Bay Village, Alaska', *Anthropological Papers of the University of Alaska*, vol. 1, no. 1, pp. 47–91
Parker, M. L. (1970) 'Dendrochronological Techniques Used by the Geological Survey of Canada' in J. H. G. Smith and J. Worrall (eds.), 'Tree-Ring Analysis with Special Reference to Northwest America', *The University of British Columbia Faculty of Forestry Bulletin*, no. 7, pp. 55–66
—— and Henoch, E. S. (1971) 'The Use of Engelmann Spruce Latewood Density for Dendrochronological Purposes', *Canadian Journal of Forest Research, 1* (2), pp. 90–8
—— and Meleskie, K. R. (1970) 'Preparation for X-Ray Negatives of Tree-Ring Specimens for Dendrochronological Analysis', *Tree-Ring Bulletin, 30*, pp. 11–22
Pearson, G. W., Pilcher, J. R., Baillie, M. G. L. and Hillam, J. (1977) 'Absolute Radiocarbon Dating Using a Low Altitude European Tree-Ring Calibration', *Nature, 270*, pp. 25–8
Pilcher, J. R. (1973) 'Tree-Ring Research in Ireland', *Tree-Ring Bulletin, 33*, pp. 1–5
—— and Baillie, M. G. L. (1980) 'Six Modern Oak Chronologies from Ireland', *Tree-Ring Bulletin, 40*, pp. 23–34

—— Baillie, M. G. L., Schmidt, B. and Becker, B. (1984) 'A 7,272-year Tree-Ring Chronology for Western Europe', *Nature, 312*, pp. 150–2

—— Hillam, J., Baillie, M. G. L. and Pearson, G. W. (1977) 'A Long Sub-Fossil Oak Tree-Ring Chronology from the North of Ireland', *New Phytology., 79*, pp. 713–29

Polge, H. (1970) 'The Use of X-Ray Densitometric Methods in Dendrochronology', *Tree-Ring Bulletin, 30*, pp. 1–10

—— (1978) 'The Contribution of Wood Density to Dendrochronology and Dendro-climatology' in J. Fletcher (ed.), 'Dendrochronology in Europe', *British Archaeological Reports, International Series 51*, pp. 77–87

Ralph, E. K. and Michael, H. N. (1967) 'Problems of the Radiocarbon Calendar', *Archaeometry, 10*, pp. 3–11

——, Michael, H. N. and Han, M. C. (1973) 'Radiocarbon Dates and Reality', *MASCA Newsletter, 9* (1), pp. 1–20

Renfrew, C. (1973) *Before Civilization: The Radiocarbon Revolution and Prehistoric Europe*, Alfred A. Knopf, New York

Robinson, W. J. (1967) *Tree-Ring Materials as a Basis for Cultural Interpretation*, doctoral dissertation, The University of Arizona, University Microfilms, Ann Arbor

—— and Evans, R. (1980) 'A Microcomputer-Based Tree-Ring Measuring System', *Tree-Ring Bulletin, 40*, pp. 59–64

Rose, M. R., Dean, J. S. and Robinson, W. J. (1981) 'The Past Climate of Arroyo Hondo, New Mexico, Reconstructed from Tree Rings', *Arroyo Hondo Archaeological Series*, vol. 4, School of American Research, Santa Fé

Schulman, E. (1937) 'Some Early Papers on Tree Rings, II. J. C. Kapteyn', *Tree-Ring Bulletin, 3* (4), pp. 28–29

—— (1956) *Dendroclimatic Changes in Semiarid America*, The University of Arizona Press, Tucson

—— (1958) 'Bristlecone Pine, Oldest Known Living Thing', *National Geographic Magazine, 113* (3), pp. 355–72

—— and Baldwin, G. C. (1939) 'The Chronology in OL-12, A Dissected Ponderosa', *Tree-Ring Bulletin, 6* (2), pp. 13–16

Schweingruber, F. H., Bräker, O. U. and Schär, E. (1978) 'X-Ray Densitometric Results for Subalpine Conifers and Their Relationship to Climate' in J. Fletcher (ed.), 'Dendrochronology in Europe', *British Archaeological Reports, International Series 51*, pp. 89–100

——, Fritts, H, C.,Bräker, O. U., Drew, L. G. and Schär, E. (1978) 'The X-Ray Technique as Applied to Dendroclimatology', *Tree-Ring Bulletin, 38*, pp. 61–91

Scott, S. D. (1966) 'Dendrochronology in Mexico', *Papers of the Laboratory of Tree-Ring Research*, no. 2, The University of Arizona Press, Tucson

Siebenlist-Kerner, V., Schove, D. J. and Fletcher, J. (1978) 'The Barn at Great Coxwell, Berkshire' in J. Fletcher (ed.), 'Dendrochronology in Europe', *British Archaeological Reports, International Series 51*, pp. 295–302

Smiley, T. L. (1955) 'The Geochronological Approach' in T. L. Smiley (ed.), 'Geochronology: With Special Reference to Southwestern United States', *University of Arizona Bulletin Series*, vol. 26, no. 2, *Physical Science Bulletin*, no. 2, pp. 15–28

—— (1958) 'The Geology and Dating of Sunset Crater, Flagstaff, Arizona', in R. Y. Anderson and J. W. Harshbarger (eds.), *Guidebook of the Black Mesa Basin, Northeastern Arizona*, New Mexico Geological Society, pp. 186–190

—— (1961) 'Evidences of Climatic Fluctuations in Southwestern Prehistory', *Annals of the New York Academy of Sciences*, vol. 95, art.1, pp. 697–704

——, Stubbs, S. A. and Bannister, B. (1953) 'A Foundation for the Dating of Some Late Archaeological Sites in the Rio Grande Area, New Mexico: Based on Studies in Tree-Ring Methods and Pottery Analysis', *University of Arizona Bulletin*, vol. 24, no. 3, *Laboratory of Tree-Ring Research Bulletin*, no. 6

Smith, L. P. and Stockton, C. W. (1981) 'Reconstructed Stream Flow for the Salt and Verde Rivers from Tree-Ring Data', *Water Resources Bulletin, 17*, pp. 939–47

Spencer, D. A. (1964) 'Porcupine Population Fluctuations in Paste Centuries Revealed by Dendrochronology', *The Journal of Applied Ecology, 1* (1), pp. 127–49

Stahle, D. W. (1979) 'Tree-Ring Dating of Historic Buildings in Arkansas', *Tree-Ring Bulletin, 39*, pp. 1–28

Stallings, W. S. Jr. (1933) 'A Tree-Ring Chronology for the Rio Grande Drainage in Northern New Mexico', *Proceedings of the National Academy of Sciences, 19* (9), pp. 803–6

—— (1937) 'Some Early Papers on Tree-Rings, I. J. Kuechler', *Tree-Ring Bulletin, 3* (4), pp. 27–28

Stockton, C. W. (1975) 'Long-Term Streamflow Records Reconstructed from Tree Rings, *Papers of the Laboratory of Tree-Ring Research*', no. 5, The University of Arizona Press, Tucson

—— and Fritts, H. C. (1971) 'Augmenting Annual Runoff Records Using Tree-Ring Data' in 'Hydrology and Water Resources in Arizona and the Southwest, Vol. 1', *Proceedings of the 1971 Meetings of the Arizona Section — American Water Resources Association and the Hydrology Section — Arizona Academy of Science, Tempe, 1971*, pp. 1–12

Stokes, M. A. and Dieterich, J. H. (eds.) (1980) 'Proceedings of the Fire History Workshop, October 20–24, 1980, Tucson, Arizona', *General Technical Report RM-81*, Rocky Mountain Forest and Range Experiment Station, Forest Service, US Department of Agriculture, Fort Collins

Stokes, M. A. and Smiley, T. L. (1968) *An Introduction to Tree-Ring Dating*, The University of Chicago Press, Chicago

Studhalter, R. A. (1955) 'Tree Growth: I. Some Historical Chapters', *The Botanical Review, 21* (1–3), pp. 1–72

Stuiver, M. and Suess, H. E. (1966) 'On the Relationship Between Radiocarbon Dates and True Age Samples', *Radiocarbon, 8*, pp. 534–40

Suess, H. E. (1965) 'Secular Variations of the Cosmic-Ray-Produced Carbon 14 in the Atmosphere and Their Interpretations', *Journal of Geophysical Research, 70*, pp. 5937–52

—— (1967) 'Bristlecone Pine Calibration of the Radiocarbon Time Scale from 4100 BC to 1500 BC' in *Radioactive Dating and Methods of Low-Level Counting*, International Atomic Energy Agency, Vienna, pp. 143–50

—— (1970) 'Bristlecone-Pine Calibration of the Radiocarbon Time-Scale 5200 BC to the Present' in I. U. Olsson (ed.), *Nobel Symposium 12: Radiocarbon Variations and Absolute Chronology*, Almqvist and Wiksell, Stockholm, pp. 303–9

—— (1978) 'La Jolla Measurements of Radiocarbon in Tree-Ring Dated Wood', *Radiocarbon, 20* , pp. 1–18

—— and Strahm, C. (1970) 'The Neolithic of Auvernier, Switzerland', *Antiquity, 44*, pp. 91–9

Thompson, M. W. (1967) *Novgorod the Great: Excavations at the Medieval City Directed by A. V. Artsikhovsky and B. A. Kolchin*, Frederick A. Praeger, New York

Van Stone, J. W. (1953) 'Notes on Kotzebue Dating', *Tree-Ring Bulletin, 20*(1), pp. 6–8

Vins, B. (1970) 'Methods and Use of Tree Ring Analyses in Czechoslovakia' in J. H. G. Smith and J. Worrall (eds.), 'Tree-Ring Analysis with Special Reference to Northwest America', *The University of British Columbia Faculty of Forestry Bulletin*, no. 7, pp. 67–73

Vogel, J. C., Casparie, W. A. and Munaut, A. V. (1969) 'Carbon-14 Trends in Subfossil Pine Stubs', *Science, 166*, pp. 1143–5

Warren, W. G. (1980) 'On Removing the Growth Trend from Dendrochronological Data', *Tree-Ring Bulletin, 40*, pp. 35–44

—— and MacWilliam, S. L. (1981) 'Test of a New Method for Removing the Growth Trend from Dendrochronological Data', *Tree-Ring Bulletin, 41*, pp. 55–66

Weakly, H. E. (1946) 'A Preliminary Report on the Ash Hollow Charcoal' in J. L. Champe, 'Ash Hollow Cave: A Study of Stratigraphic Sequence in the Central Great Plains', *University of Nebraska Studies, New Series*, no. 1, pp. 105–10

—— (1949) 'Dendrochronology in Nebraska' in 'Proceedings of the 5th Plains Conference for Archaeology', *Notebook, Number 1*, Laboratory of Anthropology, University of Nebraska, pp. 111–114

—— (1950) 'Dendrochronology and its Climatic Implications in the Central Plains' in 'Proceedings of the Sixth Plains Conference for Archaeology, 1949', *University of Utah Anthropological Papers*, no. 11, University of Utah Press, Salt Lake City, pp. 90–4

Weakly, W. F. (1971) 'Tree-Ring Dating and Archaeology in South Dakota,' *Plains Anthropologist*, vol. 16, pt. 2, *Memoir 8*

Will, G. F. (1946) 'Tree-Ring Studies in North Dakota', *Agricultural Experiment Station, North Dakota Agricultural College, Fargo, Bulletin 338*

Young, R. S. (1958) 'The Gordion Campaign of 1957: Preliminary Report', *American Journal of Archaeology, 62* (2), pp. 139–54

b. MUMMIFICATION STYLES AS A GUIDE TO THE DATING OF EGYPTIAN MUMMIES

Michael R. Zimmerman

Mummies are defined as bodies preserved either naturally, by drying or freezing, or artificially. Naturally preserved bodies have been found in bogs and in arctic and arid areas. The Egyptian practice of artificial mummification developed from the natural preservation of unclothed bodies buried in the hot, dry sands of the desert in predynastic times (Dawson and Gray, 1968). These bodies were not embalmed but simply wrapped in linen, hides or reed mats and placed in a flexed position in a shallow grave in the desert. The excellent preservation of these bodies may have had a role in the development of the belief in life after death and, indeed, small objects, apparently intended to serve the dead, are often found in the graves.

By the beginning of the Dynastic Period, about 3000 BC, it had become customary to provide the deceased with food and funerary furniture. The graves became larger and tombs, both above and below ground, were developed, all of which served to separate the bodies from the desiccating action of the sand and allowed decomposition. It became necessary to develop artificial techniques of mummification, based on removal of the fluid components of the body. The preserved body, considered the dwelling place for the deceased individual's eternal spiritual elements, was an essential component of the Egyptian theological belief system.

Deceased Egyptians were mummified until the Christian era, about 200 to 400 AD. Initially restricted to royalty, mummification became a more general practice, with a gradual refinement in technique, over the millennia (although in all periods the poor were less carefully mummified).

During the First and Second Dynasties, bodies were closely wrapped in linen bandages, eventually with separate bandaging of the extremities, and placed in wooden coffins. It was soon found out that these bodies decomposed and that evisceration was necessary. As early as the Third Dynasty, the viscera were removed through an incision in the left flank. Desiccation was attempted by immersion in a natron solution (primarily sodium carbonate and sodium choloride), but this technique is relatively ineffective. Old Kingdom and Middle Kingdom mummies were not thoroughly desiccated and have mostly crumbled to dust inside their wrappings. It was not until Dynasty XVIII that the body was completely desiccated, by the heat of either the sun or a fire, after a period of maceration in *dry* natron. By this time, the brain was also removed, by fracturing the ethmoid or sphenoid bone. At this time, the use of resin also became common. The resin was applied in a hot liquid state into the thoracic and abdominal cavities, from which the resin penetrated through the entire body. The resin is usually found by modern investigators to have dried into a glass-like substance, but it is occasionally found in a semi-solid tarry state, confirming that it had been applied as a liquid (Zimmerman and Zimmerman, 1980).

The viscera were divided into four parts, the liver, stomach, lungs and intestine, and embalmed separately. The brain was discarded and the heart and kidneys left in place. The embalmed organs were placed either in four canopic jars, a compartmented chest, or miniature coffins. Old Kingdom jars are described as roughly shaped limestone jars with a convex lid (Peck, 1980). By the Middle Kingdom, the jars had acquired human heads. During the new Kingdom, one jar had a human head and the other three were capped by the head of a falcon, jackal and baboon, respectively.

After Dynasty XXI, the viscera were returned to the body in linen parcels but empty jars continued to be associated with some mummies. By the Roman period, bodies were no longer eviscerated, being dried with the organs in place, but solid, dummy 'jars' were still provided. Another innovation of Dynasty XXI was subcutaneous packing of the trunk and limbs with mud, sand and sawdust, in an attempt to more closely approximate the natural contours of the body. The use of resin to coat the body and wrappings was introduced sometime during the New Kingdom and continued to be used on some mummies through the Roman Period.

After desiccation and treatment of the viscera, the remaining space in the body cavities was packed with sand, sawdust and linen. The flank incision was brought together, occasionally sutured, and covered with a plate of wax, metal or gilded wood inscribed with the Horus-eye (as a protective amulet). Elaborate bandaging completed the process of mummification.

The actual techniques of mummification were described in antiquity by Herodotus and Diodorus Siculus (Harris and Weeks, 1973; Peck, 1980) and reconstructed in the twentieth century by Iskandar (1980). Three types of mummification are described. The most expensive of these procedures involved removal of the brain and viscera and desiccation of the body by packing in dry natron for a period of 70 days. The dry body was then wrapped in linen bandages, coated in resin, and placed in a coffin. The second method was less expensive and utilised dissolution of the viscera *in situ* with cedar oil injected through the anus and retained for the 70-day period of desiccation. The oil was then removed, bringing with it the viscera. The body was returned to the family without wrapping. The third type of mummification was the least expensive, simply involving an intestinal purge followed by desiccation.

While these three different techniques were certainly used in ancient Egypt, it has become clear that the style of mummification was based on economic factors, particularly the status and wealth of the family of the deceased. Mummies examined by the Paleopathology Association and its individual members have shown classical evisceration through a left lower quadrant incision with organ parcels placed in the body (Cockburn *et al.*, 1975) or organs left *in situ* and decomposed (Zimmerman, 1974) or preserved (Reyman, Zimmerman and Lewin, 1977). Other mummies have shown packing of the abdomen with linens through the vagina and rectum, with the organs compressed against the diaphragm (Zimmerman and Goldberg, 1978), or evisceration through the anus with packing of the empty body cavities with sawdust (Reyman and Zimmerman, 1976). These studies have led the author to believe that variations in mummification styles provide only very general guidelines for the separation of mummies from the predynastic period and the Old, Middle and early and late New Kingdoms, as indicated in Table 4.1.

Table 4.1: Dating of Egyptian Mummies

Period	Dynasties	Dates	Mummification features
Predynastic	-------	Prior to 3100BC	Naturally preserved mummies in desert graves
Early Dynastic	I–II	3100–2686BC	Poorly preserved uneviscerated mummies; linen wrappings
Old Kingdom	III–VI	2686–2181BC	Evisceration, canopic jars with plain lids; poor preservation
First Intermediate Period	VII–X	2181–2050BC	
Middle Kingdom	XI–XIII	2050–1750BC	Canopic jars with human heads; poor preservation
Second Intermediate Period	XIII–XVII	1750–1567BC	
New Kingdom	XVIII–XX	1567–1080BC	Canopic jars with animal heads; resin in body cavities
Late New Kingdom	XXI–XXIV	1080–715BC	Viscera returned to body; packing of body with sawdust
Late Period	XXV–XXXI	715–332BC	As above
Ptolemaic Period	-------	332–30BC	As above
Roman Period	-------	30BC — 200–400AD	No evisceration; dummy 'jars'
Christian Period	-------	200–400AD — 641AD	Termination of mummification practices

References

Cockburn, A., Barraco, R. A., Reyman, T. A. *et al.* (1975) 'Autopsy of an Egyptian Mummy', *Science, 187*: pp. 1, 115–60

Dawson, W. R. and Gray, P. H. K. (1968) *Catalogue of Egyptian Antiquities in the British Museum: Mummies and Human Remains*, Trustees British Museum, London

Harris, J. E. and Weeks, K. (1973) *X-raying the Pharoahs*, Scribner's, New York

Iskandar, Z. (1980) 'Mummification in Ancient Egypt: Development, History and Techniques' in J. E. Harris and E. F. Wente (eds.), *An X-ray Atlas of the Royal Mummies*, University of Chicago Press, Chicago, pp. 1–51

Peck, W. H. (1980) 'Mummies of Ancient Egypt' in A. Cockburn and E. Cockburn (eds.), *Mummies, Disease and Ancient Cultures*, Cambridge University Press, Cambridge, pp. 11–28

Reyman, T. A. and Zimmerman, M. R. (1976) 'PUM-IV: Autopsy Report', *Paleopathology Newsletter No. 16*, pp. 5–6

—— Zimmerman, M. R. and Lewin, P. K. (1977) 'Autopsy of an Egyptian Mummy: Histologic Investigation', *Canadian Medical Association Journal, 117*, pp. 470–2

Zimmerman, M. R. (1974) 'PUM-I: Autopsy Report', *Paleopathology Newsletter No. 7*, pp. 6–7

—— and Goldberg, R. (1978) 'PUM-III: Final Anatomic Diagnosis', *Paleopathology Newsletter No. 21*, p. 10

—— and Zimmerman, B. H. (1980) 'The Paleopathology of the Human Remains from Nebwenenef's Tomb — The 1979 Season', *Newsletter No. 31*, 15–16

c. TATTOOS AND THE DATING OF MUMMIES

Michael R. Zimmerman and Jens Peder Hart Hansen

The examination of tattoos has been useful in the estimation of dating for mummies. Patterns and artistic motifs found in tattoos can be compared to those found in artefacts of known date. In the case of more recent mummies, comparison may be to historical records, such as descriptions by early explorers, or to drawings, photographs, or even living individuals. It must be stressed that the value of the study of tattoos is limited by the accuracy of dating of the comparative material. In the few cases where the technique has been applicable, the result has been an approximate date only.

The pigment used for ancient tattoos was generally soot, applied either directly by pricking the skin with a needle dipped in the pigment or by soaking a thread in liquid pigment and using a needle to draw the thread through the skin. The result is a dark blue or black tattoo, easily seen in the living but difficult to discern in the skin of a mummy, which has usually turned dark brown. As blue objects are darkened when photographed with infrared film, the use of infrared photography has proven invaluable in the detection and evaluation of tattoos in mummies.

As tattoos have not been found in Egyptian mummies, the use of tattoos for dating has been confined to northern climates, where preserved bodies with tattoos have been found. Central Siberia has yielded up frozen mummies of Scythians, one of the great tribes of nomadic horsemen of the steppes of Eurasia. The mummies of tribal chiefs, interred in burial mounds over 2,000 years ago, have been preserved by freezing (Artamonov, 1965; Rudenko, 1970). The burial chambers are deep pits covered by earth, logs and boulders. As the covering is not airtight, the chambers became filled with dense cold winter air and remained natural refrigerators over the short Siberian summers, preserving the bodies of the chiefs and their entombed entourage and trappings.

The bodies had been eviscerated and embalmed and, despite being disturbed by the inevitable tomb robbers, the skin remained in excellent condition. There were many elaborate animal tattoos, produced by the pricking method; these included various mythical felines and realistic depictions of herbivores such as mountain sheep. Personal jewellery had been looted, but coffins, clothing and a few jewellery pieces showing similar animal representations remained. Fortunately, other areas in the mounds, containing the chiefs' horses, had not been disturbed and yielded a treasure trove of artifacts, again decorated in a similar style.

The matching of the tattoos with the motifs on the artifacts is a strong indication that the bodies are indeed those of the original occupants of the tombs. (Egyptian mummies are often not in their own coffins but in reused, older coffins.) As the tombs were constructed partially of wood, dendrochronologic studies could establish a relative age range for the tombs of about 500 years. Absolute dating by radiocarbon seems to have been less specific, with a range from the seventh century BC to the first century AD. Artamonov (1965) considered stylistic grounds as well in arriving at a date of the fifth to the third century BC. Again, the tattoos are useful only in establishing a relative date.

The other studies involving tattoos as an aid to dating have been on Eskimo mummies. In 1971, beach erosion on St Lawrence Island, Alaska, exposed the frozen, naturally mummified body of a middle-aged Eskimo woman. The body was kept frozen until 1974, when an autopsy revealed

Figure 4.8: Pattern of Tattooing on the Right Forearm of a 1,600-year-old Frozen Body from St Lawrence Island, Alaska

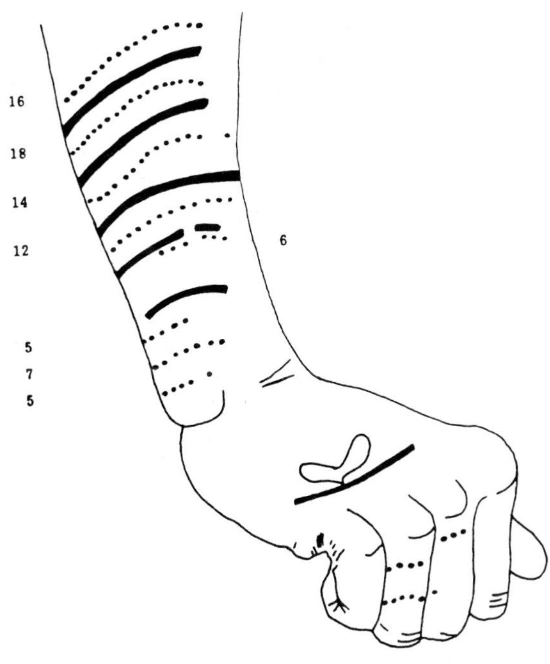

death to have been due to suffocation after accidental burial, probably in a house cave-in or landslide (Zimmerman and Smith, 1975).

Soft tissue from the body was radiocarbon-dated at two separate institutions: the Smithsonian Institution Laboratory date was AD 405 ± 70 years; the University of Pennsylvania Laboratory dates were AD 337 ± 79 years and AD 380 ± 100 years. This dating allows the body to be placed within the Old Bering Sea phase on St Lawrence Island, about AD 200–500 (Larsen and Rainey, 1948; Birket-Smith, 1959; Bandi, 1967).

During the thawing of the body, tattooing of the right arm was noted. Tattoos on the left arm were barely visible, and infrared photography was found to be necessary in order to decipher completely the tattoos.

The tattooing was seen on the dorsal aspect of the forearms, hands and

Figure 4.9: Pattern of Tattooing on the Left Forearm of the St Lawrence Island Eskimo Body

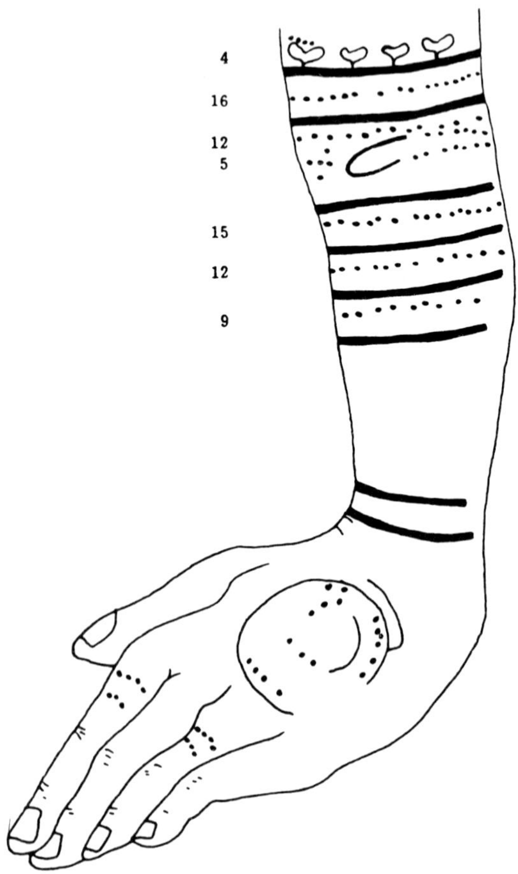

fingers. The basic pattern was one of alternating solid lines and rows of dots on the forearms, designs on the hands, and dots on the fingers (Figure 4.8). The pattern on the back of the right hand was a 'flanged heart' design attached to a solid line.

The left arm showed more elaborate tattooing (Figure 4.9), with four flanged hearts on the upper forearm, similar to but smaller than that on the right hand. An irregular oval, or ovals, was detected — with difficulty — on the hand, and the line and dot pattern was repeated on the forearm.

Further examination of the body failed to reveal any other tattoos, although it was customary on St Lawrence Island for Eskimo women to have tattoos on the chin and/or cheeks (Nelson, 1899). The practice, discontinued only recently, was described by Geist (1928):

> Some of the St Lawrence Island Eskimo women and girls have beautifully executed tattoo marks. These are made free hand although sometimes an outline is traced before the tattooing takes place. The pigment is made from the soot of seal oil lamps which is taken from the bottom of tea kettles or similar containers used to boil meat and other food over the open flame. The soot is mixed with urine, often that of an older woman, and applied with steel needles. Two methods of tattooing are practiced. One method is to draw a string of sinew or other thread through the eye of the needle. The thread is then soaked thoroughly in the liquid pigment and drawn through the skin as the needle is inserted and pushed under the skin for a distance of about a thirty-second of an inch when the point is again pierced through the skin. A small space is

Figure 4.10: Decorative Motif of the Old Bering Sea Style 2 Type, As Found on a Gorget-Like Ornament

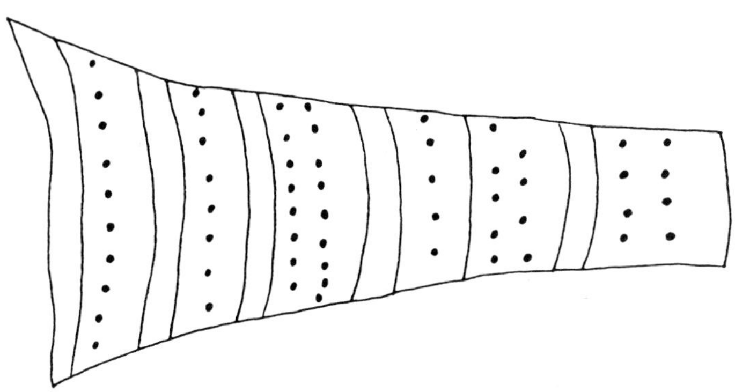

left without tattooing before the process is again repeated. The other method is to prick the skin with the needle which is dipped in the pigment each time.

Collins (1937) illustrates an Old Bering Sea phase gorget-like ornament with a motif very similar to the tattooing design found on the arms (Figure 4.10). Collins also shows four designs that are similar to the flanged hearts (Figure 4.11), also from the Old Bering Sea phase. Thus the artistic motifs of the tattooing seem to correlate with the radiocarbon dates in placing this mummy within the Old Bering Sea phase of Alaskan prehistory (Smith and Zimmerman, 1975).

Figure 4.11: Decorative 'Flanged Heart' Designs of the Old Bering Sea Style 2

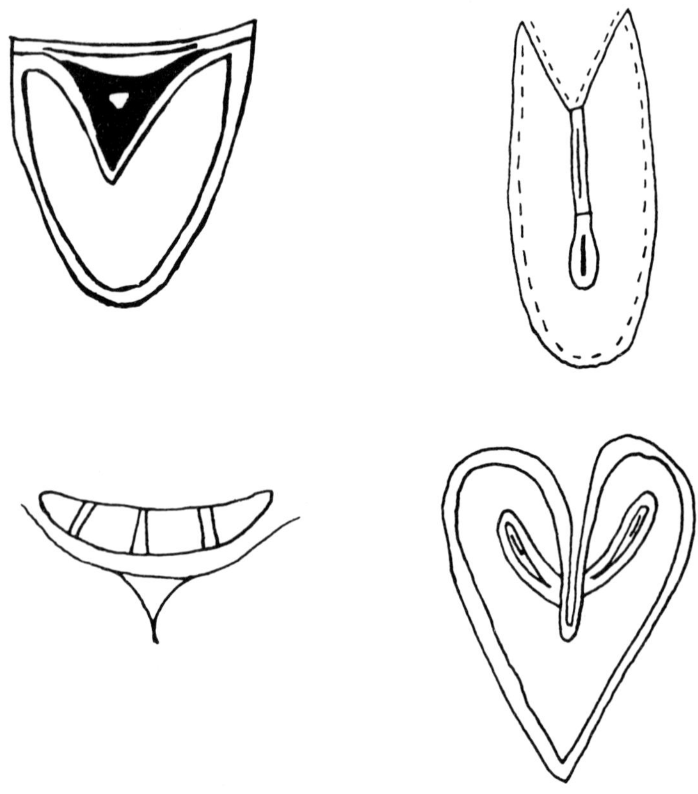

The most recent application of the study of tattoos to dating was in the examination of a group of desiccated Eskimo bodies found near the abandoned settlement of Qilakitsoq in northwestern Greenland (Hart Hansen, Nordqvist and Meldgaard, 1985). The group was composed of two children and six adult women. The faces of five of the women showed tattoos on their foreheads. Four were also tattooed on the cheeks and three on the chin. Only the youngest woman, between 18 and 22 years of age, and the children showed no tattoos. The find was radiocarbon dated at AD 1475 ± 50 years.

Unlike the St Lawrence Island mummy, the tattoos of the Greenland Eskimos were of no real importance in dating the find. Eskimo artifacts up to 1,500 years older than these Eskimos show similar motifs. Identical tattoos on Eskimo female faces have also been recognised on one of the earliest known European drawings of Eskimos, dating to AD 1566, as well as on seventeenth century paintings. Tattooing persisted in northwestern Greenland till the middle of the eighteenth century, and even later in the southern and eastern parts of the country. This 2,000-year time span for Eskimo artistic motifs makes tattoos of little value in dating human remains from this geographic area.

In summary, under rare circumstances bodies may be preserved with intact tattoos, which may be helpful in dating. Similarity of motifs between tattoos and artifacts can provide some reassurance that both belong to the same general time period. The utility of this approach is limited, first, by the accuracy of the dating of the artifacts and, second, by the innate conservatism of aboriginal artists, resulting in the use of similar patterns and motifs for hundreds or even thousands of years.

References

Artamonov, M. I. (1965) 'Frozen Tombs of the Scythians', *Scientific American, 212*, no. 7, pp. 101–9

Bandi, H.-G. (1967) *Preliminary Report on the 'St Lawrence Island Archeological Field Project 1967'*, University of Berne, Switzerland and the University of Alaska, Fairbanks

Birket-Smith, K. (1959) *The Eskimos* 2nd edn, Methuen, London

Collins, H. B. Jr. (1937) 'Archeology of St Lawrence Island, Alaska', *Smithsonian Institution Miscellaneous Collections 96 (1)*

Geist, O. W. (1928) *Diary*, University of Alaska Archives, Fairbanks

Hart Hansen, J. P., Nordqvist, J. and Meldgaard, J. (eds.) (1985) *Qilakitsoq*, Greenland National Museum and Christian Ejlers Forlag, Copenhagen

Larsen, H. and Rainey, F. (1948) 'Ipiutak and the Arctic whale hunting culture', *American Museum of Natural History Anthropological Papers 42*

Nelson, E. W. (1899) 'Eskimo about the Bering Strait', *18th Annual Report of the Bureau of American Ethnology to the Smithsonian Institution 1896–7*

Rudenko, S. I. (1970) *Frozen Tombs of Siberia*, trans. M. W. Thompson, University of California Press, Berkeley

Smith, G. S. and Zimmerman, M. R. (1975) 'Tattooing Found on a 1,600-year-old Frozen, Mummified Body from St Lawrence Island, Alaska', *American Antiquity*, 40, pp. 434–7

Zimmerman, M. R. and Smith, G. S. (1975) 'A Probable Case of Accidental Inhumation of 1,600 Years Ago', *Bulletin of the New York Academy of Medicine, 51*, pp. 828–37

PART TWO

AGE DETERMINATION

5 AGE AT DEATH ESTIMATED FROM THE SKELETON AND VISCERA

J. Lawrence Angel, Judy Myers Suchey, M. Yaşar Işcan and Michael R. Zimmerman

Estimation of age at death from the skeleton is of concern to many scientists. The archaeologist uses this valuable information as an integral part of his reconstruction of a prehistoric society. For the palaeodemographer and palaeopathologist, knowledge of age at death is vital to determine population composition, mortality, morbidity and fertility. Furthermore, the need to identify unknown individuals in modern forensic cases has made the estimation of age at death important for practical purposes as well (Angel, 1980, 1984; Krogman and Işcan, 1986).

The ageing process in the skeleton and soft tissue starts at the time of conception. Even in the very young, new cells are constantly being regenerated to replace those that have completed their 'lifespans'. While ageing is a continuous process, it operates under three general states, growth, maintenance and atrophy. The growth state is characterised by the beginning of ossification and ends with the union of epiphyses and eruption of adult teeth. Dentition is usually complete by 12 years (except the third molars), while epiphyseal fusion and union continue into the third decade of life.

During the growth stage, changes occur rapidly and consistently and are easy to observe with the naked eye. However, the maintenance phase of ageing commencing at maturity is characterised by changes that are much more subtle and varied, and thus harder to detect. In old age (over 55), atrophic changes, such as osteoporosis, usually becomes obvious.

In the living body, maintenance implies the preservation of the *status quo*. Unlike growth, the maintenance state is much more individualised. There is no precisely timed, universal sequence of events. Thus, two individuals might attain similar age-related morphologies at different periods. Furthermore, interpersonal and intrapersonal variations increase with age, both externally and internally. The challenge for the scientist is to be able to develop methods and find sites within the living system that could enable him to predict the effects of external factors and to estimate age more reliably.

There are basically three visual methods for the evaluation of age-related skeletal changes: (1) direct observations of bone morphologically, (2) observations of bone radiographically, and (3) histological examination of bone or soft tissue, including mummified remains. This chapter deals mainly with the morphologic assessment of bone age during both the growth (ossification, epiphysial union and dental calcification and eruption) and the maintenance and atrophy periods (bony hypertrophy near joints, then atrophy). This chapter outlines ageing from fetal life to old age. We pay special attention to the following skeletal areas, for which there are new and reliable data from modern autopsy samples: (1) sutures of skull vault; (2) costochondral junction of the rib; (3) epiphyses of the medial clavicle and the anterior iliac crest; and (4) pubic symphysial face.

In addition to skeletal ageing, age-related variation in mummified soft tissue is discussed. While the expected precision is far less than that obtained from the skeletal system, a rough estimate as early, middle or late adult stage can be made.

Outline of the Processes and Criteria of Gross Bony Age Change

At all stages of life what one notes in standardised detail in the skeleton are the changes of physiological age, i.e. biological age (physiological or real), as opposed to chronological age, or age at death in years. Yet for demographic, health and historical purposes in archaeological material, and for legal identification purposes in forensic skeletons, the latter is needed.

Each of us grows through childhood, adolescence and adulthood, at her or his own pace and own way. A girl may be skinny, underweight, smallish, a good runner and very healthy from age 0 to age 80, appearing and acting in adulthood always 5–20 years younger than her age. A brilliant man as a chubby child may grow fast, mature two years early, remain overweight after graduating from college at age 18, rise quickly in a competitive career, look a silver-haired 65 as a hypertensive organisation president and die at the age of 55 of coronary thrombosis, always being physiologically 10 to 20 years older than his age. Most of us are in between these extremes (actual examples), and age more in some systems than in others.

In our society, the cardiovascular-renal system ages faster, especially in modern American males, than it did in the past. This effect is likely to be a result of changes in nutrition, pollution (including inhaling cigarette smoke) and even increased levels of tension. By the 25th reunion of one of the author's college class (all males), 7 per cent were dead, as compared with the 10 per cent mortality expected for that age group. However, what was really striking was the contrast in the survivors between those who had aged quickly and those slow to age.

There is a temptation to assume that early maturers (who are indeed slightly fatter and tend to remain shorter) age more quickly as adults than late maturers (who are slightly taller at puberty). There is evidence both that early menarche links with late menopause (i.e. general excess vigour) and on the other hand that obesity (with slightly early puberty) links with quick ageing and early death. Constitutional variation in susceptibility to disease stress and age stress is so great that few predictions are possible. Skeletal evidence and common sense dictate that those who die young tend to be quick agers (compared to the survivors), and those who die old to be slow agers, i.e. appearing younger than they 'should'. However, we do not know if this is indeed the case. Likewise, we do not know if bony changes exactly parallel biological age, though variation in these bony traits is known to increase with age.

Age change is highly individual, both overall and in the relative rate of change between systems. The standard deviations of growth events around puberty are greater than one year (Mainland, 1945 [Appendix IV synthesising Todd's epiphyseal data]; Hurme, 1957; Garn *et al.*, 1958; Tanner, 1962): menarche 1.2 years, apex of growth spurt 1.5 (?), M2 eruption 1.6, and fusion of (elbow) epiphyses 1.5 years. This means a discrepancy of two to three years between biological and chronological age.

Furthermore, the correlation between dental age and timing of the three other events at puberty (bone age, growth spurt, menarche) is not high. The correlation coefficient is in the 0.4 range (0.2 to 0.6), probably because the adrenarche and sex hormonal revolutions causing sexual maturation and spurting plus cessation of epiphysial growth have only a minimal effect on genetically programmed teeth. Equally important (for example, Garn, 1958; Tanner, 1962) is the positively skewed distribution of almost all growth events; relative to the mode, there are more late maturers than early maturers simply because there is more living time ahead of than before any childhood or adolescent event (except in progeria). This is why wise students of growth tend to avoid means and to use actual distributions and modes. It is also one reason for the over-ageing of young skeletons.

There remains the problem of differences in the samples upon which age criteria are based. Todd (1920, 1921, 1924) and the Western Reserve growth students, Garn (1970) and the Yellow Springs growth students, Acheson (1957) and his Oxford method, McKern and Stewart (1957) and others have used middle-class US samples, except for the important pubic symphysis and suture closure data of Todd based on a dissecting room sample sometimes lacking death certificates. A modern US sample (Los Angeles autopsy room) shows that males mature more rapidly in pubic symphyseal development than Todd's standards show (Suchey, Wiseley and Katz, 1986).

Better nutrition may be the key to this difference. Modern Americans are indeed maturing about a year earlier than in the pre-World War I period when Todd's adult samples matured, or in the 1920s when his radiologic epiphysial subjects matured. Modern teenage girls show fusion about two years ahead of 'standards' (Angel's personal experience; Joe Davis, personal communication). This modern maturation shift is another reason for over-ageing young skeletons, which is important in forensic cases. But these criteria are correct for ancient populations, based on a 13 to 14 year age at menarche. In this connection, the ageing system of Acsádi and Nemeskéri (1970), especially their pubic symphysis data, can cause major overestimation of age of five or even ten years. This has led to scepticism concerning palaeodemography (Masset, 1976).

We are handicapped in having little precisely controlled data for infants (Angel, 1971a; Stewart, 1979 — for fetuses), for young children (Ubelaker, 1978), and for truly aged skeletons (people). Logically, the biological versus chronological discrepancy must increase steadily from birth to old age.

Hence our key problem is the ability to assess how much a given skeleton is accelerated or retarded in ageing. Applying the criteria summarised here will give the biological age. This is enough for general and archaeological purposes, but the palaeodemographer would like to correlate lifespan with fecundity, for example, or lifespan with economic status, as shown in grave-offerings. One will need to know the chronological age of the individual to do this. Obviously, one needs to know age as precisely as possible for forensic or historical identification. For example, the skeleton of James Smithson, founder of the Smithsonian Institution, appears very slightly younger than his 65 years in all age criteria, including osteon morphology (Ubelaker, Chapter 7, this volume). Why? Was it an effect of his physical vigour and early mountain-climbing and fencing (he was a rather small man, wiry and not heavy), or of his consumption of Italian food and wine (retarding remodelling) or just of the general trend to under-age after the fifth decade?

Age Criteria

For fetus and newborn the best available standard is size (Table 5.1), especially femoral length (Angel, 1971b; Ubelaker, 1978; Stewart, 1979). Reduction of depth of subarcuate fossa, start of knee epiphyses, gradual fusion of greater wings with body of sphenoid, of tympanic ring, and of two halves of mandible are all usable as approximations of early infancy (Terry and Trotter, 1953), but there is no precise scheme for them. Closure of metopic suture and bregmatic fontanelle occurs soon after one year. In infancy and throughout childhood, long bone lengths may still be the most useful age criteria, using the excellent radiologic measurements of Maresh (1955) as reference (allowing for X-ray magnification, of course), if all one has to study are one or two eroded long bones and a few partial ribs.

Table 5.1: Long Bone Dimensions (mm) in Late Fetal Life and Throughout Infancy

Maximum length (mm)	−1 Fetal	0 Newborn	Infancy 2 months	6 months	Childhood 12 months	24 months
Humerus	58	64	72+	88	104+	129
Radius	47	51	58	69	80	96+
Ulna	53	58	66	76+	90−	108−
Femur	65	76	86+	112	135−	170
Tibia	57	61	70−	89	108	139−
Fibula	54	58	66	85	105+	136
Stature	444	504	570	664	748	871

Note:
Fetal means are from 11 white 'full-term' stillborn fetuses, with mean body length of 444mm, studied in 1960 at the Jefferson Medical College (Angel, 1971a). Newborn bone measurements are an 11.6 per cent linear reduction of Maresh's (1955) 2-months means (N=55), based on the difference in living stature means. They approximate a 13.5 per cent increase from the late fetal means. The 504mm living stature at birth comes from Meredith (1943). This stature and the femur length exceed those given by Olivier and Pineau (1960, from Stewart, 1979) by 1 to 3 per cent. Maresh notes that since her radiographic bone lengths are uncorrected (2.3m focal to film plane distance), magnification for femur may be a little more than for other bones. The infancy to early childhood values come from her 1955 report. These are all bony lengths, excluding cartilage, and hence applicable to archaeological or forensic skeletons.

Formation and eruption of milk and of permanent teeth are more precise criteria, as listed in Table 5.2 (also see Costa, Chapter 8, this volume). Eruption of milk teeth begins with lower i1 at 7.5 months, upper i1 and upper i2 by 8.5 months, with lower i2 soon after, upper and lower m1s by 13 months, canines (cs) at 17 months, and m2s by two years, with standard deviations of about 3 months. Moorrees, Fanning and Hunt (1963) give more precision in establishing the degree of calcification and range of variation. While boys are slightly ahead in infancy (up to the age of one year), girls show earlier calcification of permanent teeth by two to three months, except for the canines, where their advantage is six months to one year.

The data charted by Schour and Massler (1944) are easy to use and their coloured chart (by a pharmaceutical company, Mead, Johnson and Company) is particularly helpful. [Angel kept it pinned on the wall of many field laboratories in Greece and Turkey.] Agreement is excellent with Garn *et al.* (1958), the huge sample of Boston children of Hurme (1957) and the Plains Indian data of Ubelaker (1978), if one remembers the different criteria used. The first molar is most useful all through childhood, with tooth bud in infancy, crown at platform stage (Garn *et al.*, 1958) at four years for females and 4.3 for males, alveolar eruption just before six years, occlusal level at 6.9 years in both sexes, and apical closure at the end of root formation at 10.6 and 10.3 years. The second molar, erupting before age 12, lasts into adolescence with its root closure. The third molar is unreliable. Since tooth crowns often survive cremation, the combined occurrence of crowns of milk molars, M1, and permanent incisors, for example, can indicate age accurately.

Table 5.2: Approximate Schedule of Tooth Events (in Months for Milk and Years for Permanent Teeth)

		Permanent teeth				σ of
	Crown		Eruption			eruption (SD or std
Tooth	complete	Alveolar	Hurme's	In line	Complete	deviation)
i 1	5mths	8−		10	18	
i 2	6mths	8		10	20+	
m 1	8mths	13	15	13	23	
c	10mths	16+		20−	30	
m 2	12mths	22		26−	34	
M 1	4.2 yrs	5.7	6.2	6.9	10.3	0.8
I 1	5.4	6.5	6.9	7.4	10.1	0.8
I 2	5.6	7.6	8.0	8.4	10.6	0.9+
P 1	7.3	9.9	10.4	10.6	12.8	1.5
P 2 ♀	8.0	10.3	10.9	11.3	13.4	1.6
P 2 ♂	8.2	11.1	11.3	12.2	14.1	
C ♀	7.8	10.0	10.4	11.0	14.4	1.3
C ♂	8.0	10.9	11.2+	12.0	15.5	
M 2 ♀	8.7	10.7	12.0	11.8	14.6	1.4
M 2 ♂	8.9	11.2	12.4	12.7	15.0	
M 3	14.9	17 (approx.)		18 (approx.)	21 (approx.)	1.7 (approx.)

Note:
This chart is a compromise based on Schour and Massler (1944), with careful attention to data of Robinow, Richards and Anderson (1942), Garn *et al.* (1958), and Hurme (1957). For milk teeth, males precede females by one to two months. For permanent teeth, females precede males by three to ten months, and mandibular eruption precedes maxillary eruption except for premolars. Complete eruption means final growth of root tip to form nerve foramen.

In early childhood, good age criteria are the fusion of the two halves of vertebral arches to form the spine, from below upward from 0.5 to 1.5 years (with axis midline fusing at 2+ years), followed by a cephalo-caudal fusing down the vertebral column of arches to bodies from the age of three years (cervical) to six-plus years (lumbar) (Krogman and Işcan, 1986). Closure of similar synchondroses around the foramen magnum (Redfield, 1970; Stewart, 1979) is useful if the mouth area is lost. Exoccipitals, carrying four-fifths of each condyle, join the main occipital at four to six years, and then join the basi-occipital at five to seven years. Closure of metopic suture and of fontanelles at almost one year is less precise.

First appearances of centres of ossification for long bone epiphyses and for carpals and tarsals (Scammon, 1953; Mainland, 1945; Stewart, 1968) are of little use, except for X-rays of newborn infants, of battered children and of mummies, since the small flakes or polygons of bone that represent the first years of growth dissolve in the ground, are overlooked by the excavator or the police, or are too amorphous to assign to the right joint. In later childhood and adolescence, the actual shape of the epiphysis, as spongy and finally lamellar bone replaces cartilage (Acheson, 1957), has great importance in ageing. This is also true of the appearance of traction epiphyses along the iliac crest and ischial tuberosity at about the age of 12.5 for females and 15 for males, and at about the same time the rings for holding the ligaments retaining the intervertebral disks. Child abuse with malnutrition in forensic skeletons and extreme disease (anaemia, tuberculosis) or starvation in ancient or archaeological skeletons slow down bone age change by one to four years. These factors also reduce bone size and produce 'lines' of growth arrest in long bone diaphyses and on enamel of milk and/or permanent teeth, and sometimes specific traumatic or pathological changes as healed fractures, cortical thinning, or bone-within-a-bone formation. It is important to see that such damaged children are younger biologically than chronologically.

Male and female bone age differs in relation to tooth formation, with females being advanced and males retarded in relation to teeth (Hunt and Gleiser, 1955). Females are about 1.5 years ahead of males in bone age, and the individual range of variation is about five years.

Epiphysial Fusion

There is no sharp bone age-marker of menarche, adrenarche, or other physiological events dividing childhood from adolescence and then sub-adulthood. The concurrent growth spurt is invisible in any single skeleton and the subsequent sequence of closure of epiphyses (six years for early, and 12 years for late maturers) also has only hormone-to-target-organ causes for correlation with other body changes. As noted earlier, menarche has a correlation coefficient of about 0.8 with epiphysial fusion, only about 0.4 with M2 eruption, and a standard deviation of 1.2 years — this is the start of clear differences between physiological age of different body systems and chronological age.

Fusion of the three bones of the innominate (hip) bone at the acetabulum occurs at the age of 13 years in females and 14.5 to 15 years in males. This is timed with the start of oestrogens influencing the pubis and sacro-iliac in females and of androgens influencing the ischium in males (Tanner, 1962). Four of the elbow epiphyses have the same timing, with the medial epicondyle fusing about a year later. All of these are only approximate markers for menarche and for male puberty.

It is helpful to remember the obvious mechanical generalisations that (a) pressure epiphyses for relatively simple gravity-stressed or work-stressed

Table 5.3: Bone Growth Changes. The 50th percentile ages from data of the Hamann-Todd subjects (Mainland, 1945, quoting Todd, Francis, Greulich and Pyle), and the Korean War dead (McKern and Stewart, 1957)

	Age	Upper extremity	♀	♂	Lower extremity	♀	♂	Axial skeleton	♀	♂
					Calcaneus, talus, knee $4\frac{1}{2}-9\frac{1}{2}$ fm					
	Birth	Humerus head	0+	0+	Cuboid					
		Capitate			Femur head	3mths	3mths	Mandible fuses at chin		
		Humerus gt. tuber.	6mths	9mths	Tibia ankle			Vertbr.arches fuse midline		
		Hamate			Fibula ankle	9mths	9mths	from below 6mths to $1\frac{1}{2}$ yrs		
	1	Radius wrist	9mths	1				Fontanelle		
		Metacarpals						at bregma		
	2				Metatarsals			Axis mid. fuses (2+ yrs)		
		Lunate						Vertebral arches fuse		
	3	Medial epicondyle	$3\frac{1}{4}$	6+	Patella			to bodies 3yrs (cervical) to		
					Gter trochanter	$2\frac{1}{2}$	$3\frac{3}{4}$	6yrs (lumbar)		
	4	Radius proximal	$3\frac{3}{4}$	$5\frac{1}{2}$	Fibula proximal					
								Occ.pars lateralis		
(a)	5	Scaphoid						(4–6yrs)		
		Ulna wrist	$5\frac{1}{2}$	$6\frac{1}{2}$	Calcaneus tuberosity	$5\frac{1}{2}$	$7\frac{1}{2}$	Axis and atlas		
	6							Occ.condyle area (5–7yrs)		
	7	Ulna proximal	$7\frac{1}{2}$	10						
	8				Talus post. tub.	8–	$9\frac{1}{2}$			
		Pisiform	8	11						
	9	Trochlea elbow	$8\frac{1}{2}$	$9\frac{1}{2}$	Tibia tuber	$9\frac{1}{2}$	$11\frac{1}{2}$			
					Lesser trochanter					
	10	Lateral epicondyle	9+	$11\frac{1}{2}$						
	11									
	12	Glenoid rim, subcoracoid			Iliac crest	$12\frac{1}{2}$	15–	Epiphyses of vertebrae		
		Med.clav.,scap.borders			Ischial tuber			and rib tubercles		
	13	Elbow	13	15	Acetabulum	13	$14\frac{3}{4}$	Talus tuberosity	10	12

								Ischio-pubic ramus		
	14	Medial epicondyle	14	15¾	Calc.tuber	13½	15	conjoins	11½	13½
	15	Metacarpals I	14	16	Metatarsals	14½	15½			
		Phalanges II V	14	16½	Ankle	15	16			
(b)	16							Spheno-occipital		
								synchondrosis	16?	17+
	17	Wrist	16	18						
		Scapular	15	17	Hip	17½	17½			
	18	epiphyses	–17	–19+	Knee	18	18½	Sacral no. 2–5	17?	18½
								Sternum corpus	18?	19–
	19				Iliac crest	19?	19½			
								Epiphyses of vertebrae,		
	20	Shoulder	19½	20				ribs, sacrum	19½?	20½
					Ischial tuber	20?	21–	except for		
	21							mid-thoracic lag	21+	
	22									
	23							Sacral no. 1–2	22?	23½
	24	Medial clavicle	24?	24½						
		(may remain open even after 30)								

(a) Appearance of centres of ossification for epiphyses (also carpals, tarsals)
Range of variation about 2 years

(b) Fusion of epiphyses from x-ray data (peripheral gap remains about 6 months later)
Range of variation about 1 to 5 years, increasing with age

joints (ankle, wrist, hip, knee, shoulder) start to ossify early and fuse late, and (b) traction epiphyses ossify late, often at puberty, and fuse late (some of the latter — ischium, rib heads, medial clavicle — also take some pressure). During fusion, pressure epiphyses (for example, knee and shoulder) anchor at the centre and after about six months they fuse the rim, except for oblique ones like the hip, where a superolateral start of union occurs (Haines *et al.*, 1967). Traction epiphyses like the iliac crest usually start at one edge. The ages in Table 5.3 indicate the start of solid fusion, like McKern and Stewart stages 2–3, rather than completed union, stage 4.

The sequence of union from the Hamann-Todd (Western Reserve) Collection and data of his collaborators (Mainland, 1945) — elbow, ankle, wrist, hip, knee, shoulder (like water rising around an elbows-low seated bather) — differs a little from Flecker's as given in Stewart (1979), and Krogman and Isçan (1986). Knee and shoulder fusion and union are crucial as marking the end of length growth. The spheno-occipital synchondrosis (occasionally miscalled basilar suture) fuses at about 17 years in males (McKern and Stewart, 1957), and presumably at 15 in females. In males, pelvic traction epiphyses fuse at 20+ years, vertebral rims at 21, and medial clavicle at about 24–25 years; females are presumably a little over a year younger at these fusions. There is a standard deviation of about 1.5 years and hence a range of over seven years, with the curve skewed toward the older ages (McKern and Stewart, 1957).

Fusion and time of maturation are influenced by stresses as well as by nutrition and class, and especially by heredity. The Hamann-Todd and Korean War dead data fit well with Flecker's and also apply to typical ancient populations, according to Amundsen and Diers (1969), who found evidence of menarche commencing at about 13+ in ancient Greece and Rome. For modern US or north-western European middle-class skeletons, where menarche is earlier by up to one year, one should subtract this from the age estimate and add a year or more for an obviously malnourished skeleton (for example, flat pelvis and low skull base).

Adulthood

Determining the age of adult skeletons is crucial for palaeodemography. Here, one can start adulthood at the fifteenth birthday in order to include all early childbirths (Angel, 1969); or at age 18 with stature completion as the criterion, as in modern times. It should be remembered that it is US and European urban practice to use the age of 21, for example, to mean all people in the year centred on age 21.5; the ages in this chapter are actual chronological ages.

The general physical ageing process centres on progressive denaturation of protein in collagen fibrils, a finite upper limit to cell divisions with 'genetic' errors occasionally during mitosis, and the death of non-replaceable neurons and other cells of the CNS. Bone collagen fibrils, both internal and in tendons and ligaments, become increasingly cross-bonded, inelastic and tight, eventually becoming almost impervious to tissue fluid flow (Montagna, 1965). The same applies to skin, quick-ageing in rural and fishing people much exposed to sunlight and cold. However, one of these probably does not predict the other.

In childhood and adolescence, extra stress on fibres of Sharpey, where tendons or ligaments tie into bone, produces deep fossae (for example, around pectoralis major insertion of a three-year-old swung by her arms, or adductor magnus insertion of an adolescent horseback rider). Before the age of 20, as seen in modern forensic skeletons, there is a changeover and such stress produces the deltoid, gluteal, popliteal crests and clear vertebral body rim — but smaller and smoother than they become from ages 40 to 70. This collagen-ageing process is striking at the intervertebral disc ligament attachments where the blocked flow of tissue fluid, produced by stress, raises the adjacent periosteum enough to produce exostoses around the long-fused epiphysial rings. Normally, such arthritic joint-lipping appears all over the skeleton at the start of middle age in the mid-thirties. Such bony outgrowths are useful age criteria.

As opposed to such ageing hypertrophy, there are two bony age changes involving atrophy. The first of these is a qualitative change in formation and calcification of successive layers ('rings') of each new osteon in cortical bone. This is useful in age estimation from internal bone structure (Kerley, 1965), and is discussed in Chapter 7 by Ubelaker. The second change is a general loss of bone endosteally, involving less trabeculation at the ends of long bones (Acsádi and Nemeskéri, 1970) and osteoclastic action on the irregular inner cortical surface of long bones. Trotter, Broman and Peterson (1960) show a linear loss of bone density with age, although the bone loss rate increases somewhat after the menopause in many females (Garn, 1970). This absolute bone loss would mean a crucial loss of bone strength, except that periosteal osteoblasts continuously lay down lamellar bone on the outside of long bone shafts where it is best placed to resist bending stress. However, denaturation of collagen does cause loss of elasticity, and fracture patterns change with increasing age (Buhr and Cooke, 1959).

Bone atrophy and loss become clear at the switch from middle to old age in the mid-fifties. Real osteoporosis begins later. These two processes of ageing, building of exostoses and of outer lamellae versus loss of inner bone and trabeculae, and appearance of atrophic surface texture, are somewhat independent of each other. Both are influenced by regular hard exercise, excessive stress (fatigue fractures, etc.), hormones, pollutants, diet and genes. Hence the gap between biological age and chronological age increases steadily and, at the end, almost logarithmically (Table 5.4).

Table 5.4: Deviation of Estimated from Actual Adult Age at Death in Years in the USA. (N=116 forensic, 25 willed, both races combined)

Age	Overall (ignoring + or − sign)			Absolute (with signs)			Percent of
(years)	Female	Male	Both sexes	Female	Male	Both sexes	sample
16–19	0.7 (7)	1.0 (7)	0.9 (14)	+0.7 (7)	+1.0 (7)	+0.9 (14)	9.9
20–29	1.6 (20)	1.8 (24)	1.7 (44)	+1.2 (20)	+1.5 (24)	+1.4 (44)	31.2
30–39	3.7 (9)	4.8 (16)	4.4 (25)	+0.1 (9)	+4.2 (16)	+2.7 (25)	17.7
40–49	2.3 (15)	3.0 (8)	2.6 (23)	+1.1 (15)	−0.8 (8)	+0.5 (23)	16.3
50–59	4.5 (10)	5.1 (8)	4.8 (18)	−1.5 (10)	−2.1 (8)	−1.8 (18)	12.8
60–69	3.0 (2)	7.9 (7)	6.8 (9)	+2.0 (2)	−7.9 (7)	−5.7 (9)	6.4
70–79	10.7 (3)	12.3 (3)	11.5 (6)	−10.7 (3)	−12.3 (3)	−11.5 (6)	4.3
80–89	12 (1)	13 (1)	12.5 (2)	−12 (1)	−13 (1)	−12.5 (2)	1.4
16–89 Total	3.0 (67)	4.0 (74)	3.5 (141)	−0.1 (67)	−0.3 (74)	−0.2 (141)	100.0

Notes: From the age of 16 to 49 the absolute deviation of estimated from actual age at death averages +1.4 years (106); from age 50 to 89 the absolute deviation averages −5.1 years (35). Except for the fifth decade, the overall deviation from actual age increases by one year from the age of 16 to about 55 years. After that the increase is more rapid. Note that a little of this deviation of estimated from actual age is probably technical in origin, but that most of it comes from the individual difference between physiological age and actual age in years. Statement of actual age or of birth date comes from the family of the dead person.

The scapula shows gradual lipping of the glenoid rim starting in the late 20s, the buildup of a 'plaque' on the undersurface of the acromion process (the outer margin at first) and, developing above the bursa over the rotator cuff in the 30s to 50s, 'buckling' of the whole infraspinous region in the 40s to 80s, with 'atrophic patches' in the 50s to 90s (Graves, 1922). Often, the scapula is fragmentary.

Limb joint and vertebral exostoses and cartilage change develop moderately in the 30s and 40s, and consistently in the 50s to 70s (Stewart, 1958 and Jurmain, 1975, quoted in Stewart, 1979). Spongy trabeculation at the upper end of the humerus starts to retreat in middle age, reaches the surgical neck by 52 years, the epiphysial line by 60 years, and extends up into the greater tubercle (Nemeskéri *et al.*, 1960, quoted in Stewart, 1979). All of these are usable examples of bone age change.

After the eruption of wisdom teeth in the late teens (or their suppression), the major skull changes are fusion of the spheno-occipital synchondrosis and then fusion of sutures. Since the end of skull-base growth (cartilage-based) is the precondition for suture closure, this process normally begins endocranially; in childhood and adolescence, open-skull sutures allow bending as the base expands. Except for the asterion end of the lambdoid suture, the three major vault sutures (coronal, sagittal, and lambdoid) have usually closed by the age of 40, after steadily intermittent fusing for 20 years.

Table 5.5 (Baker, 1984) gives new data illustrating the variability in these major sutures. Individuals showing open sagittal sutures endocranially are under 36 years old. Individuals showing partial union of the sagittal suture endocranially can be 19 to 79 years old. Individuals showing complete union of the sagittal suture endocranially are above 26. If the upper vault is closed ectocranially, the individual is above 35.

Table 5.5: Age Ranges for the Stages of Union of the Cranial Vault Sutures in a Multi-racial Sample of Modern Americans (144 Males, 51 Females) Autopsied at the Department of the Chief Medical Examiner-Coroner, Los Angeles[a]

	Ages in years					
	Endocranial			Ectocranial		
	Stage of union[b]			Stage of union		
Sutures	O	P	C	O	P	C
Sagittal	−35	19–79	26+	−87	19–83	34+
Left lambdoid	−70	19–74	23+	−84	24–84	23+
Right lambdoid	−70	19–74	23+	−84	22–84	23+
Left coronal	−70	22–79	26+	−84	28–89	36+
Right coronal	−70	23–79	26+	−84	24–89	36+

Notes:
a. Modified from Baker (1984);
b. For the stages of union, O stands for open, P for partial closure and C for complete.

During the 40s usually, the external aspect of sutures at the pterion start and almost reach closure. The masto-occipital closes after the 50s, with the stress on bony exostoses and the start of external shrinkage; in the 70s the parietomastoid closes, and not before the 80s or 90s does the squamous close (Todd and Lyon, 1924; Cobb, 1952; McKern and Stewart, 1957). A very frequent developmental 'error' is lapsed union of all or, usually, a part of a suture: this results in an accumulation of active and rather vascular bone next to a trench of open suture. Premature closure is genetic and much rarer in its distribution in populations. It usually influences skull form and can be recognised as pathological. In order to age a skull, one has to read past these natural 'errors'. Even so, there is incredible variation and an inexperienced person should not use suture closure to estimate age. (The same caution applies to ageing from the tooth root cementum.)

The skull vault loses density with age just like the rest of the skeleton. There is considerable thickening through endocranial growth (Angel, 1971b), so that in old age, especially past the age of 65, the meningeal vessel grooves are countersunk. From the age of 45 to 50 and onward, Pacchionian depressions for arachnoid granulation bulges increase in depth, even to the point of penetrating the outer table of the vault.

There have been very few studies testing use of the foregoing adult age criteria, for two reasons: it takes a generation to accumulate reliable data and people fear that deviations will be seen as mistakes.

Angel's own study is based on 25 willed white skeletons in the Terry Collection plus 116 forensic skeletons (27 black and 89 white) identified as known people and studied between the autumn of 1962 and January 1983, as seen in Table 5.4. Other forensic data came from about 600 skeletons submitted for 'identification', including children, native Americans and animals (not considered here). Two kinds of deviation are important: overall deviation of estimated from actual chronological age, irrespective of + or − sign, and absolute deviation, taking account of sign. Sample sizes for each age group range from two in the 80 to 89 year group, six in the eighth decade, and 44 in the third decade: there are only 35 people over 50 years, as opposed to 106 from the age of 16 to 49 years old. Hence, one has to expect considerable chance variation of means.

The results are clear. From under 20 to the age of 55, there is an approximately linear increase in overall random age variation from one to five years, or one year per decade. From 55 to 85 years the increase is twice as steep, from five to 12 years. A larger sample might show a phase of instability in the fourth decade and stability in the fifth. The mean absolute variation is −0.2 years, suggesting that age estimation for middle age is quite accurate. But the mean absolute deviation under the age of 50 is +1.4 (106), while above 50 years it is −5.1 (35). As predicted both theoretically (those who die young may age more quickly) and in terms of lower- and middle-class age criteria samples, there is a tendency to overestimate age in young skeletons and a striking tendency to underestimate age in truly old people.

Recently, skeletal age changes have been studied in autopsy room settings where age data are reliable. These samples should be fairly representative of the general population since the cases autopsied involve homicides, suicides, accidents and unexpected natural deaths. In the following studies the individuals autopsied were born throughout the United States and Mexico. There is racial and economic diversity within this sample. In the section following, standards have been developed for the costochondral junction of the rib, anterior iliac crest epiphysis, medial clavicular epiphysis and pubic bone.

Costochondral Junction of the Rib in Males and Females

Age-related metamorphosis in the sternal extremity of the rib was first detected radiologically (Michelson, 1934; Fischer, 1955; Semine and Damon, 1975; McCormick and Stewart, 1983) and later observed through direct morphological examination of the bone itself by Kerley (1970). In 1984, a new technique for age estimation from the adult skeleton, based on the lifelong changes at the costochondral junction, was established by Işcan, Loth and Wright for white males (1984a, 1984b) and later for white females (1985).

The phase technique of Isçan, Loth and Wright was developed through the analysis of 230 ribs removed from white males (N=144) and white females (N =86) of known age from death certificates (1.5 to 90 years), and known sex and race autopsied at the medical examiner's office in Ft Lauderdale, Florida.[1] They used only the sternal end of the right fourth rib. The entire range of age- related changes observed in this bone was divided into nine phases (0–8) using separate standards for males and females. They summarised the general progression of the metamorphic process in the sternal end of the rib as follows (Işcan *et al.*, 1984b, p. 1096):

> This metamorphosis begins with the formation of an indentation (pit) in the medial articular surface. Special attention is paid to the depth and shape of the pit along with the walls and rim surrounding it. Initially, the pit is merely an amorphous but noticeable indentation in the once almost flat, billowy endplate. As the pit deepens, the indentation between the anterior and posterior walls takes on a V-shaped appearance that gradually widens into a U as the walls become thinner. With increasing age the pit becomes wider and deeper. Along with further pit development, the rim progresses from a regular, rounded border to a scalloped, but still fairly regular edge, and over the years, grows increasingly sharp and irregular. The overall texture and quality of the bone itself, dense, smooth and solid in youth, deteriorate until the bone becomes very thin, brittle and porous in the elderly.

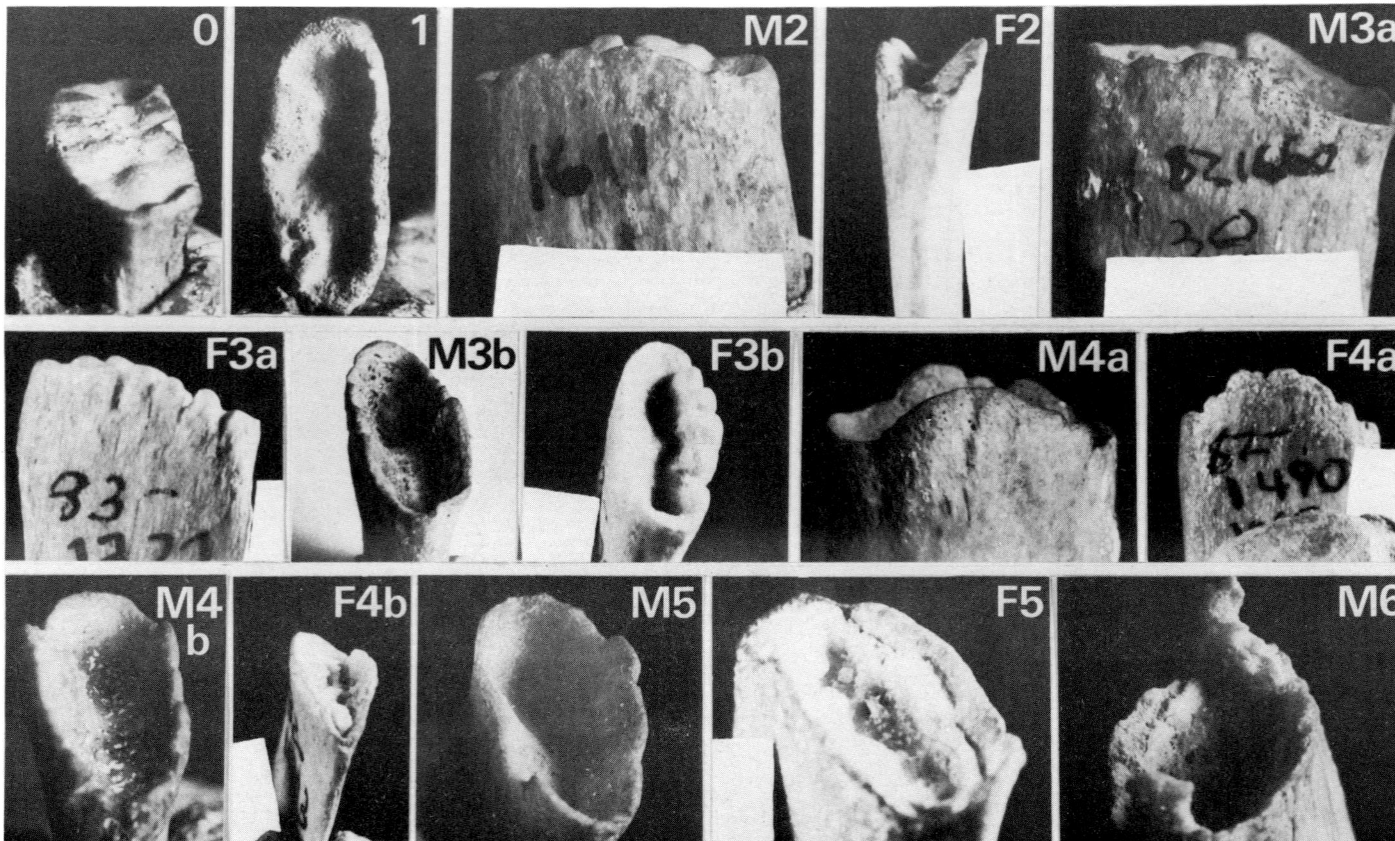
0
1
M2
F2
M3a
F3a
M3b
F3b
M4a
F4a
M4 b
F4b
M5
F5
M6

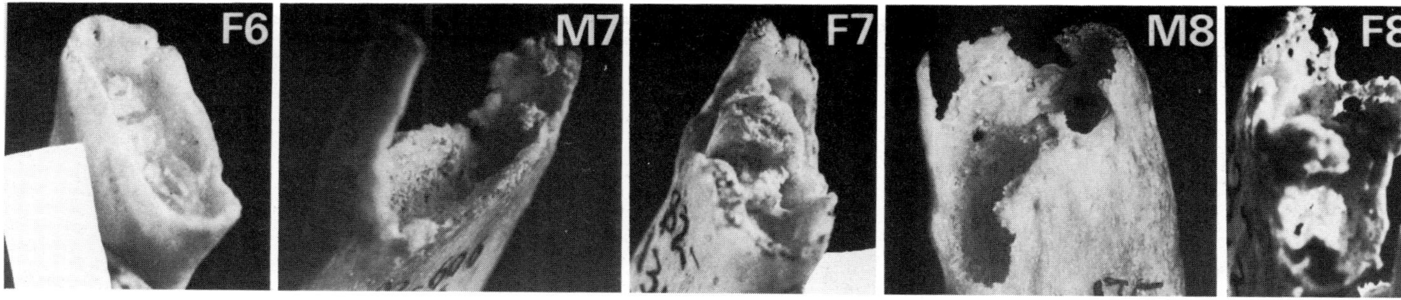

Figure 5.1: **Phase 0** — The Smooth, Regular, Rounded Rim View (0) is Typical of the Adolescent Rib (0). Note the billowy articular surface with no pit formation. **Phase 1** — Rim is still smooth and rounded, but is slightly wavier (1). The initial indentation of the pit, along with some billowing still present on the articular surface. **Phase 2** — Male: M2 shows the scalloped rim with smooth rounded eges first seen in this phase. Female: The rounded, wavy rim is first beginning to show some scallops forming at the edge. A side view of the now V-shaped pit surrounded by thick, smooth walls is seen in F2. **Phase 3** — Male: The rim is becoming more irregular with only a little scalloping remaining (M3a). The deepening pit has taken on a narrow U-shape with fairly thick walls and rounded edges (M3b). Female: The rounded rim now exhibits a pronounced, regular scalloping pattern (F3a). The still V-shaped pit has widened as the walls flare and thin slightly, but there is only a modest, if any, increase in depth (F3b). **Phase 4** — Male: Regular scalloping pattern is gone from the increasingly irregular rim (M4a). M4b shows the moderately wide U-shaped pit with slightly thinner walls whose edges are still rounded. Female: Scallops remain at the still rounded rim, but the divisions are not as pronounced and the edges look somewhat worn down (F4a). The noticeably deeper, flared V or U-shaped pit has again widened as the walls become thinner (F4b). There is also a small plaque-like deposit beginning to form in the pit. **Phase 5** — Male: Rim is slightly more irregular and shows evidence of porosity and some deterioration of bone inside the pit (M5). Note the deep, moderately wide U-shaped pit with thinner walls and sharper edges. Female: No regular scalloping remains at the now sharpening edge of the increasingly irregular rim (F5). Note the smooth plaque-like deposit covering most of the interior of the pit which is now a very flared V or U with appreciably thinner walls. **Phase 6** — Male: Note the bony projections arising from the superior and inferior borders of the rib and the noticeably deep, widely U-shaped pit, thinning walls and sharper edges (M6). Increased porosity and deterioration of bone can also be seen inside the pit. Female: F6 shows the noticeably deeper, wider U-shaped pit, and thinning walls along with porosity inside the pit. **Phase 7** — Male: M7 shows the irregular rim with long bony projections and porous, deteriorating bone in a deep, widely U-shaped pit surrounded by noticeably thin, fragile walls with sharp edges. Female: Figure F7 shows the very sharp, irregular rim. The depth of the flared U-shaped pit appears slightly shallower than in the preceding phase. Bony projections are seen arising from both the rim and floor of the pit, along with evident deterioration of the bone itself. **Phase 8** — Male: M8 shows the extremely irregular rim with sharp, brittle projections of bone with 'window' formation along with the very thin walls surrounding a very deep, porous pit. Female: F8 shows the extremely sharp, irregular rim with brittle projections of bone now prominent at the superior and/or inferior margins of the rib. The porous pit is in a wide U-shape surrounded by very thin, badly deteriorated walls with 'window' formation.

The metamorphic phases are illustrated in Figure 5.1 where M indicates males and F, females (as it also appears in parentheses at the end of each phase description). The sex designation is followed by the phase number and a letter 'a' or 'b' if more than one example of a given phase is used (a greater range of variation is presented in the photographic plates accompanying the original articles). No sex designation is made for phases 0 and 1 because sexual dimorphism in the rib is minimal in these phases. A detailed description of each phase appears below for both sexes. The age ranges were added later by Işcan and Loth (1985).

Males (Işcan et al., *1984b, pp. 1096–1099)*

Phase 0 (16 years and younger). The articular surface is flat or billowy with a regular rim and rounded edges. The bone itself is smooth, firm and very solid (0).

Phase 1 (16–19 years). There is a beginning amorphous indentation in the articular surface, but billowing may also still be present. The rim is rounded and regular. In some cases scallops may start to appear at the edges. The bone is still firm, smooth and solid (1).

Phase 2 (20–23 years). The pit is now deeper and has assumed a V-shaped appearance formed by the anterior and posterior walls. The walls are thick and smooth with a scalloped or slightly wavy rim with rounded edges. The bone is firm and solid (M2).

Phase 3 (24–28 years). The deepening pit has taken on a narrow to moderately U-shape. Walls are still fairly thick with rounded edges. Some scalloping may still be present but the rim is becoming more irregular. The bone is still quite firm and solid (M3a, M3b).

Phase 4 (26–32 years). Pit depth is increasing, but the shape is still a narrow to moderately wide U. The walls are thinner, but the edges remain rounded. The rim is more irregular with no uniform scalloping pattern remaining. There is some decrease in the weight and firmness of the bone; however, the overall quality of the bone is still good (M4a, M4b).

Phase 5 (33–42 years). There is little change in pit depth, but the shape in this phase is predominantly a moderately wide U. Walls show further thinning and the edges are becoming sharp. Irregularity is increasing in the rim. Scalloping pattern is completely gone and has been replaced with irregular bony projections. The condition of the bone is fairly good; however, there are some signs of deterioration with evidence of porosity and loss of density (M5).

Phase 6 (43–55 years). The pit is noticeably deep with a wide U-shape. The walls are thin with sharp edges. The rim is irregular and exhibits some rather long bony projections that are frequently more pronounced at the superior and inferior borders. The bone is noticeably lighter in weight, thinner and more porous, especially inside the pit (M6).

Phase 7 (54–64 years). The pit is deep with a wide to very wide U-shape. The walls are thin and fragile with sharp, irregular edges and bony projections. The bone is light in weight and brittle with significant deterioration in quality and obvious porosity (M7).

Phase 8 (65 years and older). In this final phase the pit is very deep and widely U-shaped. In some cases the floor of the pit is absent or filled with bony projections. The walls are extremely thin, fragile and brittle with sharp, highly irregular edges and bony projections. The bone is very lightweight, thin, brittle, friable and porous. 'Window' formation is sometimes seen in the walls (8M).

Females (Işcan et al., *1985),*

Phase 0 (13 years and younger). The articular surface is nearly flat with ridges or billowing. The outer surface of the sternal extremity of the rib is bordered by what appears to be an overlay of bone. The rim is regular with rounded edges, and the bone itself is firm, smooth and very solid (0).

Phase 1 (14–15 years). A beginning, amorphous indentation can be seen in the articular surface. Ridges or billowing may still be present. The rim is rounded and regular with a little waviness in some cases. The bone remains solid, firm and smooth (1).

Phase 2 (16–19 years). The pit is considerably deeper and has assumed a V-shape between the thick, smooth anterior and posterior walls. Some ridges or billowing may still remain inside the pit. The rim is wavy with some scallops beginning to form at the rounded edge. The bone itself is firm and solid (F2).

Phase 3 (20–24 years). There is only slight, if any, increase in pit depth, but the V-shape is wider, sometimes approaching a narrow U as the walls become a bit thinner. The still rounded edges now show a pronounced, regular scalloping pattern. At this stage, the anterior and/or posterior walls first start to exhibit a central, semicircular arc of the bone. The rib is firm and solid (F3a, F3b).

Phase 4 (24–32 years). There is a noticeable increase in the depth of the pit which now has a wide V or narrow U-shape with, at times, flared edges.

The walls are thinner but the rim remains rounded. Some scalloping is still present, along with the central arc; however, the divisions are not as pronounced and the edges look somewhat worn down. The quality of the bone is fairly good but there is some decrease in density and firmness (F4a, F4b).

Phase 5 (33–46 years). The depth of the pit stays about the same, but the thinning walls are flaring into a wider V or U-shape. In most cases, a smooth, hard, plaque-like deposit lines at least part of the pit. No regular scalloping pattern remains and the edge is beginning to sharpen. The rim is becoming more irregular, and the central arc is still pronounced. The bone is noticeably lighter in weight, density and firmness. The texture is somewhat brittle (F5).

Phase 6 (43–58 years). An increase in pit depth is again noted, and its V or U-shape has widened again due to pronounced flaring at the end. The plaque-like deposit may still appear but is rougher and more porous. The walls are quite thin with sharp edges and an irregular rim. The central arc is less obvious and, in many cases, sharp points project from the rim of the sternal extremity. The bone itself is fairly thin and brittle with some signs of deterioration (F6).

Phase 7 (59–71 years). In this phase, the depth of the predominantly flared U-shaped pit not only shows no increase, but actually decreases slightly. Irregular bony growths are often seen extruding from the interior of the pit. The central arc is still present in most cases but is now accompanied by pointed projections, often at the superior and inferior borders, yet may be seen anywhere around the rim. The very thin walls have irregular rims with sharp edges. The bone is very light, thin, brittle and fragile with deterioration most noticeable inside the pit (F7).

Phase 8 (70 years and older). The floor of the U-shaped pit in this final phase is relatively shallow, badly deteriorated or completely eroded. Sometimes it is filled with bony growths. The central arc is barely recognisable. The extremely thin, fragile walls have highly irregular rims with very sharp edges and, often, fairly long projections of bone at the inferior and superior borders. 'Window' formation sometimes occurs in the walls. The bone itself is in poor condition — extremely thin, light in weight, brittle and fragile (F8).

Table 5.6 presents descriptive statistics including the mean age per phase and its associated 95 per cent confidence interval. The n2 value at the bottom of Table 5.6 signifies that 85 per cent of the variation with age in males could be accounted for by the evaluation of metamorphoses in the rib. While the same statistic was somewhat lower for females (76 per cent),

Table 5.6: Descriptive Statistics of Phases of the Costochondral Junction[a]

	Males				Females			
Phase	N	Mean age	SD	95 percent confidence interval	N	Mean age	SD	95 percent confidence interval
1	4	17.3	0.50	16.5–18.0	1	14.0		
2	15	21.9	2.13	20.8–23.1	5	17.4	1.52	15.5–19.3
3	17	25.9	3.50	24.1–27.7	5	22.6	1.67	20.5–24.7
4	12	28.2	3.83	25.7–30.6	10	27.7	4.62	24.4–31.0
5	14	38.8	7.00	34.4–42.3	17	40.0	12.22	33.7–46.3
6	17	50.0	11.17	44.3–55.7	18	50.7	14.93	43.3–58.1
7	17	59.2	9.52	54.3–64.1	16	65.2	1.24	59.2–71.2
8	12	71.5	10.27	65.0–78.0	11	76.4	8.83	70.4–82.3
Total	108	41.0	7.51	39.6–42.4	83	47.8	11.00	45.4–50.2
n2	0.85				0.76			

Note:
a. Modified from Işcan *et al.* (1984b, 1985).

the sternal extremity of the rib can be considered a reliable site for the accurate determination of age in both sexes. The distribution of specimens by phase and age decade appears in Table 5.7.

In order to measure the effectiveness of the phase technique and assess inter-observer error, Işcan and Loth (1986) asked 25 physical and forensic anthropologists (with and without doctoral degrees) to determine the phase of a sample of 15 ribs. Analysis of the results indicated that inter-observer error was minimal, experience measured by educational degree played little role in the final outcome and nearly all of the participants averaged within one phase of the (chronologically) idealised phase. In conclusion, variation appeared to be greater among the ribs than the judges. Further analysis of the data in relation to cause of death, occupation and drug/alcohol abuse yielded no obvious association with age. However, it must be pointed out that the size of the test sample was not large enough to evaluate fully these antemortem conditions.

The phase technique for the determination of age from the rib assumes that the sex of the specimen is known. If this is not the case, one may use the discriminant function formulae that Işcan (1985) developed for this purpose.

Sex-specific standards for age determination at this site are necessary because, although the initial morphological changes in the rib are nearly identical in both sexes, onset occurs at different ages. Beginning pit formation was first detected at about the age of 14 in females, but not until 17 years in males, and is felt to reflect the cessation of growth at the sternal end of the rib (Işcan *et al.*, 1984a, 1985). This is not unexpected in light of the generally earlier skeletal development of females (Tanner, 1962; Krogman, 1972). Therefore, in Figure 5.1, while phases 0 and 1 are represented by only one sex, the age difference (Table 5.6) must not be ignored.

This age difference was found to continue until phase 4 when the mean age for both sexes reaches 28. From this point on, age per phase remained very close until the final phases, when females averaged a maximum of five years older than males. This turnabout most likely resulted from differences in the age distribution of the bones; the female sample averaged nearly seven years older than males.

Another factor leading to the establishment of separate standards is the existence of sex-specific patterns of ageing in the bony rib. Sexual dimorphism at the costochondral junction becomes readily observable by the middle- to late-20s.

The sternal extremity of the rib shows every indication of being a reliable anatomical site for age estimation in the adult. It is particularly advantageous in that it covers an age range extending into the 70s in both sexes, and is not directly affected by parturition trauma like the pubic symphysis in females (Angel, 1969).

Table 5.7: Frequency Distribution of Age Intervals by Phase of the Costochondral Junction[a]

Age Interval	Male phases 1	2	3	4	5	6	7	8	N	Female phases 1	2	3	4	5	6	7	8	N
14–19	4	1	1						6	1	4							5
20–29		14	13	10	1				38		1	5	9	2				17
30–39			3	2	7	3			15					7	4			11
40–49					5	5	4	1	15				1	6	8	1		16
50–59					1	6	4		11					1	1	4		6
60–69						2	8	2	12						1	6	3	10
70–79						1		7	8					1	4	2	3	10
80–90							1	2	3							3	5	8
Total	4	15	17	12	14	17	17	12	108	1	5	5	10	17	18	16	11	83

Note:
a. Modified from Işcan *et al.* (1984b, 1985).

However, several factors must be kept in mind if this technique is to be successfully applied. First, the necessity of determining the sex of the skeleton is reiterated so that the appropriate standard can be used. Second, intercostal variation must be considered since this technique was based on the fourth rib. Finally, while the phase technique does attempt to mediate the problem of individual variation in the normal ageing process, this variation is always an important factor.

Medial Clavicular Epiphysis and Anterior Iliac Crest

Standards for epiphysial union of the medial clavicular epiphysis and anterior iliac crest are available on a sample of 859 modern Americans autopsied at the Department of the Chief Medical Examiner-Coroner, Los Angeles (Owings, 1981). These bones were removed at the time of autopsy and prepared by soaking them in water in order to make direct observation. Four stages of epiphysial union were used for the classification: (1) non-union without epiphysis, (2) non-union with epiphysis, (3) partial

Table 5.8: Epiphysial Union (in percentages) of the Left Anterior Iliac Crest in a Modern Multiracial Sample of Persons Autopsied at the Department of the Chief Medical Examiner-Coroner, Los Angeles[a]

Age	No.	Males (n=522) Stage of union[b]				Age	No.	Females (n=198) Stage of union			
		1	2	3	4			1	2	3	4
11	2	100	—	—	—	11	1	100	—	—	—
13	3	67	33	—	—						
14	7	43	43	14	—	14	3	—	—	100	—
15	12	25	67	8	—	15	9	—	44	56	—
16	24	4	33	63	—	16	4	—	—	100	—
17	22	—	9	82	9	17	6	—	—	100	—
18	30	—	7	53	40	18	16	—	—	69	31
19	34	—	9	47	44	19	9	—	—	56	44
20	28	—	—	25	75	20	14	—	—	14	86
21	29	—	—	10	90	21	8	—	—	—	100
22	38	—	—	—	100	22	13	—	—	8	92
23	30	—	—	3	97	23	10	—	—	10	90
24–40	263	—	—	—	100	24–39	105	—	—	—	100

Notes:
a. From Owings (1981, p. 63);
b. Stages of union: (1) non-union without epiphysis, (2) non-union with epiphysis, (3) partial union, (4) complete union.

union, and (4) complete union. Figure 5.2 illustrates these stages on the anterior iliac crest; Figure 5.3 illustrates these stages on the medial clavicle. Table 5.8 presents the data for the left anterior iliac crest for both males and females.

As can be seen from Table 5.8, crests with no union (stages 1 and 2) can be seen in males 19 years and younger and in females 15 years and younger. Partial union occurs in both males and females from the age of 14

Figure 5.2: Anterior Iliac Crest in Four Stages of Epiphysial Union. From top to bottom: complete union; partial union; non-union with separate epiphysis; non-union, no epiphysis

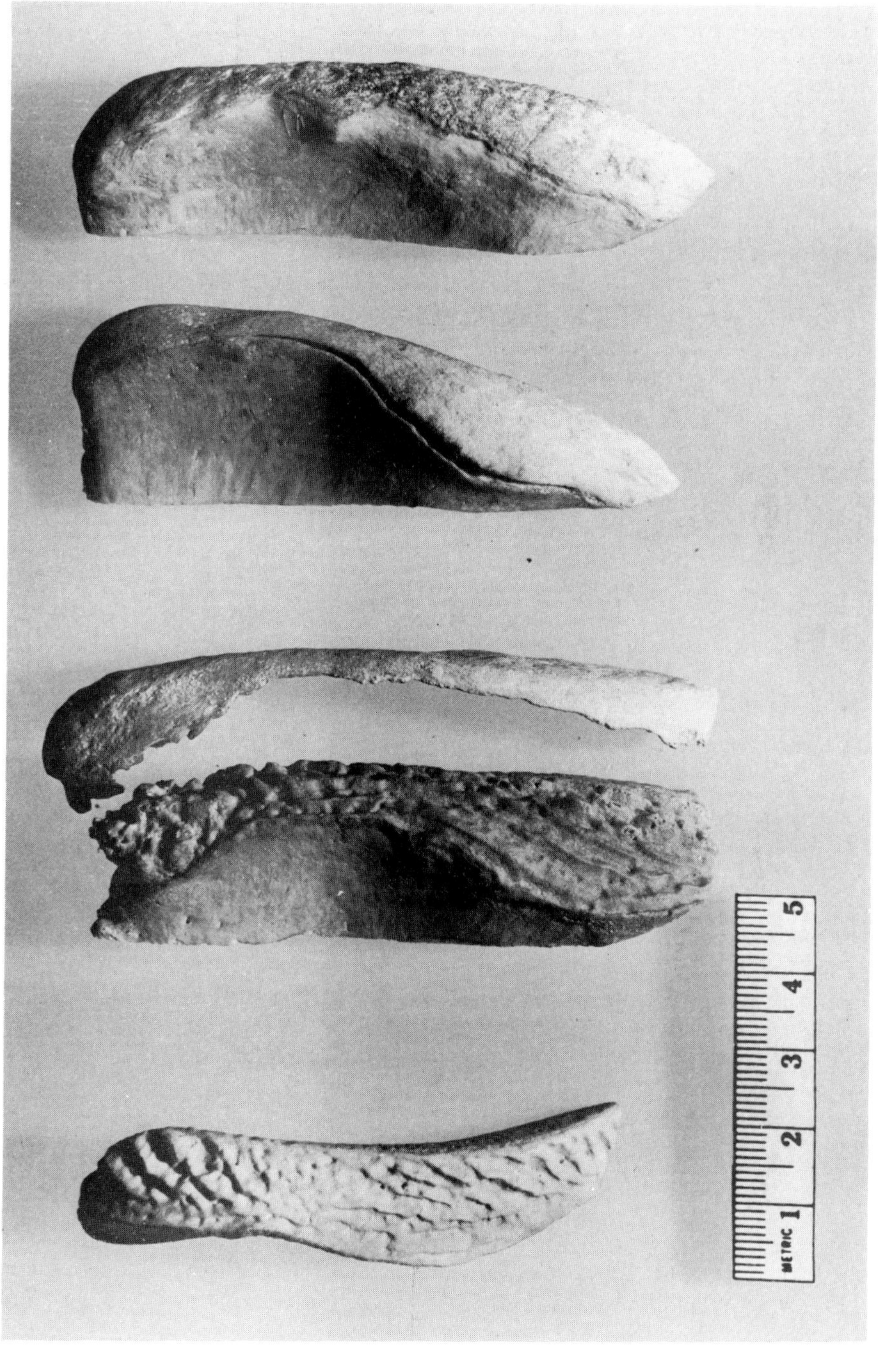

Figure 5.3: Medial Clavicle in Four Stages of Epiphysial Union. From left to right: non-union, no epiphysis; non-union with separate epiphysis; partial union; complete union

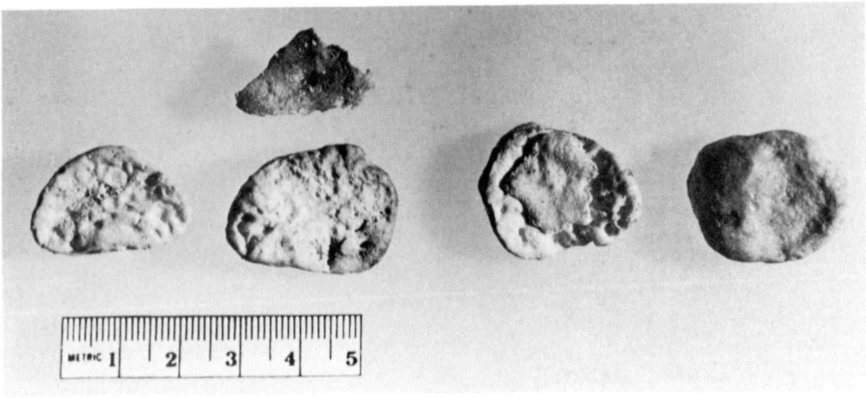

to 23. Complete union occurs as early as the age of 17 in males and 18 years in females. Table 5.9 presents the data for the left medial clavicular epiphysis for males and females. Clavicles with no union (stage 1 and 2) can be seen in males 25 years or younger and in females 21 years or younger. Partial union can be seen in males aged 17 to 30, and in females aged 16 to 33. Complete union is seen as early as the age of 21 in males and 20 years in females.

Pubic Symphysis — Male

The pubic symphysis is a good age indicator in males aged 17 to 30 years (Suchey, Wiseley and Noguchi, 1980; Suchey, Wiseley and Katz, 1985), and less useful from the age of 30 to 40 years. The ageing standards presented here are derived from research on modern Americans autopsied at the Department of the Chief Medical Examiner-Coroner, Los Angeles.[2] The sample is the largest of its kind, with 739 males ranging in age from 14 to 92. The age distribution of the sample is presented in Table 5.10, where it can be seen that the majority of specimens came from individuals less than 60 years old. Age data are precise, being based on death certificates. These data ultimately originate from the family or from those persons taking responsibility for the disposition of the remains.

The sample is fairly representative, since the cases autopsied involved homicides, suicides, accidents and unexpected natural deaths. Former studies used samples with major limitations. Todd's (1920, 1921) Western Reserve (Hamman-Todd) Collection lacked adequate representation of the younger ages and most of the individuals were in the older age ranges, after the major symphysial changes are completed. Also, the accuracy of his stated ages was in question, as Todd (1920, p. 289, p. 314) was careful

Table 5.9: Epiphysial Union (in percentages) of the Left Medial Clavicle in a Modern Multiracial Sample of Persons Autospied at the Department of the Chief Medical Examiner-Coroner, Los Angeles[a]

Age	No.	Males (n=597) Stage of union[b]				Age	No.	Females (n=247) Stage of union			
		1	2	3	4			1	2	3	4
11	2	100	—	—	—	11	1	100	—	—	—
13	3	100	—	—	—	13	1	100	—	—	—
14	7	100	—	—	—	14	2	100	—	—	—
15	12	100	—	—	—	15	9	100	—	—	—
16	25	92	8	—	—	16	6	66	17	17	—
17	23	83	—	17	—	17	7	43	14	43	—
18	33	52	18	30	—	18	16	19	31	50	—
19	33	55	21	24	—	19	11	18	18	64	—
20	27	30	15	55	—	20	15	7	—	86	7
21	33	9	3	85	3	21	9	11	11	78	—
22	40	12	10	70	8	22	15	—	—	80	20
23	32	6	—	78	16	23	12	—	—	83	17
24	29	3	—	62	35	24	16	—	—	50	50
25	38	3	—	45	52	25	7	—	—	43	57
26	21	—	—	38	62	26	16	—	—	37	63
27	35	—	—	37	63	27	9	—	—	33	67
28	22	—	—	23	77	28	19	—	—	5	95
29	22	—	—	5	95	29	11	—	—	—	100
30	30	—	—	7	93	30	11	—	—	9	91
31–40	130	—	—	—	100	31	10	—	—	—	100
						32	10	—	—	10	90
						33	12	—	—	17	83
						34–39	22	—	—	—	100

Notes:
a. From Owings (1981, pp. 54–55);
b. Stages of union: (1) non-union without epiphysis, (2) non-union with epiphysis, (3) partial union, (4) complete union.

to note. Hence, certain skeletons were deleted if they did not fit existing standards for skeletal development, perhaps falsely reducing sample variability. Then, the municipal records were found not always to be reliable. There are notable peaks of stated age at 35, 40, 45, 50 and 60

Table 5.10: Description of the Pubic Bone Samples as to Age Autopsied at the Department of the Chief Medical Examiner-Coroner, Los Angeles[a]

Age range	Males	Females
10–19	81	48
20–29	192	130
30–39	120	69
40–49	117	44
50–59	101	41
60–99	128	37
Total N	739	369

Note:
a. Males are from Suchey *et al.* (1980) and females from Suchey *et al.* (1981).

years in the white male sample; the black male sample shows this feature to a lesser degree. These peaks may come simply from rounding off, but perhaps from a greater problem in accuracy of vital statistics. Of the males, Brooks (1984) notes that 'many of them were transients, with no known age data. Their age estimation was made by Todd or one of the anatomists when the cadaver was being prepared for the WRU Medical School dissection classes'. Thus, a number of skeletons in Todd's sample may have stated ages which are more than a couple of years wrong. McKern and Stewart's (1957) accurate war dead sample was limited almost entirely to young white males.

Racial composition of the Los Angeles sample is primarily white, black and Mexican. In the majority of cases, race was determined by morphological observations of the body before autopsy. Death certificate information was utilised in the remaining cases.

Todd System. Todd's system (1920, 1921) is based on ten phases. These phases are illustrated in Figure 5.4 and are described as follows (1920):

I. First post-adolescent phase — age 18 to 19 years. Symphysial surface rugged, traversed by horizontal ridges separated by well marked grooves; no ossific (epiphysial) nodules fusing with the surface; no definite delimiting margin; no definition of extremities (p. 301).

II. Second post-adolescent phase — age 20 to 21 years. Symphysial surface still rugged, traversed by horizontal ridges, the grooves between which are, however, becoming filled near thiphysial) nodules fusing with upper symphysial face may occur; dorsal limiting margin begins to develop; no delimitation of extremities; foreshadowing of ventral bevel (pp. 302–3).

III. Third post-adolescent phase — age 22 to 24 years. Symphysial face shows progressive obliteration of ridge and furrow system; commencing formation of the dorsal plateau; presence of fusing ossific (epiphysial) nodules; dorsal margin gradually becoming more defined; bevelling as a result of ventral rarefaction becoming rapidly more pronounced; no delimitation of extremities (p. 304).

IV. Fourth phase — age 25 to 26 years. Great increase of ventral bevelled area; corresponding diminution of ridge and furrow formation; complete definition of dorsal margin through the formation of the dorsal plateau; commencing delimitation of lower extremity (p. 305).

V. Fifth phase — age 27 to 30 years. Little or no change in symphysial face and dorsal plateau except that sporadic and premature attempts at the formation of a ventral rampart occur; lower extremity, like the dorsal margin, is increasing in clearness of definition; commencing formation of upper extremity with or without the intervention of a bony (epiphysial) nodule (p. 306).

VI. Sixth phase — age 30 to 35 years. Increasing definition of extremities;

Figure 5.4: Diagram of Pubic Bones Illustrating Todd's Ten Typical Phases (after G. Neumann). All represent the left side

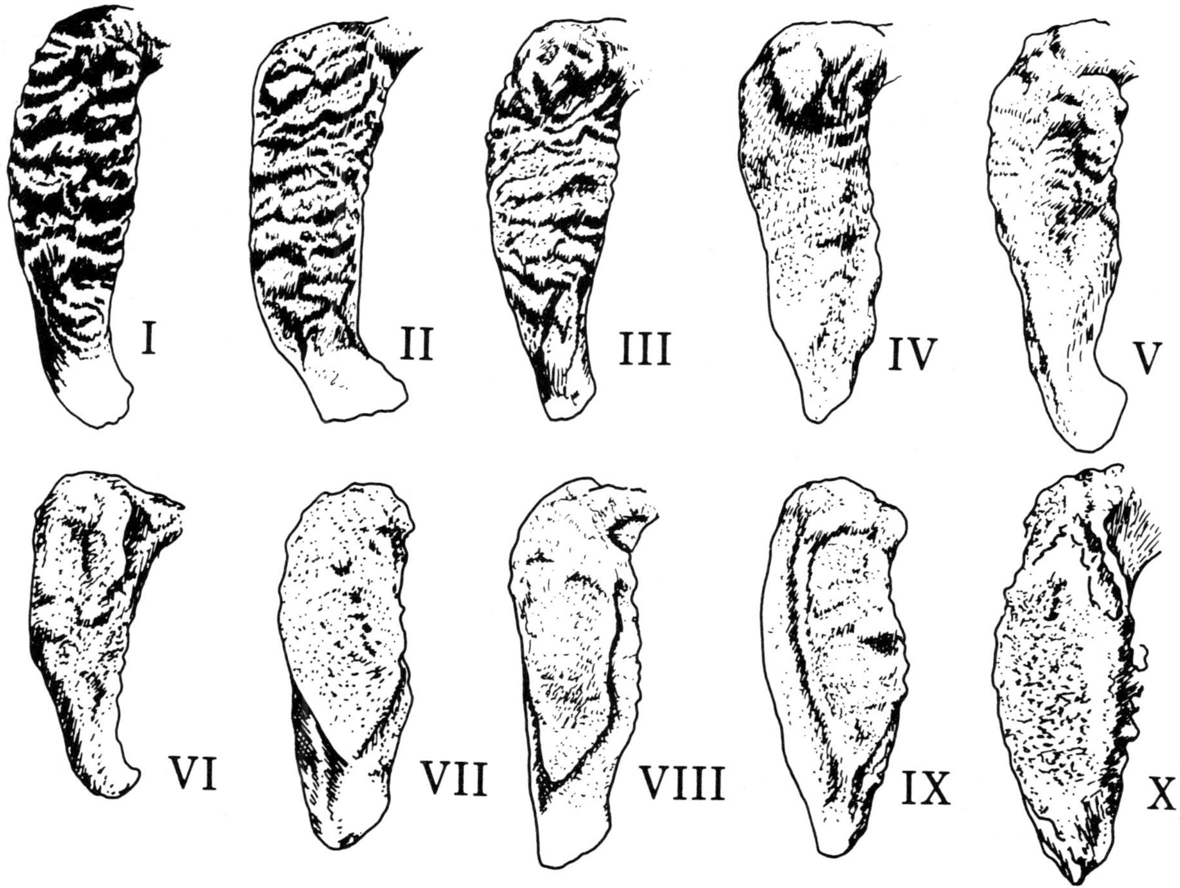

development and practical completion of ventral rampart; retention of granular appearance of symphysial face and ventral aspect of pubis; absence of lipping of symphysial margin (p. 308).
VII. Seventh phase — age 35 to 39 years. Changes in symphysial face and ventral aspect of pubis consequent upon diminishing activity; commencing bony outgrowth into attachments of tendons and ligaments, especially the gracilis tendon and sacro-tuberous ligament (p. 310).
VIII. Eighth phase — age 39 to 44 years. Symphysial face generally smooth and inactive; ventral surface of pubis also inactive; oval outline complete or approximately complete; extremities clearly defined; no distinct 'rim' to symphysial face; no marked lipping of either dorsal or ventral margin (p. 311).
IX. Ninth phase — age 45 to 50 years. Symphysial face presents a more or less marked rim; dorsal margin uniformly lipped; ventral margin irregularly lipped (p. 312).
X. Tenth phase — age 50 and upward. Symphysial face eroded and showing erratic ossification; ventral border more or less broken down; disfigurement increases with age (p. 313).

Using Todd's descriptions and the illustrations of the ten phases, the male pubic bones in the Los Angeles sample could be easily classified. The statistics based on the Los Angeles sample are presented in Table 5.11. Comparing these to Todd's ranges, it can be seen that the Todd system tends to over-age individuals in phases I to VIII. In these phases the Los Angeles mean falls below the lowest part of the age range. It is suggested that current researchers who utilise the Todd system replace his age ranges with these developed from the Los Angeles sample.

Table 5.11: Statistics of the Todd Ageing System when Applied to the Los Angeles Sample of Male Pubic Bones[a]

Todd's phases	Todd's age range	Los Angeles sample size	Los Angeles range	Los Angeles mean	Los Angeles standard deviation
I	18–19	19	14–19	16.68	1.46
II	20–21	50	14–23	18.12	1.74
III	22–24	52	17–24	20.38	1.90
IV	25–26	30	20–36	24.83	3.72
V	27–30	51	19–45	24.57	4.58
VI	30–35	43	22–51	28.81	5.89
VII	35–39	56	20–64	34.04	9.47
VIII	39–44	97	23–71	38.33	9.43
IX	45–50	241	21–87	51.00	13.64
X	50+	100	26–92	62.74	12.40

Note:
a. From Todd (1920, p. 313–314) and Suchey *et al.* (1980).

McKern-Stewart System. McKern and Stewart (1957) divided the symphysial face into three major components (the dorsal plateau, ventral rampart and symphysial rim) and developed a six-stage system for each component. This system allows individuals falling outside the typical developmental scheme to be aged. Figure 5.5 illustrates the phases of these three components. The descriptions of the phases are as follows (1957, pp. 75–9):

Component I. Dorsal Plateau

0. Dorsal margin absent.
1. A slight margin formation first appears in the middle-third of the dorsal border.
2. The dorsal margin extends along entire dorsal border.
3. Filling-in of grooves and resorption of ridges to form a beginning plateau in the middle-third of the dorsal demi-face.
4. The plateau, still exhibiting vestiges of billowing, extends over most of the dorsal demi-face.
5. Billowing disappears completely and the surface of the entire demi-face becomes flat and slightly granulated in texture.

Component II. Ventral Rampart

0. Ventral bevelling is absent.
1. Ventral bevelling is present only at superior extremity of ventral border.
2. Bevel extends inferiorly along ventral border.
3. The ventral rampart begins by means of bony extensions from either or both of the extremities.
4. The rampart is extensive but gaps are still evident along the earlier ventral border, most evident in the upper two-thirds.
5. The rampart is complete.

Component III. Symphysial Rim

0. The symphysial rim is absent.
1. A partial dorsal rim is present, usually at the superior end of the dorsal margin; it is round and smooth in texture and elevated above the symphysial surface.
2. The dorsal rim is complete and the ventral rim is beginning to form. There is no particular beginning site.
3. The symphysial rim is complete. The enclosed symphysial surface is finely grained in texture and irregular or undulating in appearance.
4. The rim begins to break down. The face becomes smooth and flat and the rim is no longer round but sharply defined. There is some evidence of lipping on the ventral edge.

Figure 5.5: Photograph of the Model Casts Used in Conjunction with the McKern-Stewart Ageing System (I Dorsal Plateau, II Ventral Rampart, III Symphysial Rim). All represent the left side except for II 1, 2, 4, 5

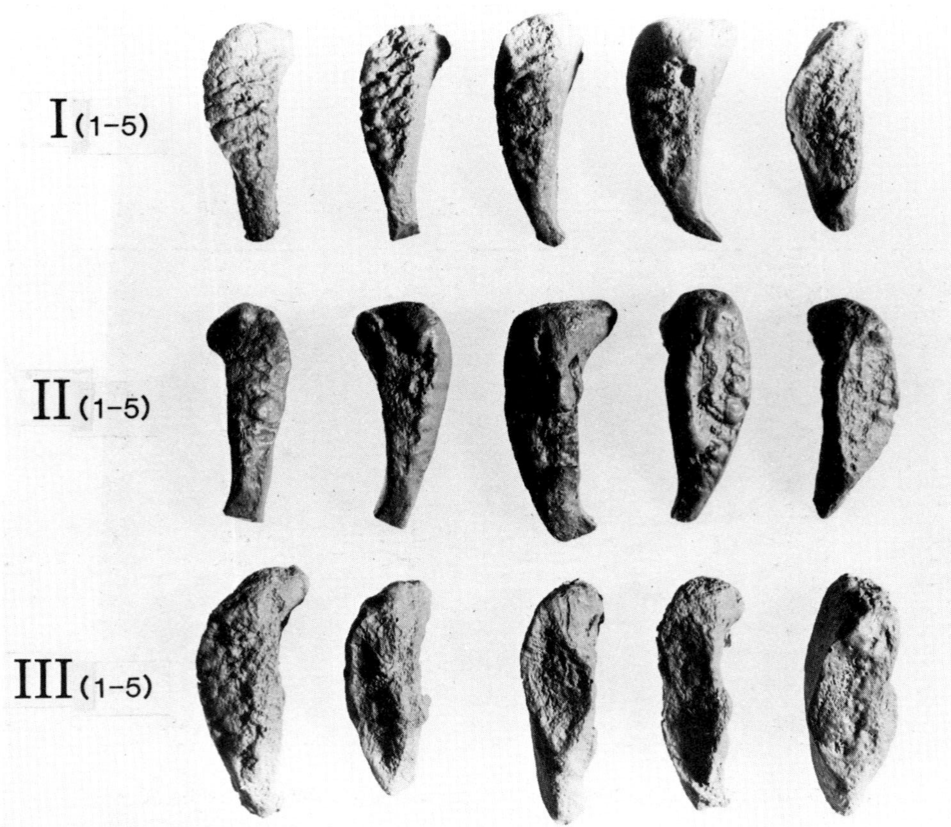

Table 5.12: Statistics of the McKern-Stewart System When Applied to the Los Angeles Sample of Male Pubic Bones[a]

	McKern-Stewart sample (N=349)					Los Angeles sample (N=739)			
Total score	No.	Age range	Mean age	Standard deviation	Total score	No.	Age range	Mean age	Standard deviation
0	7	0–17	17.29	0.49	0	2	14–19	16.50	3.54
1–2	76	17–20	19.04	0.79	1	14	14–23	17.50	2.14
					2	3	16–17	16.33	0.58
3	43	18–21	19.79	0.85	3	26	15–23	18.08	1.98
4–5	51	18–23	20.84	1.13	4	7	16–18	17.00	0.82
					5	23	16–23	18.83	1.95
6–7	26	20–24	22.42	0.99	6	35	17–25	20.46	2.13
					7	5	21–26	22.60	2.30
8–9	36	22–28	24.14	1.93	8	48	18–37	23.73	4.56
					9	0	–	–	–
10	19	23–28	26.05	1.87	10	54	19–49	27.30	6.52
11,12,13	56	23–39	29.18	3.33	11	45	20–78	35.91	12.48
					12	26	20–59	33.27	9.94
					13	93	21–71	36.26	9.96
14	31	29+	35.84	3.89	14	201	24–80	47.00	12.16
15	4	36+	41.00	6.22	15	157	25–92	61.27	13.89

Note:
a. From McKern and Stewart (1957, p. 85) and Suchey *et al.* (1980).

5. Further breakdown of the rim (especially along superior ventral edge) and rarefaction of the symphysial face. There is also disintegration and erratic ossification along the ventral rim.

Examining Tables 5.11 and 5.12 it is seen that there is a great deal of variability in male pubic bone patterns when compared to age. 'Younger patterns' are not seen in the older males but 'advanced patterns' can be seen in young males. Thus, the higher scores in the McKern-Stewart system and the advanced patterns in the Todd system occur in males with very broad age ranges.[3] Evaluating the male pubic ageing data, the McKern-Stewart system is found to be complicated and not usable by

Table 5.13: Modification of the Todd System of Pubic Symphysial Ageing. Reduction to five phases (A, B, C, D, E)[a] using age data from the male sample (n=739) autospied at the department of the Chief Medical Examiner-Coroner, Los Angeles

New phases	Todd's phases	Ageing rule	Assessment of rule
Todd A	I, II, III	24 years or under	Covers all of sample variability
Todd B	IV, V	19–30 years	Leaves out six advanced cases from 30–45; 93% sample variability is included
Todd C	VI, VII	22–50 years	Leaves out 13 advanced cases; 94% of sample variability is included
Todd D	IX	30 years or older	Leaves out 13 young cases; 95% of sample variability is included
Todd E	X	45 years or older	Leaves out five young cases; 95% of sample variability is included

Note:
a. N=739 males autopsied at the Department of the Chief Medical Examiner-Coroner Los Angeles.
Source: Suchey, Wiseley and Katz (1986) in Reichs (ed.) (in press).

everyone. The variability found in male pubic symphysial metamorphosis does not warrant the complexities of a three-component system. It is recommended that this system be avoided. Todd's phases can be used in conjunction with the age ranges from the Los Angeles study. A modification of the Todd system is presented in Table 5.13. Here, Todd's phases I, II and III are combined to form a Todd-A with the age range of 24 years or under. Todd's phases IV and V are combined to form a Todd-B with the age range of 19 to 30 years. Todd-C consists of phases VI, VII and VIII with a range of 22 to 50 years. Todd-D (IX) indicates 30 years or older; Todd-E (X) indicates 45 years or older. The available data do not warrant an ageing system any more definitive than this.

Pubic Symphysis — Female

The female pubic symphysis does not follow regular patterns in ageing as does that of the male. This conclusion has been arrived at after careful study of a sample of 500 modern American females autopsied at the Department of the Chief Medical Examiner-Coroner, Los Angeles. A preliminary study (Suchey, 1979) showed that a series of eleven pubic bones of known age could not be aged consistently by expert forensic anthropologists.

Study began on a large sample of female pubic bones to pursue this problem. The first focus was on the effect of pregnancy on pubic morphology (Suchey *et al.*, 1979, pp. 517–540). The sample consisted of 486 females for whom the following variables were known: (1) age, (2) number of pregnancies, and (3) interval since last pregnancy. A statistical association was found between the number of full- term pregnancies and the degree of apparent dorsal pitting. However, the correlation is not strong; 17 females reported as being nulliparous have 'medium to large' dorsal changes, whereas 22 females having from one to five full-term pregnancies have an absence of apparent dorsal changes. Females who had their last child 15 or more years before death are more apt to have 'medium to large' dorsal changes than are females who have more recently given birth. This effect appears to be independent of the number of full-term pregnancies but is probably related to age.

When the data were subjected to multiple regression analysis, it was found that age and number of pregnancies are most important in predicting the degree of pitting and that the effect of interval on pitting is not significant. Age was found to be an important variable independent of the number of full-term pregnancies. In nulliparous women, an absence of dorsal pitting is far more frequently found in females younger than 30 years old than in those over 30.

The next part of the study involved age changes on the symphysial face. The literature on female pubic changes was searched (Todd, 1920; Brooks, 1955; Gilbert and McKern, 1973; Hanihara and Suzuki, 1978) for aids to the identification of age-related traits. A sample of 369 females was studied using all traits that were easy to identify (Suchey *et al.*, 1981). This sample is predominantly white and black, with an emphasis on young individuals (Table 5.10). Only females with birth certificates available were included in this sample in order to eliminate any possible errors in the age factor. Extreme variability was found in female pubic morphology, which made the formation of an ageing system virtually impossible.

Certain traits are useful in the ageing of specific female pubic bones and these general rules are helpful in those cases:

Figure 5.6: Female Pubic Bones Illustrating Major Ageing Rules (A — Symphysial Face Shows Deep Distinct Ridges, B — Ventral Rampart has Begun its Formation But Is Not Complete, C — Ossific Nodules Are Present). All represent the left side

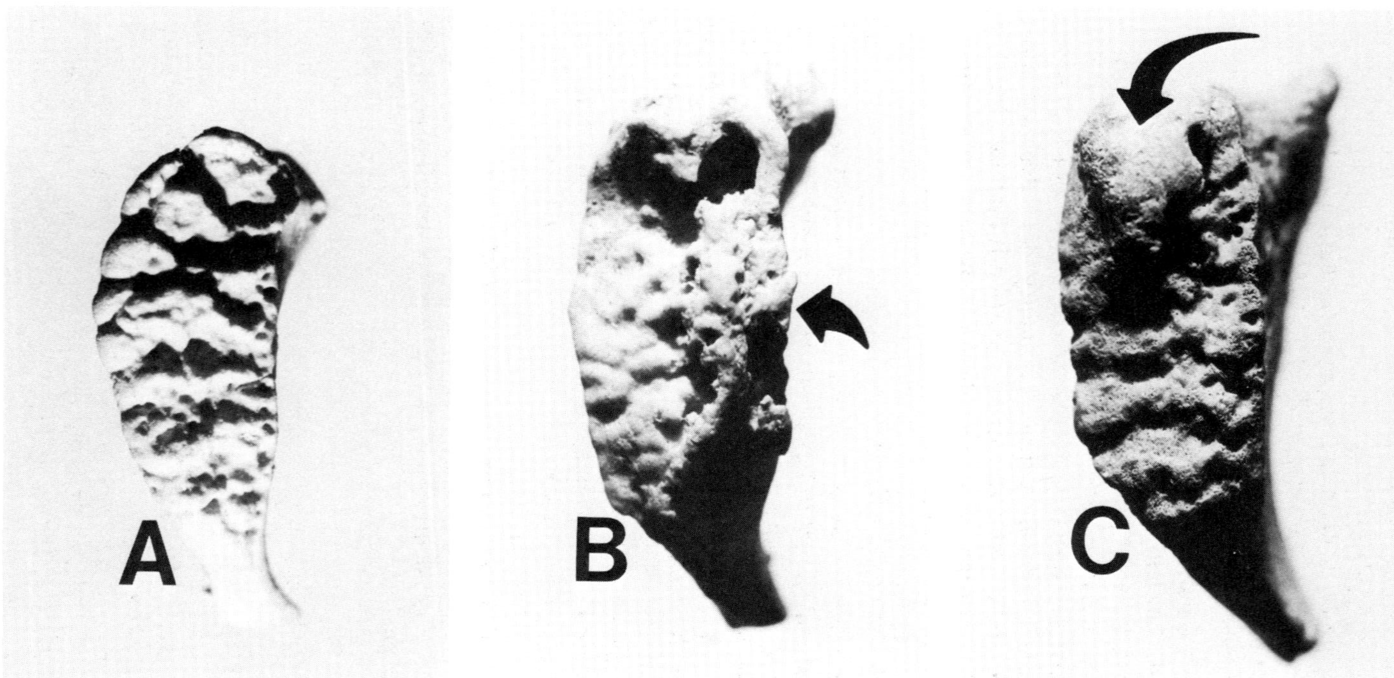

1. Deep, distinct ridges on both the dorsal and ventral demi-face. Less than 24 years old (Figure 5.6A).
2. The ventral rampart has begun its formation, but it is not complete. The symphysial rim has begun its development but is not complete. Range of 20 to 49 years (Figure 5.6B).
3. Ossific nodules are present. Less than 28 years old (Figure 5.6C).

Many female pubic bones will not exhibit any of the above conditions. Instead, they will be characterised by the completion of a ventral rampart. The ventral rampart is seen to be complete as early as 20 years old in some females. In the Los Angeles sample, 41 of the 118 females in their 20s had a complete ventral rampart. Erratic ossification of the pubic symphysis can also occur in this age period. Pregnancy alone does not appear to explain this variability. An alternative suggestion is that use of oral contraceptives may be causing some females to advance earlier than others; this hypothesis has not been tested rigorously to date.

Skeletal Ageing Conclusions

In the infant, child and adolescent to sub-adult age range, a variety of skeletal and dental indicators serve as reliable means of age estimation. The selection of the indicators depends on the completeness and kind of sample under study, as well as the available methods. In adults, age indicators span broader ranges. Recent research in autopsy room samples has allowed re-evaluation of certain of the classic indicators and development of new techniques. The costochondral junction of the rib, epiphyses of the anterior iliac crest and medial clavicle, and the pubic symphysis in young males are found to be reliable age indicators. The pubic symphysis in females and cranial vault sutures are less reliable, but are still useful indicators in certain cases. It is recommended that a variety of age indicators be used for adults. No one area can always be depended upon to supply the best answer. Furthermore, one area may provide a good lower end while another supplies an upper end to the age range.

Visceral Changes with Ageing

Ageing changes in the various organ systems present a continuum that shows great individual and intersexual variation, but are nonetheless of some value in estimating the age at death of mummified human remains. Skeletal and dental criteria are generally best in evaluating children and adolescents, so visceral morphological criteria are best applied to adults. Classification is quite rough, generally being on the order of young (20 to 40 years), middle-aged (40 to 60 years) and old (over 60 years).

The cardiovascular system is most useful for this process. In modern Western populations, ageing is accompanied by enlargement of the heart and advancing arteriosclerosis, the deposition of cholesterol and other lipids in the walls of the arteries. In mummies, changes in heart weight generally cannot be appreciated, as the degree of desiccation is variable from organ to organ and for the body as a whole. The weight of the mummified heart therefore has little relation to the true weight of the organ at the time of death.

On the other hand, mummies have shown a rough but consistent correlation between age at death and degree of arteriosclerosis. Modern adult men show a progression from mild disease, beginning in the third decade (or earlier), to severe arteriosclerosis by the sixth or seventh decade. In women, the process is delayed, with the onset of arteriosclerotic disease in the perimenopausal years (45 to 55 years). This correlation has held true in mummies from such diverse areas as Egypt and Alaska.

Arteriosclerosis was used as an aid to age estimation as early as 1909, when Shattock noted severe disease in the aorta of the XXIst dynasty Pharaoh Merneptah. Shattock's conclusion was that Merneptah was 'an old man'. Historical evidence indicates that he was 57 to 70 years old at death (Wente, 1980), but radiological findings yield a slightly younger age, 45 to 50 years (Krogman and Baer, 1980).

Other Egyptian mummies have shown atherosclerosis, as reviewed by Sandison (1980). Of particular interest was the study of the Lady Teye by Long (1931), showing arteriosclerosis involving the aorta and coronary arteries, and arteriosclerosis of the kidneys with fibrosis in the heart. The age had been estimated at 50 years, but Long suggested that she had been somewhat older at death. Studies such as these make it clear that arteriosclerosis was present in ancient Egypt, and that the modern investigator can consider mummies showing the disease to be of advanced age.

The few frozen, mummified human bodies found in Alaska have been intact, allowing for age determination by a variety of techniques. The presence of arteriosclerosis has resulted in the assignment of two Aleuts (Zimmerman *et al.*, 1971; 1981) and two Eskimo mummies to middle age. The age of the second Aleut (Zimmerman *et al.*, 1981) was confirmed at 51 + 6 by osteon analysis. In the case of one Eskimo, found on St Lawrence Island (Zimmerman and Smith, 1975), and dated to about 400 AD, the mummy was initially estimated to be in her 50s. Subsequent osteon analysis yielded an age estimate of 50 to 59 years (Stout and Tietelbaum, 1976) and the application of the amino acid racemisation technique described in this volume by Masters (Chapter 2) resulted in an age of 53 ± 5 years (Masters and Zimmerman, 1978).

The other Eskimo mummy, found near Barrow, Alaska, and dated at

1600 to 1800 AD (Zimmerman and Aufderheide, 1984) also showed coronary and aortic arteriosclerosis, leading to a similar initial estimate of death in the sixth decade. However, histologic examination of the breasts revealed evidence of lactation, suggesting an earlier age, and osteon analysis and amino acid racemisation studies were confirmatory, both yielding an age in the early 40s. Calcified arteries in her uterus also indicated a woman near the end of her active reproductive years, but it is clear that such changes antedate the postmenopausal period.

The absence of atherosclerosis can also be used as a clue to age. Another Eskimo woman found at the Barrow site was estimated to be in her 20s, an assessment confirmed by osteon analysis (24 to 27 years) and amino acid racemisation (27 + 2 years). Parenthetically, while osteoporosis is a postmenopausal phenomenon in modern western society, this criterion cannot be applied to modern or ancient Eskimos; both of these women showed severe osteoporosis.

While pigments accumulate in body tissues throughout life, this process is not very useful for age determination. Heavy deposits of anthracotic pigment have been seen in the lungs of an ancient Egyptian child (Zimmerman, 1977). Lipofuscin, a degenerative pigment, accumulates in the heart and liver, and the St Lawrence Island Eskimo did show this brown pigment in her liver. Histologic diagnosis involves excluding iron or bile as the pigment, by specific stains, but lipofuscin is a non-specific indicator and can, in general, only be taken as evidence that the individual is over 50 years old.

Other changes in the organs, such as atrophy of the genitalia, kidneys, brain and other organs, are equally non-specific and have proven difficult to evaluate. These findings should only be considered at the time of autopsy in concert with other features, such as dental attrition, to arrive at a rough estimate, awaiting the more specific tests outlined in this and other chapters.

Acknowledgements

We are grateful to medical examiners of Broward County, Florida, and Los Angeles County, California, and their staff for the collection of data; and to Marjorie Wolf for word-processing the manuscript. We also thank Susan R. Loth for her assistance in proof-reading the manuscript.

Notes

1. This study was supported by Florida Atlantic University Sponsored Research Grant No. 121210016.
2. This research was supported by the National Science Foundation under Grant BN S78–13025.
3. For further information on the distribution of the 739 males in the Los Angeles sample according to the McKern-Stewart system or the Todd system, contact Suchey directly for the data in histogram form. Data from regression analyses are also available.

References

Acheson, R. M. (1957) 'The Oxford Method of Assessing Skeletal Maturity', *Clinical Orthopedics, 10*, pp. 19–39

Acsádi, G. and Nemeskéri, J. (1970) *History of Human Life Span and Mortality*, Akademiai Kiado, Budapest

Amundsen, D. W. and Diers, C. J. (1969) 'The Age of Menarche in Classical Greece and Rome', *Human Biol., 41*, pp. 125–32

Angel, J. L. (1969) 'The Bases of Paleodemography', *Am. J. Phys. Antrhopol., 30*, pp. 427– 37

—— (1971a) *The People of Lerna*, Smithsonian Institution Press, Washington DC

—— (1971b) 'Skull Vault Thickness Variation', *Am. J. Phys. Anthropol., 35*, p. 272 (abstract)

—— (1980) 'Physical Anthropology: Determining Sex, Age, and Individual' features in Cockburn, A. and E. (eds.), *Mummies, Disease and Ancient Cultures*, Cambridge University Press, Cambridge, pp. 241–57

—— (1984) 'Variation in Estimating Age at Death of Skeletons', *Coll. Anthropol., 8*, pp. 163–8

Baker, R. K. (1984) *The Relationship of Cranial Suture Closure and Age Analyzed in a Modern Multi-Racial Sample of Males and Females*, MA Thesis, California State University, Fullerton

Brooks, S. T. (1955) 'Skeletal Age at Death: The Reliability of Cranial and Pubic Age Indicators', *Am. J. Phys. Anthropol., 13*, pp. 567–97

—— (1984) Personal communication

Buhr, A. J. and Cooke, A. M. (1959) 'Fracture Patterns', *The Lancet, 1*, pp. 531–6

Cobb, W. M. (1952) 'Skeleton', in A. I. Lansing (ed.)' *Cowdry's Problems of Ageing*, Williams and Wilkins, Baltimore, pp. 791–856

Fischer, E. (1955) 'Verkalkungsformen der Rippenknorpel', *Fortschritte auf dem Gebiete der Rontgen Strahlen und der Nuklear Medizin, 82*, pp. 474–81

Garn, S. M. (1970) *The Earlier Gain and Later Loss of Cortical Bone*, Thomas, Springfield

——,Lewis, A. B., Koski, K. and Polachek, D. L. (1958) 'The Sex Difference in Tooth Calcification', *J. Dent. Res., 37*, pp. 561–7

Gilbert, B. M. and McKern, T. W. (1973) 'A Method for Ageing the Female Os Pubis', *Am. J. Phys. Anthropol., 38*, pp. 31–8

Graves, W. W. (1922) 'Observations on Age Changes in the Scapula', *Am. J. Phys. Anthropol., 5*, pp. 21–33

Haines, R. W., Mohiuddin, A., Opka, F. T. and Viega-Pines, J. A. (1967) 'The Sites of Early Epiphyseal Union in the Limb Girdles and Major Long Bones of Man', *J. Anat., 101*, pp. 823–31

Hanihara, K. and Suzuki, T. (1978) 'Estimation of Age from the Pubic Symphysis by Means of Multiple Regression Analysis', *Am. J. Phys. Anthropol.*, *48*, pp. 233–40
Hunt, E. E. Jr., and Geilser, I. (1955) 'The Estimation of Age and Sex of Preadolescent Children from Bones and Teeth', *Am. J. Phys. Anthropol.*, *13*, pp. 479–87
Hurme, V. O. (1957) 'The Human Dentition in Forensic Medicine: Symposium. Time and Sequence of Tooth Eruption', *J. Forensic Sci.*, *2*, pp. 377–88
Işcan, M. Y. (1985) 'Osteometric Analysis of Sexual Dimorphism in the Sternal End of the Rib', *J. Forensic Sci.*, *30*, (4), pp. 1090–9
—— and Loth, S. R. (1985) 'Estimation of Age and Determination of Sex from the Sternal Rib', in K. J. Reichs (ed.), *Forensic Osteology: Advances in Techniques of Human Identification*, Thomas, Springfield, in press
—— (1986) 'Determination of Age from the Sternal End in White Males: A Test of the Phase Method', *J. Forensic Sci.*, *31*, in press
—— and Wright, R. K. (1984a) 'Metamorphosis at the Sternal Rib End: A New Method to Estimate Age at Death in White males', *Am. J. Phys. Anthropol.*, *65*, pp. 147–56
—— (1984b) 'Age Estimation From the Rib by Phase Analysis: White Males', *J. Forensic Sci.*, *29*, pp. 1094–1104
—— (1985) 'Age Estimation From the Rib by Phase Analysis: White Females', *J. Forensic Sci.*, *30*, pp. 853–63
Kerley, E. R. (1965) 'The Microscopic Determination of Age in Human Bone', *Am. J. Phys. Anthropol.*, *23*, pp. 149–63
—— (1970) 'Estimation of Skeletal Age after About Age 30', in T. D. Stewart (ed.), *Personal Identification in Mass Disasters*, National Museum of Natural History, Washington DC, pp. 57–70
Krogman, W. M. (1972) *Child Growth*, University of Michigan Press, Ann Arbor
—— and Baer, M. J. (1980) 'Age at Death of Pharaohs of the New Kingdom', in J. E. Harris and E. F. Wente, (eds.), *An X-Ray Atlas of the Royal Mummies*, University of Chicago Press, Chicago, pp. 188–233
—— and Işcan, M. Y. (1986) *The Human Skeleton in Forensic Medicine*, 3rd edn, Thomas, Springfield
Long, A. R. (1931) 'Cardiovascular Renal Disease: Report of a Case Three Thousand Years Old', *Arch. Pathol.*, *12*, pp. 92–4
Mainland, D. (1945) *Anatomy as a Basis for Medical and Dental Practice*, P. B. Hoeber, New York
Maresh, M. M. (1955) 'Linear Growth of Long Bones of Extremities from Infancy Through Adolescence', *Am. J. Dis. Children, 89*, pp. 725–42
Masset, C. (1976) 'Sur Quelques Fâcheuses Méthodes, de Determination de Lage des Squelettes', *Bull. et Mem. de la Société d'Anthropol. de Paris, 13*, (serie no. 3), pp. 329–36
Masters, P. M. and Zimmerman, M. R. (1978) 'Age Determination of an Alaskan Mummy: Morphologic and Biochemical Correlation', *Science, 201*, pp. 811–12
McCormick, W. F. and Stewart, J. H. (1983) 'Ossification Patterns of Costal Cartilages as an Indicator of Sex', *Arch. Pathol. Lab. Med.*, *107*, pp. 206–10
McKern, T. W. and Stewart, T. D. (1957) 'Skeletal Age Changes in Young American Males, Analyzed from the Standpoint of Age Identification', *Headqu. QM Res. and Dev. Command, Tech. Rep. EP-45*, Natick, MA
Meredith, H. V. (1943) *Physical Growth from Birth to Two years: I. Stature. University of Iowa Studies: Studies in Child Welfare, No. 19*, University of Iowa Press, Iowa City
Michelson, N. (1934) 'The Calcification of the First Costal Cartilage Among Whites and Negroes', *Human Biol.*, *6*, pp. 543–57
Montagna, W. (ed.) (1965) *Advances in Biology of Skin. Vol. 6.*, Pergamon Press, Oxford
Moorrees, C. F. A., Fanning, E. A. and Hunt E. E. (Jr.) (1963) 'Formation and Resorption of Three Deciduous Teeth in Children', *Am. J. Phys. Anthropol.*, *21*, pp. 205–13
Owings, P. A. (1981) *Epiphyseal Union of the Anterior Iliac Crest and Medial Clavicle in a Modern Multi-Racial Sample of Males and Females*, MA Thesis, California State University, Fullerton
Redfield, A. A. (1970) 'A New Aid to Aging Immature Skeletons: Development of the Occipital Bone', *Am. J. Phys. Anthropol.*, *33*, pp. 207–20
Robinow, M., Richards, T. W. and Anderson, M. (1942) 'The Eruption of Deciduous Teeth', *Growth, 6*, pp. 127–33

Sandison, A. T. (1980) 'Diseases in Ancient Egypt', in A. Cockburn (ed.), *Mummies, Disease and Ancient Cultures*, Cambridge University Press, Cambridge, pp. 29–44

Scammon, R. E. (1953) 'Developmental Anatomy', in J. P. Schaeffer (ed.), *Morris' Human Anatomy*, Philadelphia, Blakiston, pp. 11–62

Schour, I. and Massler, M (1944) *Development of the Human Dentition* (chart), American Dental Association

Semine, A. A. and Damon, A. (1975) 'Costochondral Ossification and Aging in Five Populations', *Human Biology, 47*, pp. 101–116

Shattock, G. S. (1909) 'Report on the Pathology of King Menephtah', *Lancet, 1* (1), p. 319

Stewart, T. D. (1968) 'Identification by the Skeletal Structures', in F. E. Camps (ed.), *Gradwohl's Legal Medicine*, Bristol, John Wright and Sons, pp. 123–54

—— (1979) *Essentials of Forensic Anthropology: Especially as Developed in the United States*, Thomas, Springfield

Stout, S. D. and Teitelbaum, S. L. (1976) 'Histomorphometric Determination of Formation Rates of Archaeological Bone', *Calc. Tiss. Res., 21*, pp. 163–9

Suchey, J. M. (1979) 'Problems in the Aging of Females Using the Os Pubis', *Am. J. Phys. Anthropol., 51*, pp. 467–70

——, Wiseley, D. V., Green, R. F. and Noguchi, T. T. (1979) 'Analysis of Dorsal Pitting in the Os Pubis in an Extensive Sample of Modern American Females', *Am. J. Phys. Anthropol., 51*, pp. 517–40

——, Wiseley, D. V. and Noguchi, T. T. (1980) *Aging the Male Os Pubis*, paper presented at the 32nd Annual Meeting of the American Academy of Forensic Sciences, New Orleans

—— (1981) *Age Changes in the Female Os Pubis in an Extensive Sample of Modern American Females*, paper presented at the 33rd Annual Meeting of the American Academy of Forensic Sciences, Los Angeles

——, Wiseley, D. V. and Katz, D. (1986) 'Evaluation of the Todd and McKern-Stewart Methods for Aging the Male Os Pubis' in K. J. Reichs (ed.), *Forensic Osteology: Advances in the Identification of Human Remains*, Thomas, Springfield, in press

Tanner, J. M. (1962) *Growth at Adolescence*, Blackwell, Oxford

Terry, R. J. and Trotter, M. (1953) 'Osteology' in J. P. Schaeffer (ed.), *Morris' Human Anatomy*, Philadelphia, Blakiston, pp. 89–286

Todd, T. W. (1920) 'Age Changes in the Pubic Bone: I. The Male White Pubis', *Am. J. Phys. Anthropol., 3*, pp. 285–334

—— (1921) 'Age Changes in the Pubic Bone: II. The Pubis of the Male Negro-White Hybrid; III. The Pubis of the White Female; IV. The Pubis of the Female Negro-White Hybrid', *Am. J. Phys. Anthropol., 4*, pp. 1–70

—— and Lyon, D. W. (Jr.)(1924) 'Cranial Suture Closure, Its Progress and Age Relationship. Parts I and II', *Am. J. Phys. Anthropol., 7*, pp. 325–84

Trotter, M., Broman, G. E. and Peterson, R. R. (1960) 'Densities of Bone of White and Negro Skeletons', *J. Bone and Joint Surgery, 42A*, pp. 50–8

Ubelaker, D. H. (1978) *Human Skeletal Remains: Excavation, Analysis, Interpretation*, Aldine, Chicago

Wente, E. F. (1980) 'Age at Death of Pharaohs of the New Kingdom, Determined from Historical Sources' in J. E. Harris and E. F. Wente (eds.), *An X-Ray Atlas of the Royal Mummies*, University of Chicago Press, Chicago, pp. 234–85

Zimmerman, M. R. (1977) 'The Mummies of the Tomb of Nebwenenef: Paleopathology and Archaeology', *J. Am. Res. Center in Egypt, 14*, pp. 33–6

—— and Aufderheide, A. C. (1984) 'The Frozen Family of Utkiavik: The Autopsy Findings', *Arctic Anthropol., 21*, pp. 53–64

——and Smith, G. S. (1975) 'A Probable Case of Accidental Inhumation of 1,600 Years Ago', *Bull. New York Acad. Med., 51*, pp. 828–37

——, Trinkaus, E., LeMay, M., *et al.* (1981) 'The Paleopathology of an Aleutian Mummy', *Arch. Pathol. and Lab. Med. 105*, pp. 638–41

——, Yeatman, G., Sprinz, H. and Titterington, W. P. (1971) 'Examination of an Aleutian Mummy', *Bull. New York Acad. Med., 47*, pp. 80–103

6 THE USE OF HAND-WRIST RADIOGRAPHS FOR AGE ASSESSMENT IN DECEASED INDIVIDUALS

Evelyn J. Bowers

For individuals who are not fully mature, an X-ray or radiograph showing the hand and wrist, including the distal ends of the radius and ulna, is a standard tool for estimating a biological age. Maturation involves changes in function — development, by definition — as well as growth or increase in size (Krogman, 1972, Roche, 1978). An X-ray that shows the presence and form of calcification of the carpals and the epiphyses of the long bones of the hand and wrist, contains a large number of sites that can serve as markers of maturation. These bones gradually become calcified in the course of maturation. Ultimately, the epiphyses of the long bones fuse with their shafts, and linear growth is no longer possible.

Although there is some variation in both the timing and the sequence of the developmental changes, methods have been devised that evaluate the portion of this process that has been completed. In any decedent in whom the wrist is not disarticulated — a mummy, for example — an X-ray can be obtained and a bone age derived by one of the techniques described below. Attempts to derive a bone age directly from skeletal material where the bones of the hand and wrist have become disarticulated is inadvisable except by the most experienced. Disarticulation alters the orientation of the carpals, which introduces a source of error in observing their shape, and makes highly likely the loss of the tiny bits of bone that are unfused epiphyses. However, for material in which an articulated wrist and hand remain, the estimation of biological age from a standard hand-wrist radiograph should be not only feasible but of appreciable utility.

The Nature of Biological Age

A biological age, derived from whatever method, is, of course, not exactly comparable to an age in calendar years or chronological age. This is due to variation in the rate of maturational change, a phenomenon pointed out by Boas (1892a, 1892b). An X-ray of the hand-wrist is a useful tool for examining this variation (Boas, 1912). Tanner (1962) points out that the variation seen in skeletal age differs with chronological age. Plus and minus two standard deviations encompasses over 97 per cent of the distribution.

This amounts to eight months in a two-year-old but two years in a 16-year-old in a modern population.

The difference between skeletal age and chronological age is not equally likely to be positive or negative. An individual's skeletal age is younger than the chronological age more frequently than it is older. This is due to the nature of the factors that impinge on maturation. Biological variation may result in either advancement or retardation of skeletal age with respect to chronological age. Comfort (1979) has pointed out that the primary timing mechanism(s) for development are not yet known. However, development is at least partly dependent on growth, which is conditioned by environmental factors such as nutrition and stress. Inadequate nutrition and excessive stress can retard growth and development from their innate timing, but no environmental factors are established as being able to accelerate this innate development schedule before full adulthood (unless one counts exogenous androgens). At least partial genetic control is indicated by the occurrence of familial clustering of early and late maturers (Garn and Bailey, 1978). The acceleration of the timing of maturation in Western populations over the last 100 years or so, known as secular trend, is thought to be due to the removal of inhibiting factors such as malnutrition and childhood illness, rather than an acceleration of the basic innate rate of maturation (Tanner, 1962).

The individual differences in the timing of maturation are sometimes large. The normal time of maturation ranges from eight to 16 years in girls, and from ten to 18 in boys (Tanner, 1962). Some of this variation seems to come from idiosyncratic familial patterns, while some appears to be more common in one group or another (Eveleth, 1978). Moreover, the environmentally caused variations in the timing of maturation are more frequent in some groups than in others. All these factors need to be considered when arriving at a probable age for a decedent. The sources of variation are reviewed below.

Techniques

Garn, Rohmann and Silverman (1967) have presented tables of the timing of the appearance of bones and epiphyses on X-ray. These have some use in evaluating skeletal age in certain ranges of chronological ages. Such maturational markers were used in studies of war dead by McKern and Stewart (1957). The markers are, however, widely dispersed through the body and require a large number of radiographs for their examination, but there are, as shown in Figure 6.1, a considerable number of potential maturation markers in the hand and wrist. The human hand contains 14 phalanges and five metacarpals: each of these has an epiphysis during some part of development. Each wrist contains eight carpals, and the radius and ulna both have distal epiphyses. In the course of development, these features undergo changes in shape as well as increases in size and changes in proportions (Michelson, 1946). Figures 6.2 to 6.6 show individuals who have completed differing portions of maturation. In the course of these changes, the epiphyses, for example, appear, enlarge, differen-

Figure 6.1: Bones of the Hand of an Adolescent and Their Features. The numerous pseudoepiphyses are unusual. The X-ray from which this was drawn is shown in Figure 6.5

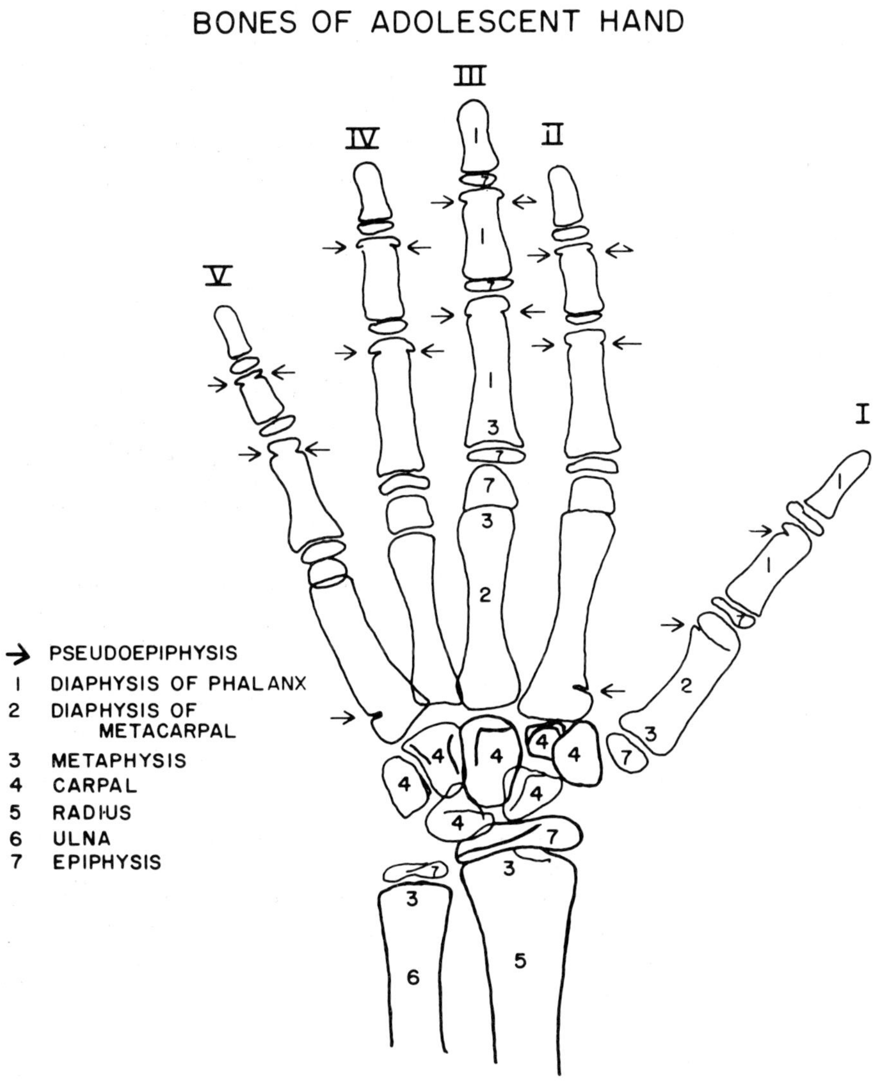

Figure 6.2: A Fully Mature Hand Belonging to a Small Caucasoid Female from North America

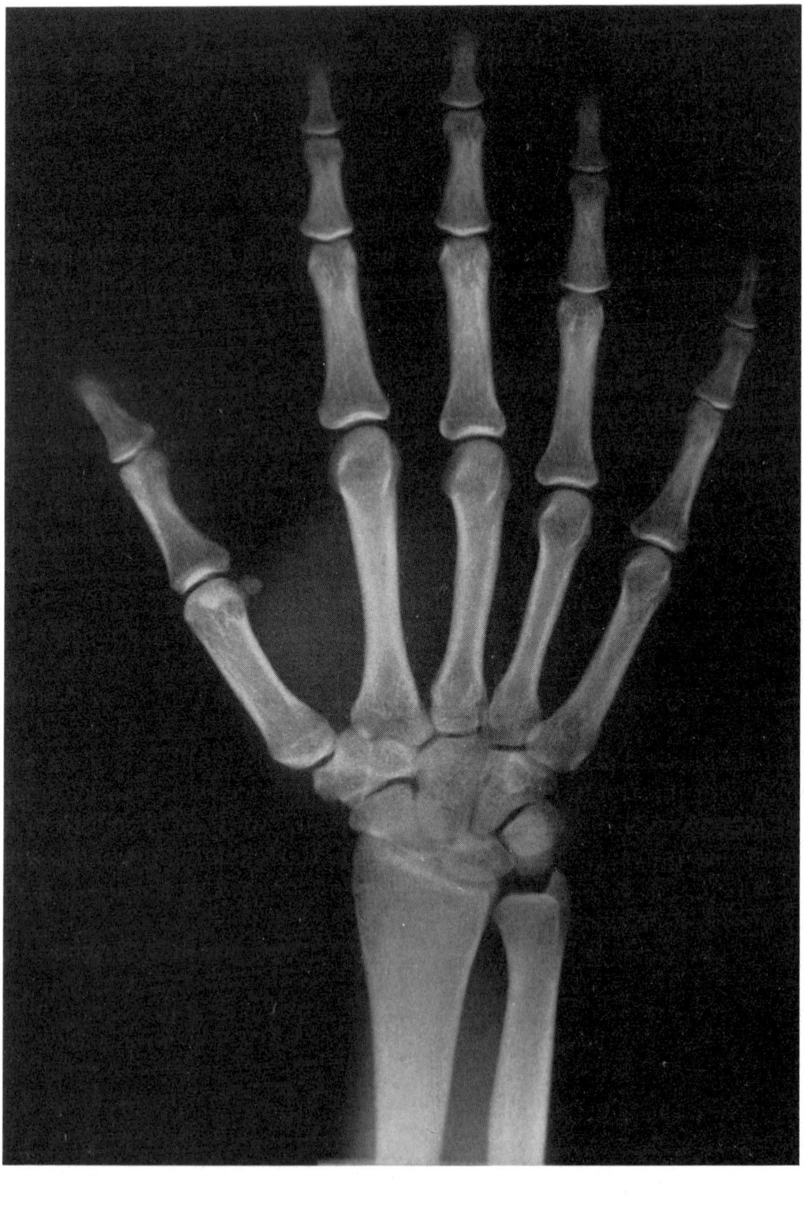

Figure 6.3: Hand of a Caucasoid Male Who is Small for Age. Chronological age four years; skeletal age three years six months by the Greulich-Pyle atlas method. Tanner-Whitehouse-2 Total bone age 3.8 years; RUS age 4.2 years; carpal age 3.1 years

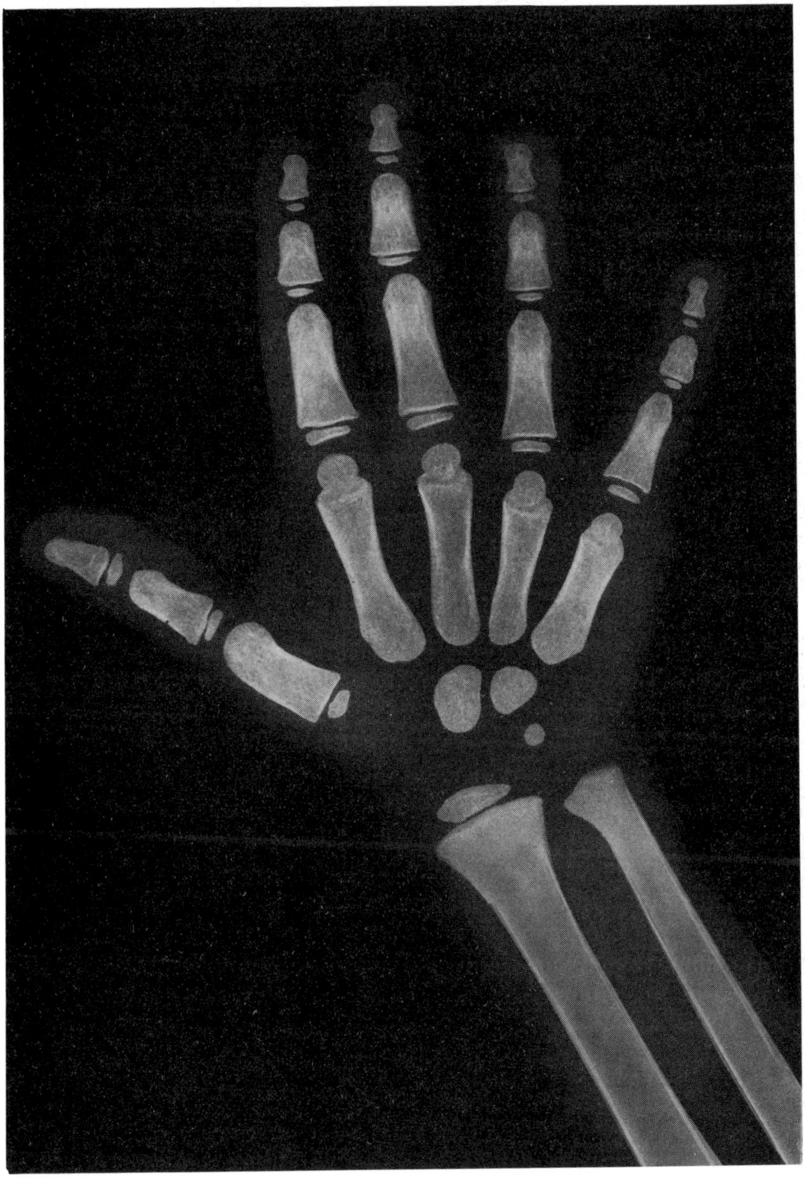

Figure 6.4: Hand of a Caucasoid Male, SA = 9:0 (GP)

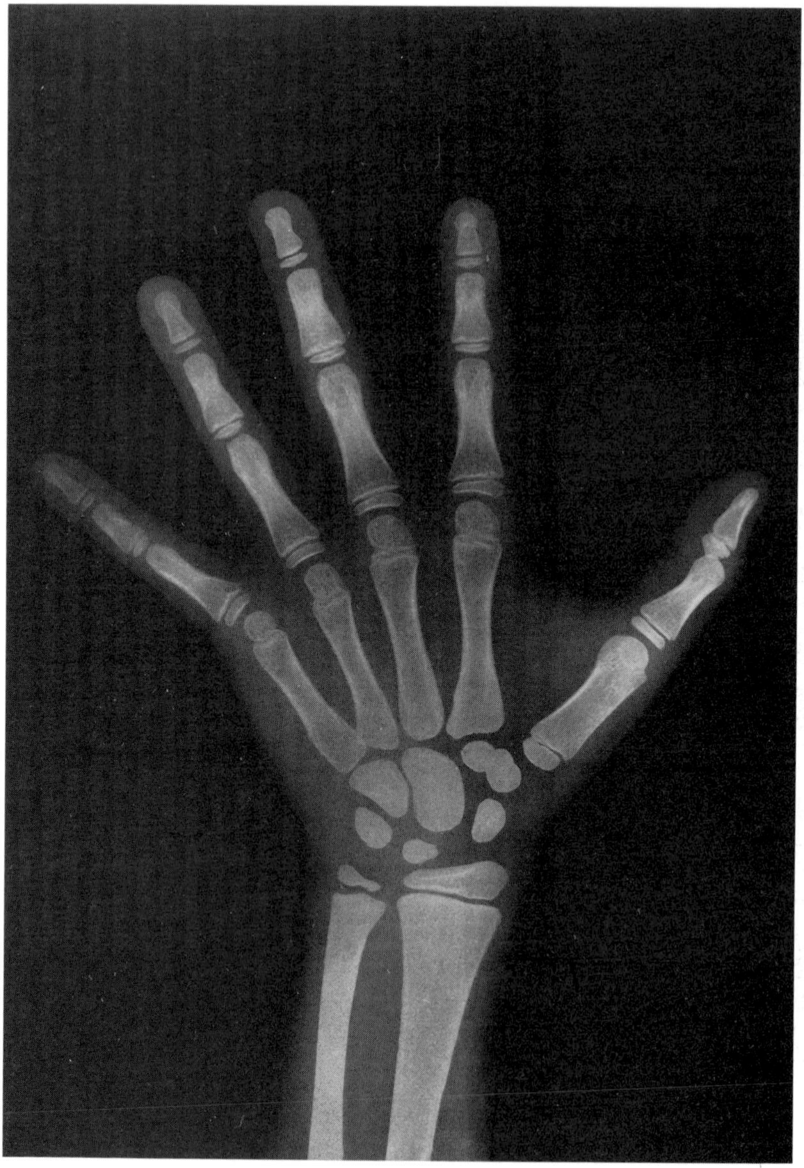

Figure 6.5: Hand of a North American Male of African Ancestry (Predominantly Congoid), SA = 11.4 (TW-2). Figure 6.1 was drawn from this X-ray

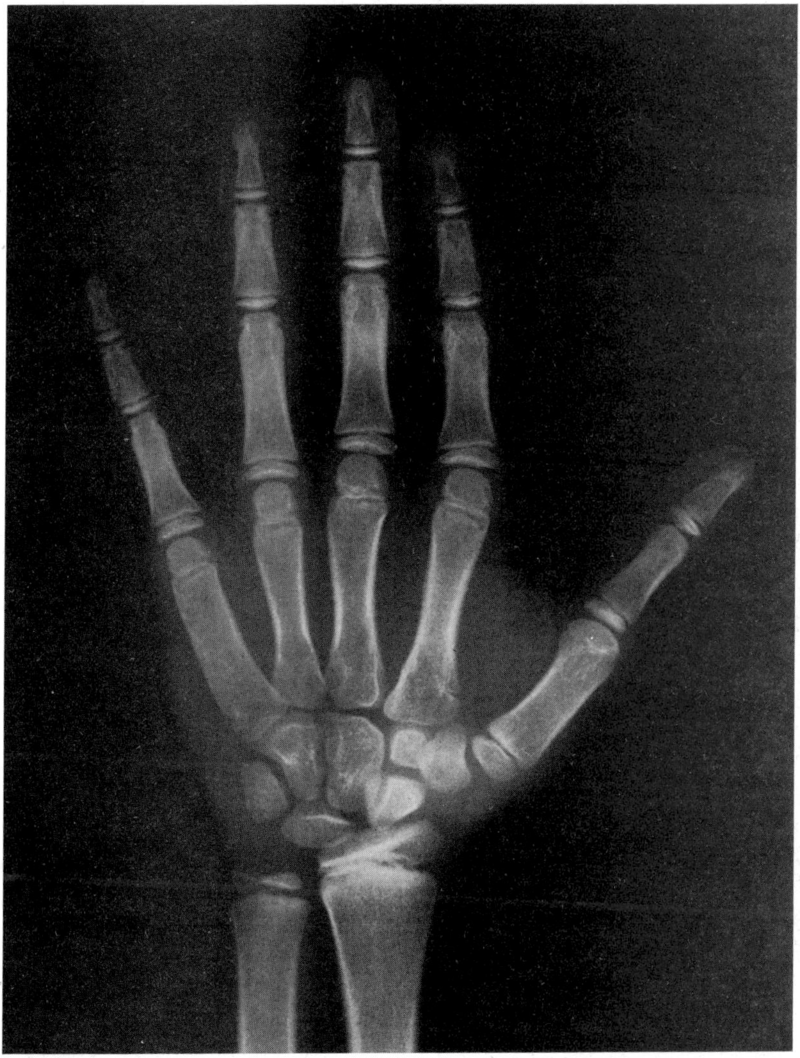

Figure 6.6: Hand of North American Female of Chinese Ancestry (Major Racial Group, Mongoloid), SA = 11:0 (GP). Note that this child has completed a greater share of her total maturation, as indicated, for example, by the capping of a number of epiphyses, than the boy in Figure 6.5

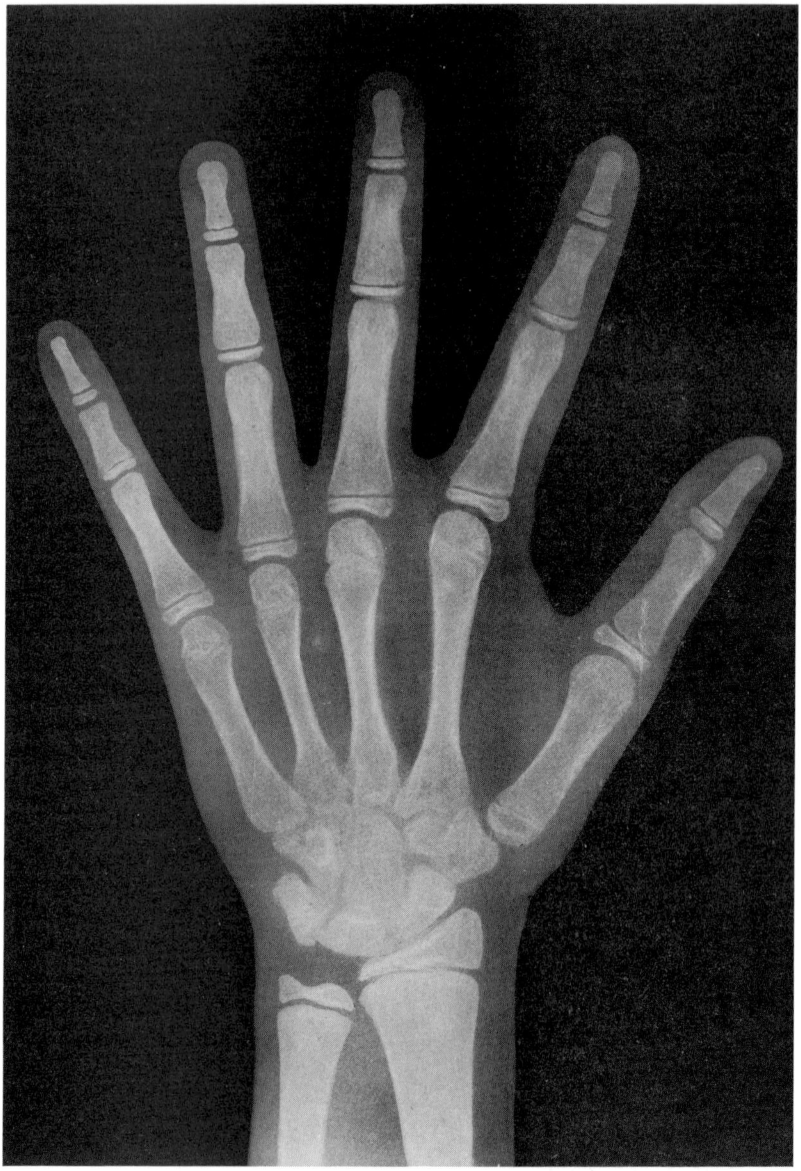

tiate, cap, and fuse with their metaphyses. The carpals appear and then undergo a series of changes in shape. Although change is a continuous process, the achievement of a certain shape can be used as a developmental marker.

There are two major techniques presently in use for the analysis of biological age from a hand-wrist radiograph: the Greulich-Pyle atlas method (Greulich and Pyle, 1959) and the Tanner-Whitehouse-2 method (Tanner *et al.*, 1975). Both of these have appreciable developmental histories (see below), and are quite reliable when used by well-trained individuals. Although it would be preferable to use standards derived from the population from which the individual being investigated derives, both the Greulich-Pyle and the Tanner-Whitehouse techniques have been used satisfactorily on many thousands of children from various parts of the world.

The Greulich-Pyle method uses an atlas of plates showing the bones of the hand and wrist at different stages of development. In the usual application of this technique, a radiograph is compared to the atlas plates, and assigned the age corresponding to the plate that is considered the best match (Greulich and Pyle, 1950, 1959). Alternatively, individual bones can be compared. Each bone is then assigned an age individually, and these are averaged to derive a biological age (Pyle, Waterhouse and Greulich, 1971; Roche, Roberts and Hamill, 1973a and 1973b).

The atlas approach has an appreciable degree of subjectivity to it since one can only pick the plate that one thinks matches best, whether for the whole hand or a single centre. If one averages the centres, then should they be weighted and, if so, how? Considerable inaccuracy may be introduced into biological age estimates if there are appreciable differences in the maturation of different parts of the hand, as is likely when illness and malnutrition are present. Pyle and her colleagues had developed what is termed the 'red graph' method of individually scoring bones for use in clinical settings (Pyle *et al.*, 1948) in order to accommodate dysharmonic maturation. Such variation is sometimes of diagnostic significance (Poznanski, 1974).

The Tanner-Whitehouse-2 system takes better account of differences in maturation level within a hand since it scores maturity indicators individually and then sums the scores (Marshall, 1966). It looks at a particular set of ossification centres in the wrist and the bones of the hand. From the degree of maturation these show, based on a sequence of precise marker criteria, each associated with a maturity score, a summed rating is derived. The rating can be considered a percentage of maturation accomplished. Only subsequently is an age associated with a score. Both the ratings for a particular indicator and the ages associated with a sum score are different for boys and girls.

The Tanner-Whitehouse-2 system was designed so that it produces three scores, because of the difference in the timing of the development of the carpus and the rest of the hand: a total bone score (TBS), which gives an age comparable to the Greulich-Pyle rating; a RUS score, based on the development of the radius, ulna and the short bones (metacarpals and phalanges) of the I, III and V digital rays; and a carpal bone score. The II and IV rays are omitted based on considerations of redundancy. The RUS score is particularly useful for adolescents where the carpus is fully mature (Tanner *et al.*, 1975). In a young child, it would be possible to assess a biological age from the carpals alone, if need be. This extends only to the age of 13 years in girls and 15 years in boys, but may prove useful in some circumstances.

The Tanner-Whitehouse-2 and the Greulich-Pyle atlas methods both give reasonable assessments of biological age if used with care. The Greulich-Pyle atlas may be. more readily applied, while the Tanner-Whitehouse-2 system, which requires more training to use, may be marginally more accurate and allows greater appreciation of dysharmonic maturation.

History

The earliest work on the developmental anatomy of the hand and wrist as indicated by radiographs had been thought to be that of Pryor (1905, 1907) and Rotch (1909). Pryor drew attention to sex differences in the timing of maturation, while Rotch was interested in child welfare and education. Recently, however, the orthopaedist John Poland's *Skiagraphic Atlas* (1898) has resurfaced (Himes, 1984). Apparently it was based on cross-sectional data, that is X-rays of different children at various ages, rather than X-rays of one child at different ages, as was most early work. Cross-sectional data can, of course, be accumulated rapidly, but they tend to obscure the abruptness with which developmental change sometimes takes place (Tanner, 1962). Later atlases were based on data with a longitudinal component.

Early work on the maturation of the hand and wrist developed in part under the impetus of the need for an objective method to determine the proper placement of children in educational settings (Crampton, 1908; Rotch, 1909). This influence waned with the development of IQ testing, although an IQ test does not distinguish between a child of average intelligence whose maturation is advanced and one of average maturation-for-age who is bright. The difference may or may not be relevant for school placement, but it matters greatly for expected outcome. Only recently has interest in the inter-relationships among physical, cognitive and emotional development begun to revive, aided by modern methods and approaches to multifactoral systems (Lampl, 1982).

Some early work used a method of maturation assessment based on the area ossified to derive a maturity rating. This approach, used in the Harvard Growth Studies of the 1920s (Shuttleworth, 1939), was not particularly satisfactory, since changes in shape were not clearly distinguished from changes in size (Marshall, 1966).

In order to devise a more useful system for ascertaining biological age, Todd and colleagues in Cleveland, starting in 1926, developed standards for the maturation of the hand and wrist. The same impulse to devise standards lead Rochlin to publish *Roentgenostiologie und Roentgenanthropologie* in 1936. The Cleveland standards were published as an atlas of maturation of the hand by Todd in 1937. This atlas utilised primarily X-rays from the Brush Foundation Study children,, and was semi-longitudinal in concept. It was refined and improved to form the Greulich-Pyle atlas (1950), which has, itself, been revised once (Greulich and Pyle, 1959). More recently, a special revision was prepared (Pyle, Waterhouse and Greulich, 1971) for use in the United States' Health Examination Survey (Roche, Roberts and Hamill, 1973a, 1973b). The problems with the atlas approach have been mentioned above.

The concept of maturity indicators as reflected in the hand and wrist appears to have originated with Todd's work (1937). Illustrations of these indicators were published in the back of the 1950 Greulich and Pyle atlas. Acheson seems to have introduced the concept of scores for the appearance of maturity indicators (Acheson, 1954, 1957; Acheson, Kemp and Parfit, 1955). Next, Tanner and Whitehouse (1959) developed a set of standards under the auspices of the International Children's Centre in Paris. They then published, with Healy (1962), the TW-1 system with its scores for individual maturation steps, which were summed to give percentage of maturation completed. Lastly, this was revised into the TW-2 system. This process involved some simplifications and clarifications, and the assignment of skeletal ages to the various total bone scores (Tanner *et al.*, 1975). This system has been used widely and has demonstrated conclusively its worth.

In working with non-western and non-modern material, it might be of some advantage to retain the terminal stages that were dropped for certain bones in the TW-2 method. Although, as both Pyle and Tanner point out, the change from the commencement of fusion of the radial epiphysis to its completion may be a matter of a number of years, this may not be a disadvantage in archaeological or palaeontological studies. In TW-2 full maturity is said to correspond to 18.1 years, while full maturity in the Greulich-Pyle atlas is listed as 19 years. Full fusion of the radius is the major difference.

Sources of Variation

There are, as already stated, a number of sources of variation in the skeletal ages of children of the same chronological age. One of these is sex. From infancy to adulthood, girls mature more rapidly than do boys. Thus a bone showing the appearance of some particular maturational marker will come from a younger individual if the source is a female than will be the case if it comes from a male. Pryor, the pioneer scholar of the radiographic appearance of the maturation of the hand and wrist, pointed out this fact in his early work (1907). On average, girls mature two years before boys. Moreover, girls are more frequently early maturers than boys. Krogman (1972) reported the proportions of early, average and late maturing children as 30:48:22 for females, and 9:60:31 for males, based on material from Cleveland in the 1930s. This is due to the role of the gonadal steroids in the process of maturation and the sex difference therein. Establishing the sex of an individual usually presents no major difficulties, but it should be remembered that eunuchoid individuals do not mature in the usual manner and their body proportions are altered. They may have open epiphyses in the hand and wrist in their 20s and 30s. Disproportion between the size of a hand and its degree of maturation might alert one to this possibility. Tables of sizes and proportions of the bones of the hand and wrist have been published and can be consulted where needed (Garn *et al.*, 1972). The ordinary sex differences are dealt with in the Greulich-Pyle atlas by providing separate plates for girls and boys. In the 1971 atlas, separate male and female ages for one set of plates are given. The Tanner-Whitehouse system gives separate scores for boys and girls for each maturity indicator, as well as separate tables for conversion of bone scores to biological ages.

A second source of variation is population group. The characteristics used to define the world's racial groups intergrade along clines and no characteristics are unique to any group. However, there are differences in the frequencies of a number of characteristics among racial groups. The range of variation within any of the major races exceeds the differences between them. However, human populations adapt to their environments biologically as well as culturally, so that modal characteristics occur. Awareness of these differences and the fact that the real individuals may be very different from group norms is important in accurate assessment of any biological characteristic, including skeletal age.

Both racial and social class variations in rate of maturation have been observed. The differences in maturation rates among Congoids, Caucasoids and Mongoloids appear to have both innate and environmental components (Eveleth, 1978; Eveleth and Tanner, 1976). Data on Capoids and Australoids appears to be scarce, although they would clearly be desirable since they are as distinct populations as the others. Garn and colleagues have shown that Americans of African descent (Congoids), living in modest circumstances, have skeletal ages that are advanced with respect to Caucasoid children from similar environments (Garn *et al.* 1972). Congoid children living in Africa have skeletal ages that are advanced at birth but that become retarded with respect to their chronological ages in the first two or three years of life. The contrast with data from North America indicates that the difference in rates is due to environmental factors. Whether this is the case for Capoid (Bushman-Hottentot) children is unclear. Atlases for Japanese children have been published (Sutow and Ohwada, 1953; Sugiura and Nakazawa, no date) and should be consulted for individuals whose ancestors came from Asia north of the Himalayas. Eveleth has pointed out (1978) that puberty is generally traversed more rapidly by Mongoloids than by Caucasoids.

Both the Greulich-Pyle atlas and the Tanner-Whitehouse-2 standards are based on Caucasoid children. Besides rate differences, there is an additional source of difficulty when they are used for other groups. On average, the bones of Congoids are more robust, and those of Mongoloids less robust, than those of Caucasoids. Care must be exercised to avoid confusing size differences with shape differences, and thus mis-assigning some bones or radiographs. The carpal angle is of some help in distinguishing Congoids from Caucasoids. The angle formed by tangents to the medial and lateral proximal margins of the carpus generally lies between 124.3 and 138.7 degrees. Harper, Poznanski and Garn (1974) indicate that it is about 10 degrees higher in American blacks (Congoids, predominantly) than whites (Caucasoids), and it increases between the ages of five and 14 years by between 5 and 10 degrees. It should be noted that they say this angle is also about 5 degrees higher in the right hand than the left, although that finding may be influenced by laterality.

While the racial difference is not definitive, it is helpful. In persons of mixed ancestry, it is not unreasonable to assume that if the carpus tends toward one racial form, the rest of the hand and wrist will do likewise. These differences do not, however, preclude using one of the standard methods. Malina (1968) and more recently Eveleth, Bowers and Schall (1979) have used the Tanner-Whitehouse-2 system with Philadelphia blacks, with satisfactory results. Population-specific standards would, nevertheless, be preferable. There are the two works on Japanese for orientals (Sutow and Ohwada, 1953; Sugiura and Nakazawa, no date.), but only the very important series of plates from Mackaye (1952) on Congoid children in Africa. Todd (1937) used some radiographs of American children of African ancestry, and argued that differentiation was not necessary for children in optimal circumstances.

Differences within the major races have also been observed (Eveleth and Tanner, 1976). The Tanner-Whitehouse-2 standards and the Greulich-Pyle atlas have been known to differ for some time, partly from social class differences (Roche, Davila and Eyman, 1971). It is not clear whether ethnicity within major race is a further source of variation beyond that which can be attributed to social class. There are separate standards for different ethnic groups only among Caucasoids. In addition to Rochlin's work (1936), and Schmid and Mull's atlas of German children (1960), there is also an atlas based on Dutch children (Speijer, 1950) as well as the Tanner-Whitehouse material on British children, and the Todd and Greulich-Pyle sequence for the United States. Certainly, differences do occur for the rates of ageing in the post-mature portion of the life span, resulting in differences in longevity in different areas with comparable levels of modern medical care (Comfort, 1979; Roberts, 1981). Variation in skeletal development is less clear.

Differences that probably stem from social class variation are seen when the two major skeletal ageing techniques are compared. The Greulich-Pyle atlas is based on radiographs from the Brush Foundation Study. The participants in this study were largely white children of established professionals from the middle-west of the United States. Active parental concern with their child's development was a criterion for inclusion (Todd, 1937). These children's health and nutritional status were among the best ever observed. As a result, the distinct absence of growth-retarding influences impinging upon these children allowed their genetic potential for growth and development to be fulfilled at something approaching the maximum rate and degree.

In contrast, the children from the Harpenden Growth Study, whose X-rays the Tanner-Whitehouse System-2 utilises, were from a less well circumstanced group of English children. Moreover, Tanner and colleagues were attempting to characterise skeletal development in its average form, while Todd had been interested in the optimal course. Reflecting these, the Tanner-Whitehouse standards are about six months behind Greulich-Pyle atlas plates, even though they were assembled a generation more recently.

In addition to differences between populations, there are also differences within populations that occur over time. Over approximately the last century, the average size of North American and European children has increased, and their average ages at maturity have decreased (Tanner, 1962). These changes in growth and development, termed secular trend, have been attributed to many factors, but their cause or causes are not established. Clearly, improved nutrition and reduced childhood illness are involved. Light also influences the timing of development in many species (Comfort, 1979). A role for changes in duration and luminescence of light exposure has also been postulated as influencing secular trend (Katz, 1972). In addition, changes in exercise, dress, and stressful rearing practices have also been thought likely to influence secular trend.

All of these factors need to be considered when making a biological age assessment, particularly for an individual from a population other than the ones on which the standards are based, since all these factors can exert growth retarding influences. It should be remembered that the populations that do and do not tend to be growth-retarded change with time. In the Age of Discovery, non-European children were said to develop early. Recently, it has come to be appreciated that the non-Europeans had ages at maturation that, for the ones in good circumstances, are unchanged. It was the Europeans of the Age of Discovery who were growth-retarded; the non-Europeans were not advanced.

Another factor to be considered in examining an X-ray of the hand and wrist for skeletal age in the contexts of palaeontology and archaeology is the postulate of normality; it is customary to assume that any specimen found is non-pathological. In part, this is enjoined by the scarcity of material. This postulate need not and should not be invoked for hand-wrist X-rays. Appreciable numbers of these have been examined in the nine decades since Roentgen invented radiography in 1895. A certain number of pathological conditions can be detected readily from a hand-wrist radiograph (Poznanski, 1974). Some of these, for example Turner's syndrome, are not incompatable with life but will influence the rate of maturation and so need to be considered in estimating a biological age. The hand-wrist X-ray is informative in endocrine disorders and chromosome anomalies. For example, the carpal angle is high in Turner's syndrome, and the fourth metacarpal is frequently short. In gonadal dysgenesis, disharmonic maturation of the various hand and wrist bones is to be expected since the carpals mature, but open epiphyses occur in the metacarpals and phalanges, perhaps reflecting atypical ratios or levels of androgens and oestrogens, past the age of 20 years (Kosowicz, 1965). The same might be expected in castrati. Poznanski (1974) has discussed extensively the variations seen in X-rays of the hand and wrist and their significance.

Cautions

Perhaps the major difficulty in assessing age from a hand-wrist X-ray of a mummy would be estimating what retarding effect poor environmental circumstances might have. Poor-quality bone would be one factor to suggest that growth retardation might have been present. Marked variation in the maturational level of different carpals might be another. Greulich and Pyle (1959) indicate that an illness at the time when a particular carpal was due to ossify may lead to delay in so doing, although the next in sequence may appear on time if the stress has been withdrawn. It may take several years for this delay to be made up. Another feature that indicates stress is the presence of what are termed Harris lines. These are lines of increased calcification that are sometimes seen on long bones and show up on the radius and ulna in hand-wrist films. They are probably due to the resumption of normal calcification rate before the resumption of growth, leading to dense calcification marking the time when growth resumed following its stress-induced cessation (Garn *et al.*, 1968). Presence of these stress-induced features suggest that an observed biological age will be more of an underestimate than would be the case in their absence.

Another feature that, while it is not pathological, should arouse suspicion of slowed growth and development when it is seen on a hand-wrist film, is the pseudoepiphysis. Several of these are shown in Figures 6.1 and 6.4. Pseudoepiphyses appear to be a normal feature of the way epiphyses ossify (Lee and Garn, 1967; Lee, Garn and Rohman, 1968; Bowers *et al.*, 1979). They are, however, transitory. If they last long enough that they are viewed on an X-ray, or if there are a number of them present, they probably do indicate a slow rate of growth and development. For this reason, a number of authors have considered them markers of pathology (Dreizen, Spirakis and Stone, 1965; Snodgrasse *et al.*, 1955; a and b). For the purposes of age assessment in palaeontological and archaeological material, their utility is to suggest that a biological age estimate is likely to be an underestimate.

Conclusion

This chapter has reviewed the use of radiographs of the hand and wrist as a means for the estimation of biological age. It has discussed the difference between biological age and chronological age, and has considered some of the factors that contribute to such differences. It indicates that a radiograph of the hand and wrist is a useful tool for the estimation of biological age in certain palaeontological and archaeological contexts.

Acknowledgement

To the W. M. Krogman Center for Research in Child Growth and Development of the University of Pennsylvania, Philadelphia, for the use of their facilities in preparing this chapter, and for supplying illustrative material. I also wish to thank Dr Solomon H. Katz, the Growth Center's Director, for his continuing support and encouragement.

References

Acheson, R. M. (1954) 'A Method of Assessing Skeletal Maturity from Radiographs: A Report from the Oxford Child Health Survey', *J. Anatomy, 88,* pp. 498–508

—— (1957) 'The Oxford Method of Assessing Skeletal Maturity', *Clinical Orthopaedics, 10,* pp. 19–39

—— and Kemp, F. H. and Parfit, J. (1955) 'Height, Weight and Skeletal Maturity in the First Five Years of Life', *Lancet, ii*, pp. 691–2

Boas, Franz (1892a) 'The Growth of Children', *Science, 19*, pp. 256–5, pp. 281–2

—— (1892b) 'The Growth of Children', *Science, 20*, pp. 351–2

—— (1912) 'The Growth of Children', *N.S., 5*, pp. 570–3

Bowers, E. J., Schall, J. I., Eveleth, P. B. *et al.*, (1979) 'Relationship Between Pseudo-Epiphyses and Growth in Philadelphia Adolescents, Abstract', *AJPA, 50*, p. 421.

Comfort, A. (1979) *The Biology of Senescence*, 3rd edn, Elsevier, New York

Crampton, C. W. (1908) 'Physiological Age, A Fundamental Principle', *Child Devel., 15*, pp. 1–52

Dreizen, S., Spirakis, C. N. and Stone, R. E. (1965) 'The Distribution and Disposition of Anomalous Notches in the Non-Epipyseal Ends of Human Metacarpal Shafts', *AJPA, 23,* pp. 181–8

Eveleth, P. B. (1978) 'Population Differences in Growth: Environmental and Genetic Factors' in Frank Falkner and J. M. Tanner (eds.), *Human Growth, Vol. 3: Neurobiology and Nutrition*, Plenum Press, New York, pp. 373–94

——, Bowers, E. J. and Schall, J. I. (1979) 'Secular Change in the Growth of Philadelphia Black Adolescents', *Human Biol., 51*, pp. 213–28

——and Tanner, J. M. (1976) *World Wide Variation in Human Growth*, Cambridge University Press, Cambridge

Garn, S. M. and Bailey, S. M. (1978) 'Genetics of Maturational Processes', in Frank Falkner and J. M. Tanner (eds.), *Human Growth, Vol. 1: Principles and Prenatal Growth*, Plenum Press, New York, pp. 307–30

——, Hertzog, K., Poznanski, A. K. and Nagy, J. M. (1972) 'Metacarpophalangeal Length in the Evaluation of Skeletal Malformations', *Radiology, 105*, pp. 375–81

——, Rohmann, C. G. and Silverman, F. M. (1967) 'Radiographic Standards for Postnatal Ossification and Tooth Calcification', *Med. Radiogr. Photogr. 43*, pp. 45–66

——, Sandusky, S. T., Nagy, J. M. and McCann, M. B. (1972) 'Advanced Skeletal Development in Low-Income Negro Children', *J. Pediatrics*, 80, pp. 965–9

——, Silverman, F. N., Hertzog, K. P. and Rohman, C. G. (1968) 'Lines and Bands of Increased Density, Their Implications to Growth and Development', *Med. Radiogr. Photogr., 44*, pp. 58–89

Greulich, W. W. and Pyle, S. I. (1950) *Radiographic Atlas of Skeletal Development of the Hand and Wrist*, Stanford University Press, Stanford, California

—— (1959) *Radiographic Atlas of Skeletal Development of the Hand and Wrist*, 2nd edn, Stanford University Press, Stanford, California

Harper, H. A. S., Poznanski, A. K. and Garn, S. M. (1974) 'The Carpal Angle Effects of Age, Race and Sex' in A. K. Poznanski, *The Hand in Radiological Diagnosis*, Saunders, Philadelphia

Himes, J. H. (1984) 'An Early Hand-Wrist Atlas and its Implications for Secular Change in Bone Age', *Ann. Human Biol., 11* (1), pp. 71–5

Katz, S. H. (1972) 'Biological Factors in Population Control' in B. Spooner (ed.), *Population Growth: Anthropological Implications*, MIT Press, Cambridge, Massachusetts, pp. 351–68

Kosowicz, J. (1965) 'The Roentgen Appearance of the Hand and Wrist in Gonadal Dysgenesis', *Am. J. Roentgenol., 93*, pp. 354–61

Krogman, W. M. (1972) *Child Growth*, Ann Arbor, Michigan, University of Michigan Press

Lampl, M. (1982) 'Episodic Growth Patterns in the First Year of Life', thesis, University of Pennsylvania, Department of Anthropology

Lee, M. M. C. and Garn, S. M. (1967) 'Pseudo-Epiphyses or Notches in the Non-Epiphyseal Ends of Metacarpal Bones in Healthy Children', *Anat. Rec., 159*, pp. 263–72

—— and Rohman, C. G. (1968) 'Relation of Metacarpal Notching to Stature and Maturational Status of Normal Children', *Invest. Radiol.*, *3*, pp. 96–102
Mackaye, D. H. (1952) 'Skeletal Maturation in the Hand: A Study of Development in East African Children', *Trans. R. Soc. Trop. Med. Hyg.*, *46*, pp. 135–50
Malina, R. (1968) 'Growth, Maturation and Performance of Philadelphia Negro and White Elementary School Children', thesis, Department of Anthropology, University of Pennsylvania
Marshall, W. A. (1966) 'Assessment of Skeletal Maturity' in J. J. Van der Werfften Bosch and A. Haak (eds.), *Somatic Growth of the Child*, Thomas, Springfield, Illinois, pp. 31–7
McKern, T. W. and Stewart, T. D. (1957) 'Skeletal Age Changes in Young American Males', *[U.S.] Quartermaster Res. and Devel. Ctr. Natick, MA: Tech. Rept. No. EP45*
Michelson, N. (1946) 'A Method for Assessing the Development of the Hand Skeleton', *AJPA N.S.*, 4, pp. 235–42
Poland, J. (1898) *Skiagraphic Atlas of the Bones of the Wrist and Hand*, London
Poznanski, A. K. (1974) *The Hand in Radiological Diagnosis*, Saunders, Philadelphia
Pryor, J. W. (1905) 'Development of the Bones of the Hand and Wrist as Shown by the X-ray Method', *Bull. of the State College of Kentucky*, ser. 2, no. 5, Press of Transylvania, Lexington, Kentucky
—— (1907) 'The Hereditary Nature of Variation in the Ossification of Bones', *Anat. Record, 1*, pp. 84–8
Pyle, S. I., Mann, A. W., Dreizen, S. *et al.*, (1948) 'A Substitute for Skeletal Age (Todd) for Clinical Use', *J. Pediat.*, *32*, pp. 125–36
——, Waterhouse, A. M. and Greulich, W. W. (1971) *A Radiographic Standard of Reference for the Growing Hand and Wrist*, The Press of Case-Western Reserve University, Cleveland, Ohio
Roberts, D. F. (1981) 'Culture and Disease in Britain and Western Europe' in H. R. Rothschild (ed.), *Biocultural Aspects of Disease*, Academic Press, New York, pp. 417–58
Roche, A. F. (1978), 'Bone Growth and Maturation', in Frank Falkner and J. M. Tanner (eds.) *Human Growth Vol 2: Postnatal Growth*, Plenum Press, New York, pp. 317–55
——, Davila, G. H. and Eyman, S. L. (1971) 'A Comparison Between Greulich-Pyle and Tanner-Whitehouse Assessments of Skeletal Maturity', *Radiology, 98,* pp. 273–80
——, Roberts, J. and Hamill, P. V. V. (1973a) 'Skeletal Maturity of Children 6–11 Years', *Vital and Health Statistics Ser. 11 No. 140, DHEW Pub. No. (HRA) 75–1622*, Washington DC [U.S.], National Center for Health Statistics
—— (1973b) 'Skeletal Maturity of Youths 12–17 Years', *Vital and Health Statistics Ser. 11 No. 160, DHEW Pub. No. (HRA) 77–1642*, Washington DC [U.S.], National Center for Health Statistics
Rochlin, D. G. (1936) *Roentgenosteologie und Roentgenanthropologie*, Biomedgiz, Moscow
Rotch, T. M. (1909) 'A Study of the Development of the Bones in Childhood by the Roentgen Method With the View of Establishing a Development Index for the Grading of and the Protection of Early Life', *Trans. Assoc. Am. Physicians, 24*, pp. 603–24
Schmid, F. and Mull, H. (1960) *Atlas der Normalen und Pathologischen Handskeletentwicklung*, Springer Verlag, Berlin
Shuttleworth, F. K. (1939) 'The Physical and Mental Growth of Girls and Boys Age 6 to 19 in Relation to Age at Maximum Growth', *Monogr. Soc. Res. Child Devel.*, vol. 4 no. 3
Snodgrasse, R. M., Dreizen, S., Currie, C. *et al.*, (1955a) 'The Association Between Anomalous Ossification Centers in the Hand Skeleton, Nutritional Status and Rate of Skeletal Maturation in Children 5 to 14 Years of Age', *Am. J. Roentgenol.*, *74*, pp. 1037–77
——, Dreizen, S., Parker, G. S. and Spies, T. D. (1955b) 'Serial Sequential Development of Anomalous Metacarpal and Phalangeal Ossification Centers in the Human Hand', *Growth, 19*, pp. 307–22
Speijer, B. (1950) *Betekenis en Bepaling van de Skeletleeftijd* (with a summary and atlas in English), Sijthoff, Leiden
Sutow, W. W. and Ohwada, K. (1953) 'Skeletal Maturation in Healthy Japanese Children 6 to 19 years of Age. Skeletal Age Standards', *Shonika Rinsko*, 6, p. 738
Sugiura, Y. and Nakazawa, O. (n.d.) *Roentgen Diagnosis of Skeletal Development*, Chugai-Igaku Co., Tokyo
Tanner, J. M. (1962) *Growth at Adolescence*, 2nd edn, Blackwell Scientific Publications, Oxford

—— (1978) 'A Concise History of Growth Studies from Buffon to Boas' in Frank Falkner and J. M. Tanner (eds.), *Human Growth, Vol. 3: Neurobiology and Nutrition*, Plenum Press, New York, pp. 515–93

—— and Whitehouse, R. H. (1959) *Standards for Skeletal Age*, Centre Internationale de l'Enfants, Paris

—— and Healy, M. J. R. (1962) *A New System for Estimating Skeletal Maturity From the Hand and Wrist, With Standards Derived From a Study of 2,600 Healthy British Children. Part II. The Scoring System*, Centre Internationale de l'Enfants, Paris

——, Whitehouse, R. H., Marshall, W. A., Healy, M. J. R. and Goldstein, H. (1975) *Assessment of Skeletal Maturity and Prediction of Adult Height (TW-2 Method)*, Academic Press, London

Todd, T. W. (1937) *Introduction to the Study of Skeletal Maturation, Part I: Hand*, Kimpton, London

7 ESTIMATION OF AGE AT DEATH FROM HISTOLOGY OF HUMAN BONE

Douglas Ubelaker

Research in the last few decades has documented the complexity of human bone growth and remodelling, and identified some of the physiological, mechanical, hormonal and even electrical factors involved (Ortner and Putschar, 1981; Enlow, 1963). This increased understanding of the normal processes of endochondral ossification, intramembranous ossification and cortical remodelling has had the unexpected payoff of a greater understanding of the ageing process and has led to several methods for calculating age at death from histological structures.

As early as 1936, Amprino and Bairati (1936) argued that histological structures in the midshafts of the femur showed regular age changes. Jowsey (1960) later further developed the approach in her examination of femoral midshaft cross sections from 24 individuals. Using microradiographs, she concluded that subadults show greater bone formation and resorption, suggestive of a higher rate of bone turnover. She concluded that the rate of turnover was low in young adults and then increased again with advancing age. Her data indicated an increase in overall resorption in older adults as well as in the number of osteons that are incompletely filled. Jowsey also described general increased porosity and variation in mineral density with advancing age.

Frost (1963) used tetracycline fixation to examine the formation time of human osteons. Although his sample consisted of different bones from only 19 individuals, he did quantify some of the variability in the time involved in osteon formation. He also studied sections of a variety of bones from 87 other patients to quantify possible age changes in measurements of haversian systems. Frost found that with age the rate of individual osteon formation decreases, that is more time is required for a resorption space to be transformed into an osteon. He found only a very slight, if any, decrease with age in osteon outside diameter and no change in mean haversian canal diameter.

By 1965, numerous histological structures had been correlated with age in human bone, but no study had quantified the data into a method that enabled the estimation of age at death from observations on the structures. To produce such a method, Kerley (1965) examined histological structures at the femoral midshaft in 126 ground-cross sections of the femur, tibia and fibula, all from individuals of known sex and clinical history, and with ages at death ranging from birth to nearly 100 years. The resulting method for estimating age at death examines the numbers of secondary osteons, non-haversian canals (primary osteons), fragments of secondary osteons and the percentage of remaining circumferential lamellar bone in four visual fields distributed around the circumference of the bone. The circular fields were all immediately adjacent to the periosteal edge of the bone and placed equidistantly on the anterior, medial, lateral and posterior (on the linea aspera) aspects. Numbers of osteons, fragments and non-haversian canals are then totalled from the four fields, while the percentage figures for circumferential lamellar bone are averaged. Ages at death can then be estimated by using individual regression equations for each structure or by using a series of graphs and a summary profile chart (Kerley, 1965, p. 161; 1970, p. 67).

Later, Kerley and Ubelaker (1978) noted that the original Kerley method called for using a 'ten-power objective lens combined with ten-power, wide-field oculars', which produce a circular visual field '1.25mm in diameter' (Kerley, 1965, p. 154). A survey of microscopes revealed that combination of lens and oculars could produce a field size ranging from 1.2 to 2.0mm, depending upon the structure of the microscope. To resolve the problem, new counts were made from the original slides using a calibrated field size. These new counts were then compared to the original counts made in the 1965 study to reveal that actual field size was about 1.62mm, not the published 1.25mm. Kerley and Ubelaker (1978, p. 546) conclude the following:

> In using the Kerley method, investigators should be aware of the variance in field diameters with different microscopes and should adjust the data to a field diameter of 1.62mm. The area (πr^2) of the 100 × field of the microscope used must be calculated using a stage micrometer and that field size divided into 2.06mm^2 (area of a field diameter of 1.62mm) to determine the relationship between the original field size and the field size of the individual microscope being used. All counts of osteons, fragments or non-haversian canals should then be multiplied by that factor. The estimate of percentage of lamellar bone is not affected.

In addition, the original data was reanalysed and new regression equations produced (Table 7.1). Standard errors ranged from 3.66 years for fibular osteons to 14.62 years for fibular non-haversian canals. In making the counts, it should be noted that Kerley counted 50 per cent of those osteons on the border of the field (not entirely visible) as whole osteons.

Since publication of the original method in 1965, several modifications or different approaches have appeared. In 1969, Ahlquist and Damsten offered a modification of the Kerley method that examines the percentage

Table 7.1: Regression Formulae for Estimating Age at Death (Y) from Cortical Microstructure (from Kerley and Ubelaker, 1978; p. 546)

Factor	Regression	Standard error of estimate[a]
Femoral osteons	$Y = 2.278 + 0.187X + 0.00226X^2$	9.19
Femoral fragments	$Y = 5.241 + 0.509X + 0.017X^2 - 0.00015X^3$	6.98
Femoral lamellar	$Y = 75.017 - 1.790X + 0.0114X^2$	12.52
Femoral non-haversian	$Y = 58.390 - 3.184X + 0.0628X^2 - 0.00036X^3$	12.12
Tibial osteons	$Y = -13.4218 + 0.660X$	10.53
Tibial fragments	$Y = -26.997 + 2.501X - 0.014X^2$	8.42
Tibial lamellar	$Y = 80.934 - 2.281X + 0.019X^2$	14.28
Tibial non-haversian	$Y = 67.872 - 9.070X + 0.440X^2 - 0.0062X^3$	10.19
Fibular osteons	$Y = 23.59 + 0.74511X$	8.33
Fibular fragments	$Y = -9.89 + 1.064X$	3.66
Fibular lamellar	$Y = 124.09 - 10.92X + 0.3723X^2 - 0.00412X^3$	10.74
Fibular non-haversian	$Y = 62.33 - 9.776X + 0.5502X - 0.00704X^3$	14.62

Note:
a. Square root of the mean square residual.

of remodelled bone in four fields distributed around the cortex. Their method calls for the insertion of a square-ruled reticule with 100 squares at such a point in the eyepiece so that it measures 1mm at the level of the section. This method then calls for counting the number of squares that are more than half-filled with either secondary osteons or osteon fragments, and then dividing the total in four fields by four to produce the percentage figure. As in Kerley's method, four fields adjacent to the periosteal edge of the bone and distributed around the cortex at bone midshaft were employed. The actual locations of the four sites were shifted to fall between those of Kerley in order to avoid the posterior site falling on the linea aspera. Ahlquist and Damsten felt a site on the linea aspera might contain too much non-age related variability. They tested their method on 20 ground sections taken from the femoral midshaft of individuals of known age at death. Ages at death ranged from four to about 89 years. The correlation between age at death and the percentage of remodelled bone is shown in the regression $y = 0.991 \times -4.96 \pm 6.71$ (Ahlquist and Damsten, 1969, p. 209).

Singh and Gunberg (1970) offered a third approach to the estimation of age at death from histological structures. In their study, they examined cross-sections taken from the midshafts of 33 femora and tibiae of known age at death from male individuals and from 'the posterior border of the mandibular ramus opposite the lingula' (Singh and Gunberg, 1970, p. 374) of 52 males. They utilised both ground thin sections and decalcified sections in their study. They examined two circular fields on each section, each measuring 2.0mm in diameter and located on the periosteal third of the section. Location of the fields around the circumference of the bone was selected at random. They recorded: (1) the total number of secondary osteons in the two fields; (2) the average number of lamellae per osteon, determined by counting all osteons in the two fields; and (3) the average minimum diameter of each haversian canal in the two fields. Table 7.2 presents their resulting regression equations. In the equations, X_1 refers to the number of osteons, X_2 to the average number of lamellae, and X_3 to the average haversian canal diameter.

Table 7.2: Regression Equations for Estimating Age at Death from Histological Measurements (from Singh and Gunberg, 1970; p. 377)

Selection	Dependent variable: age in years Regression equations	Multiple R^a	Standard error of estimates
	Mandible		
I	$20{\cdot}82 + 0{\cdot}85\,X_1 + 0{\cdot}87\,X_2 - 0{\cdot}22\,X_3$	0·979	2·55
II	$-18{\cdot}99 + 1{\cdot}13\,X_1 + 1{\cdot}76\,X_2$	0·976	2·69
III	$32{\cdot}23 + 0{\cdot}92\,X_1 - 0{\cdot}30\,X_3$	0·978	2·58
IV	$74{\cdot}73 + 1{\cdot}52\,X_2 - 0{\cdot}45\,X_3$	0·969	3·04
V	$-28{\cdot}24 + 1{\cdot}68\,X_1$	0·969	3·02
VI	$5{\cdot}31 + 5{\cdot}00\,X_2$	0·950	3·83
VII	$103{\cdot}99 - 0{\cdot}63\,X_3$	0·966	3·16
	Femur		
I	$27{\cdot}65 + 0{\cdot}65\,X_1 + 0{\cdot}78\,X_2 - 0{\cdot}26\,X_3$	0·958	3·24
II	$-14{\cdot}69 + 1{\cdot}13\,X_1 + 1{\cdot}11\,X_2$	0·948	3·55
III	$29{\cdot}59 + 0{\cdot}79\,X_1 - 0{\cdot}28\,X_3$	0·957	3·25
IV	$61{\cdot}25 + 1{\cdot}74\,X_2 - 0{\cdot}44\,X_3$	0·949	3·52
V	$16{\cdot}10 + 1{\cdot}38\,X_1$	0·945	3·60
VI	$2{\cdot}00 + 5{\cdot}16\,X_2$	0·889	5·01
VII	$89{\cdot}01 - 0{\cdot}62\,X_3$	0·937	3·82
	Tibia		
I	$43{\cdot}52 + 0{\cdot}291\,X_1 + 1{\cdot}47\,X_2 - 0{\cdot}34\,Xd3$	0·964	3·02
II	$-\ 3{\cdot}40 + 0{\cdot}67\,X_1 + 2{\cdot}27\,X_2$	0·936	3·93
III	$48{\cdot}61 + 0{\cdot}53\,X_1 - 0{\cdot}38\,X_3$	0·957	3·22
IV	$54{\cdot}79 + 2{\cdot}19\,X_2 - 0{\cdot}4\,x_3$	0·960	3·12
V	$-\ 4{\cdot}76 + 1{\cdot}15\,X_1$	0·919	4·33
VI	$5{\cdot}10 + 4{\cdot}88\,X_2$	0·908	4·59
VII	$91{\cdot}32 - 0{\cdot}64\,X_3$	0·935	3·88

Note:
a. Significant as assessed by F-test for regression at 5 per cent level of significance.

A final method of estimating age from histological structures has been proposed by Thompson (1979). Thompson examined a core of bone 0.4cm in diameter removed from the following sites in 166 cadavers ranging in known age at death from 30 to 97 years: anterior midshaft of femur, medial midshaft of tibia; midshaft humerus, medial to deltoid tuberosity; and lateral ulna, one third of the distance from the distal end. Thompson first measured cortical thickness, weight and mineral content and then prepared a ground thin section from each core. He used a grid eyepiece micrometer to quantify aspects of secondary osteons and haversian canals. He developed regression equations for the following 19 variables: (1) cortical thickness measured from intact core with calipers; (2) core weight, grams of wet weight of unfinished core; (3) cortical bone density, grams of core per unit cubic centimeter of core; (4) mineral content, grams per centimeter as measured by bone mineral analyser; (5) mineral index, variable 4 divided by refinished core length; (6) percentage of area of fields containing osteon lamellae; (7) percentage of area of fields containing canals; (8) percentage of area of fields containing osteon lamellae plus haversian canal area; (9) number of secondary osteons in a field; (10) number of canals and primary osteons in a field; (11) osteon lamellae area divided by secondary osteon number; (12) haversian canal area divided by haversian canal number; (13) total osteon perimeter length in a field; (14) total haversian canal perimeter length in a field; (15) aggregate osteon perimeter divided by osteon number; (16) aggregate canal perimeter divided by canal number; (17) aggregate haversian canal area divided by aggregate secondary osteon lamellae area; (18) aggregate haversian canal perimeter divided by aggregate osteon perimeter; (19) individual canal perimeter divided by individual osteon perimeter.

A large number of regression equations are available to estimate age from the above variables for 28 categories of unknown specimens (different combinations of bones, sex and extent of disease present). The lowest standard error (6.2 years) is for an equation for calculating age from the total osteon perimeter length in a field from the left humerus. Thompson (1979) should be consulted for additional details of the method and the complete formulae.

The application of the histological methods described above is considerable in both forensic cases (Thompson, 1979) and demographic reconstruction from archaeologically recovered skeletons (Ubelaker, 1974, 1978). In general, they offer more precise and objectively determined age estimates and are effective for older adults, in contrast to most methods using gross skeletal morphology. Nevertheless, there are general limitations to them, and each has its particular advantages and disadvantages. All of the methods assume adequate bone preservation and sufficient visual clarity to identify the structures. As Stout (1978) has pointed out, under ideal conditions preservation of such structures can be maintained for thousands of years. Frequently, however, even in relatively recent bone, much of the structure is lost or is obscured by fungus, soil or other contaminants. All of the methods use sites adjacent to the periosteal surface of the bone. Since this surface is most in contact with the environment, it is usually the first to be weathered away or invaded by structure-obscuring contaminants. If this surface has been eroded or invaded by contaminants, then the accuracy of all of the available methods may be markedly reduced. Under adverse preservation conditions the Ahlquist-Damsten method may be advantageous since it does not require osteons to be distinguished from osteon fragments. The Thompson approach also may be useful since particular regression equations can be selected depending upon the structures visible.

The Singh and Gunberg regression equations offer the lowest standard error, but the age range of their sample was very restricted. Their use of both decalcified and undecalcified sections further complicate a standard use of the technique.

Bouvier and Ubelaker (1977) compared the Kerley and Ahlquist-Damsten techniques. Kerley's original sample of documented sections was obtained, the age at death calculated using both techniques and the results compared to the known ages. It was found that ages estimated from the Kerley method were closer to actual age, but the methods were comparable in precision and objectivity. The difference in accuracy was attributed to the more restricted age range of the Ahlquist-Damsten sample, as well as methodological differences.

Stout and Gehlert (1980) carried this study one step further in applying the two methods to sections taken from 13 documented dissecting room skeletons. They argued that the Bouvier and Ubelaker study may have been biased in favour of Kerley's technique, since Kerley's original sections were used in the comparison. Their own comparison of the techniques on the small but independent sample produced similar results. They did find, however, that the Kerley-Ubelaker regressions for femoral osteons were more accurate than those derived from the fibula, in contrast to that suggested by the standard errors of the regressions. They also found that the most accurate estimates were obtained by averaging the ages estimated from each of the regressions.

The published standard errors of the Thompson regression equations are slightly higher than those of Kerley and Ubelaker. However, they are based on a large, if not somewhat old, sample and have the great advantage of not requiring a complete cross-section of bone. The coring technique allows the bone to be left intact and thus minimises damage to museum collections. The success of this technique in estimating age at death in the eight forensic cases described by Thompson (1979, p. 913) is most impressive.

The major problem in utilising existing histological techniques is one of sampling. Each regression equation or other method attempts to place the unknown specimen within the database from which the method is derived. One assumes that age changes in the base sample are identical to those in the population represented by the unknown specimen. Of course, one not only does not know how the unknown specimens relate to the documented variation in the base sample, but also the extent of differences in the basic rate of ageing between populations. The Kerley method was derived from a sample of mostly white, healthy individuals collected at the Armed Forces Institute of Pathology in Washington DC. The Thompson sample consists mostly of elderly whites from New England. How applicable are these data to the theoretical forensic case of a young black from urban Los Angeles with a history of drug abuse and malnutrition, or the skeletons recovered from a prehistoric American Indian cemetery in South America? The applicability is simply not known since so few studies have been done, and most of those have focused on modern, predominantly white, elderly samples.

In 1974, collaborators at the Museo del Hombre Dominicano in Santo Domingo, Dominican Republic, worked with the author to assemble from local city cemeteries a collection of 158 skeletons of known age at death and sex. The skeletons were removed in response to space problems in the cemeteries, and represent poor urban blacks either from the city of Santo Domingo or migrants from the surrounding countryside or from neighbouring Haiti. Data concerning each individual were established from cemetery records and verified, when possible, with city records. All were classified as black or mulatto.

Undecalcified ground thin sections were prepared from those individuals whose bones were well-preserved and showed no evidence of disease. Ages at death were estimated using the Kerley profile chart on sections from 114 femora, 108 tibiae and 102 fibulae of 55 males and 59 females, whose confirmed age at death ranged from 15 years to more than 90 years.

Error of the estimated ages ranged from 0 to 47 years, with an average error for the entire sample of over 10 years; 42 per cent of the estimated ages were within five years of actual age, 62 per cent within ten years, and 95 per cent within 28 years. Average error was much lower for females (eight years) than for males (13 years). Average error was also less for individuals younger than 50 years (six years) than for those older than 50 (13 years).

Obviously, the rate of age change in the Santo Domingo sample is somewhat different than the sample used in the original Kerley study. These differences can lead to substantial errors in individual estimates. Fortunately, the impact of these individual errors is minimal in a demographic study of a large sample. In the Santo Domingo study, the actual average age at death was about 59 years, while the average of the estimates was about 55 years. Actual and estimated average values for females were almost identical.

More research is needed to develop improved, objective histological techniques based upon large and diverse samples. The effects of such variables as disease, activity, trauma and diet on the histological ageing process need to be clarified through careful research. It is difficult to imagine two populations differing more than those represented by the Santo Domingo and Kerley samples, yet additional histological studies of other groups should clarify the actual variability involved. This information, coupled perhaps with new objective counting methods, should push histological ageing techniques to new levels of accuracy.

Acknowledgements

The author thanks Jeanne Mahoney of the Department of Anthropology, Smithsonian Institution, Washington DC, for her assistance in typing this manuscript.

References

Ahlquist, J. and Damsten, O. (1969) 'A Modification of Kerley's Method for the Microscopic Determination of Age in Human Bone', *Journal of Forensic Sciences, 14*, 2, pp. 205–12

Amprino, R. and Bairati, A. (1936) 'Processi de Ricostruzione e di Reassorbimento nella Sostanza Compatta delle Ossa dell'Uomo', *Zeitschrift fur Zellforschung und Mikroskopische Anatomie, 24*, pp. 439–511

Bouvier, M. and Ubelaker, D. H. (1977) 'A Comparison of Two Methods for the Microscopic Determination of Age at Death', *American Journal of Physical Anthropology, 46*, 3, pp. 391–4

Enlow, D. H. (1963) *Principles of Bone Remodeling*, Charles C. Thomas, Springfield

Fros, H. M. (1963) 'Mean Formation Time of Human Osteons', *Canadian Journal of Biochemistry and Physiology, 41*, 5, pp.1307–10

Jowsey, J. (1960) 'Age Changes in Human Bone', *Clinical Orthopaedics, 17*, pp. 210–17

Kerley, E. R. (1965) 'The Microscopic Determination of Age in Human Bone', *American Journal of Physical Anthropology, 23*, pp. 149–64

—— (1970) 'Estimation of Skeletal Age: After about Age 30' in T. D. Stewart (ed.), *Personal Identification in Mass Disasters*, Smithsonian Institution, Washington, pp. 57–70

Kerley, E. R. and Ubelaker, D. H. (1978) 'Revisions in the Microscopic Method of Estimating Age at Death in Human Cortical Bone', *American Journal of Physical Anthropology, 49*, 4, pp. 545–6

Ortner, D. J. and Putschar, W. G. J. (1981) 'Identification of Pathological Conditions in Human Skeletal Remains', *Smithsonian Contributions to Anthropology, 28*, Washington

Singh, I. J. and Gunberg, D. L. (1970) 'Estimation of Age at Death in Human Males from Quantitative Histology of Bone Fragments', *American Journal of Physical Anthropology, 33*, 3, pp. 373–82

Stout, S. D. (1978) 'Histological Structure and Its Preservation in Ancient Bone', *Current Anthropology, 19*, 3, pp. 601–4

—— and Gehlert, S. J. (1980) 'The Relative Accuracy and Reliability of Histological Aging Methods', *Forensic Science International, 15*, pp. 181–90

Thompson, D. D. (1979) 'The Core Technique in the Determination of Age at Death in Skeletons', *Journal of Forensic Sciences, 24* (4), pp. 902–15

Ubelaker, D. H. (1974) 'Reconstruction of Demographic Profiles from Ossuary Skeletal Samples', *Smithsonian Contributions to Anthropology, 18*, Washington

—— (1978) *Human Skeletal Remains*, Taraxacum, Washington

8 DETERMINATION OF AGE AT DEATH: DENTITION ANALYSIS

Raymond L. Costa, Jr.

It is sometimes necessary to ascertain the age at death of archaeologically derived skeletal material or the remains of individuals who have been burned, drowned, desiccated or decomposed to the extent where normal means of identification are no longer possible. Teeth are very good for this purpose since they are hard, very durable and do not melt or burn, other than in exceptional instances. Since teeth continue to change from approximately 20 weeks *in utero* until the last tooth is lost or its owner dies, age changes in the dentition can often be used to provide an age at death when this information is not known.

The methods used by archaeologists to gain age-at-death information are somewhat different than those used by forensic pathologists, but both rely on the relationship between chronological age and age-related changes in the dentition. These changes can be divided into: (1) changes that occur while the deciduous and permanent teeth are forming and erupting into occlusion; (2) degenerative changes, which are the results of tooth usage and dental pathologies that occur during the functional life of the tooth; and (3) histological changes, which are related to the age of the tooth, but seen only in section.

It should be noted at the outset that developmental methods seem to work best and, despite more than a century of research in forensic odontology, it is not possible to age precisely an individual after the crowns of the teeth have finished developing (although analysis of annulation rings in cementum holds the promise of rectifying this situation). For this reason, dental ageing should always be accompanied by other skeletal methods whenever possible.

Calcification of the Crown and Root

Calcification of the crowns of deciduous teeth begins at 20 to 28 weeks *in utero*. Permanent teeth begin to calcify just before birth and continue until the crowns of the third molars are completed between the ages of nine and 15 years (Cameron and Sims, 1973). Both enamel and dentin develop simultaneously, growing outward from the amelo-dentinal junction. The rate of enamel apposition has been estimated at 4μ per day (Massler and Schour, 1946), although there is considerable variation. Enamel prisms grow along their long axes, starting at cusp tips or incisal lines and spreading outwards. Due to diurnal growth variations, each day's growth can be seen as optically distinct variations in these prisms. The variations have the appearance of alternating light and dark bands. When the prisms cease forming, the crown is complete.

It is possible to section a still-forming tooth, count the daily bands and arrive at the number of day's growth of enamel past the point at which the tooth began to form. While it is quite possible to arrive at the age of death by counting growth bands for fetuses or children in whom the deciduous dentition is not fully formed, there is such variation in the timing of calcification initiation for the permanent dentition that this method gives inaccurate results. Garn *et al.* (1958) identified two stages of tooth development, which can be easily seen in radiographs, namely, full follicle development and full calcification of the crown. Yamaji (1958) and Hayakawa (1959) use six stages and have prepared tables of calcification timing for permanent teeth (reprinted in Furuhata and Yamamoto, 1967).

Around the time of birth, and for about a week thereafter, growth of enamel prisms is markedly slowed, resulting in an observable 'neonatal line' (Schour and Massler, 1937). If such a line exists, the tooth is from an individual who has survived the birth process. Since the crowns of the molar deciduous teeth and the first permanent molars are being formed at the time of birth, young infants and children can be aged by counting the daily enamel bands formed after the neonatal line. Unfortunately, the neonatal line is not always readily observable in all individuals, so that its absence does not always imply an individual born dead. Indeed, since it takes between several days and two weeks for the neonatal line to form, a tooth from an individual who died shortly after birth may not display a neonatal line.

In addition to the neonatal line, systemic illnesses that occur during crown development also slow the apposition of enamel. This results in the formation of Wilson bands, which are similar to the neonatal line in appearance and occur on all teeth that are developing at the time of the illness (Kreshover, 1940). Wilson bands can be used to help establish identity by matching an individual's medical record of illnesses with effects of these illnesses on the dentition. Wilson bands have been used archaeologically to determine the palaeoepidemiology of some childhood diseases (Rose, Armelagos and Lallo, 1978).

Since enamel and dentin grow at similar rates in most children, Stack (1960) developed a simple way of assigning age at death for fetuses and very young individuals, providing all the teeth in one half of one jaw are available. The teeth in one quadrant are carefully extracted and weighed. The weight can be correlated with age, yielding the age at death. It is not known if this method has any advantages over counting diurnal growth bands.

Root formation begins before teeth start to erupt and continues after the teeth have come into occlusion. Generally, roots are one-half to two-thirds formed by the time they reach the occlusal table. The final third of root development, and closure of the root apex, can take a variable amount of time depending on the tooth and the individual. This takes up to four years for first permanent molars, and six years for second molars (Furuhata and Yamamoto, 1967).

From approximately six months of age, when the first deciduous teeth appear, to the emergence of the third molars at 18 to 20 years, dental eruption timing can provide a means of estimating age at death. Eruption timing has the advantage that no laboratory methods are needed to acquire data; one need only examine the mouth of the individual to see which teeth have erupted. Saunders (1837) compiled early systematic tooth eruption data as a response to the child labour laws then being enacted in England. It was not possible to enforce these laws without some impartial observational indicator of a child's age, and eruption timing seemed to provide this information.

Many studies of tooth eruption timing were undertaken in Europe and the United States in the nineteenth and early twentieth century for various reasons. These are summarised by Cattell (1928), and have in common a lack of precision and inadequate statistical analysis. Most of these studies simply compared the teeth observable in the mouth to the stated age of the individual. This approach begs two questions: what exactly is meant by the term 'eruption', and exactly how old is the individual in question?

Garn *et al.* (1958) suggest five levels of tooth development, including three stages of eruption that can be observed in both living and skeletonised subjects. The first stage is the eruption of the tooth crown through the alveolar bone, which is directly observable in skeletal material, and can be seen on radiographs of living people. The second stage is the eruption of the tooth crown through the gingiva, so that the crown can be observed in the oral cavity. This is difficult to ascertain in dried skeletal material. The final stage occurs when the tooth crown reaches the occlusal plane. Depending on which tooth is in question, the eruptive process can take anything from a few weeks for anterior deciduous teeth, to more than a year for permanent molars.

To allow compatibility, it is of utmost importance to be specific about which of these criteria is being used in an eruption study. Many early authors, such as Cattell (1928), were not precise in this regard, thus limiting the value of their contribution. More recent workers have been more specific, but have not agreed on which phase of eruption should be considered the one of choice. Eruption through the alveolar bone and attainment of the occlusal plane are useful when studying dried skeletal material, whereas gingival and occlusal eruption are most suited to living subjects.

Some statistical problems exist in the study of tooth eruption. Both sample selection and sample size hampered Logan and Kronfield (1933), who studied histologically the dental development and eruption timing of 30 young individuals, many of whom had died from systemic diseases. Almost all other workers have used very large samples of normal individuals selected by population (Dahlberg and Menegez-Bock, 1958), country of origin (Boesen, Eriksen and Helm, 1976), or sex (Hunt and Gleiser, 1955). Dahlberg and Menegez-Bock note that while some authors have gathered data from cross-sectional samples of children, others have used longitudinal samples, leading to small but statistically important systematic differences. These errors suggest that only longitudinal or cross-sectional data should be compared. Dahlberg provides a method for equilibrating cross-sectional and longitudinal data.

Gleiser and Hunt (1955) and Hunt and Gleiser (1955) studied sex differences in eruption and crown calcification timing of permanent first molars. They found that females erupt these teeth when they are approximately 5 per cent younger than males, which results in females erupting first permanent molars approximately three months earlier. Demisch and Wartmann (1956) noted the same different in third molar development and eruption. Garn *et al.* (1958) found that females erupted permanent teeth at an average age advanced approximately 3 per cent over males, and while the difference in months increased with age, the percentage different did not. The average difference was 0.6 years for third molars. Additionally, Garn *et al.* (1958) state that this difference in timing was not due to females erupting teeth at an earlier stage of dental development. Fanning (1962) also noticed a 0.6 year difference between males and females in third molar emergence, but placed males earlier than females. Barret, Brown and Cellier (1964) observed a similar male-female difference in eruption timing in Australian aborigines.

Differences in eruption timing have been examined in many different populations. Dahlberg and Menegez-Bock (1958) studied Pima Indian children and compared them to several earlier studies, including American blacks (Steggerda and Hill, 1942), African Zulu (Suk, 1919), Maya and Navajo Indians (Steggerda and Hill, 1942), American whites (Cattell, 1928; Cohen, 1928; Fulton and Price, 1954; Hellman, 1943; Hurme, 1948; Klein, Palmer and Kramer, 1938; Steggerda and Hill, 1942) and English whites (Ainsworth, 1925; Clements, Davies-Thomas and Pickett, 1953; Stones *et al.*, 1951). Dahlberg found that the European and American whites erupted teeth on a different schedule than the American Indians or African blacks, whose median age for eruption of premolars and molars was earlier by one-half to two-thirds of a year.

Moorrees (1957) notes that whites have delayed eruption of teeth compared to Mongoloids and blacks, and shows that although his sample of Aleuts had consumed an essentially modern soft food diet for 15 years before his study, their eruption timing was still advanced over the same American whites used for comparison by Dahlberg and Menegez-Bock (1958). Boesen, Eriksen and Helm (1976) studied two Greenland Eskimo samples and found that their teeth generally erupted earlier than white comparison groups, but that the pattern varied from tooth to tooth and no clear trend could be established. Mayhall, Belier and Mayhall (1978) compared Canadian Inuit Eskimo to French Canadians and found that although there were no significant variations in timing of incisor eruption, canines and posterior teeth erupted earlier in the Eskimo. In summary, whites seem to have later eruption timing than other population groups, especially in posterior teeth, and this difference amounts to between one-half and three-quarters of a year.

Most eruption studies have not included third molars due to their high rates of attrition and impaction and their large range of eruption-timing variation (Mucci, 1982). Fanning (1962) studied white Bostonians and found the median age of eruption was 20.5 years, with males erupting these teeth at a median age of 19.8 years, and females at 20.4 years. Nine per cent of these Bostonians were missing one or more third molars. Fanning compared these results to Chagula's (1960) study of East African black university students. In the 13- to 16-year age range, blacks erupted third molars approximately 2.5 years earlier than whites, while in the 17- to 20-year age range, this difference dropped to 1.25 years. These differences may well be within the range of variation for Boston whites, which is not given by Fanning. Begg (1954) suggests that populations eating high-attrition diets that cause significant interstitial wear will erupt their third molars at an earlier age than those who do not. Several studies substantiate Begg's position (Lombardi, 1982; Richardson, 1980).

Helm (1969) compared physical maturation with dental eruption timing for Danish schoolchildren surveyed in 1913 (Degerbol, 1931) and 1965. He found that while there was a discernible secular trend in physical maturation (the 1965 sample matured approximately one year earlier than the 1913 sample), dental eruption timing remained essentially constant.

Composite Charts

The composite data for dental development and eruption can best be summarised by diagrammatic or pictorial charts. Massler and Schour published the first pictorial chart of human dental development in 1941, based in large part on the work of Logan and Kronfeld (1933), whose sample contained only 30 individuals, many of whom died of chronic, debilitating diseases. Dental development, however, is under strong genetic control (Glasstone, 1963; 1964).

This chart shows cut away views of the developing dentition at 21 different

Figure 8.1: Composite Chart of Dental Formation and Eruption From Ubelaker (1978), Similar to Massler and Schour's (1941) Chart Distributed by the American Dental Association. Ubelaker's ages and variations are a summary of works by Anderson, Thompson and Popovich (1976), Banerjee and Mukherjee (1967), Christensen and Kraus (1965), Coughlin and Christensen (1966), Dahlberg and Menegez-Bock (1958), Demisch and Wartmann (1956), Gilster, Smith and Wallace (1964), Hurme (1948), Moorrees (1956), Moorrees, Fanning and Hunt (1963a, 1963b), Nolla (1960), Robinow, Richards and Anderson (1942) and Steggerda and Hill (1942)

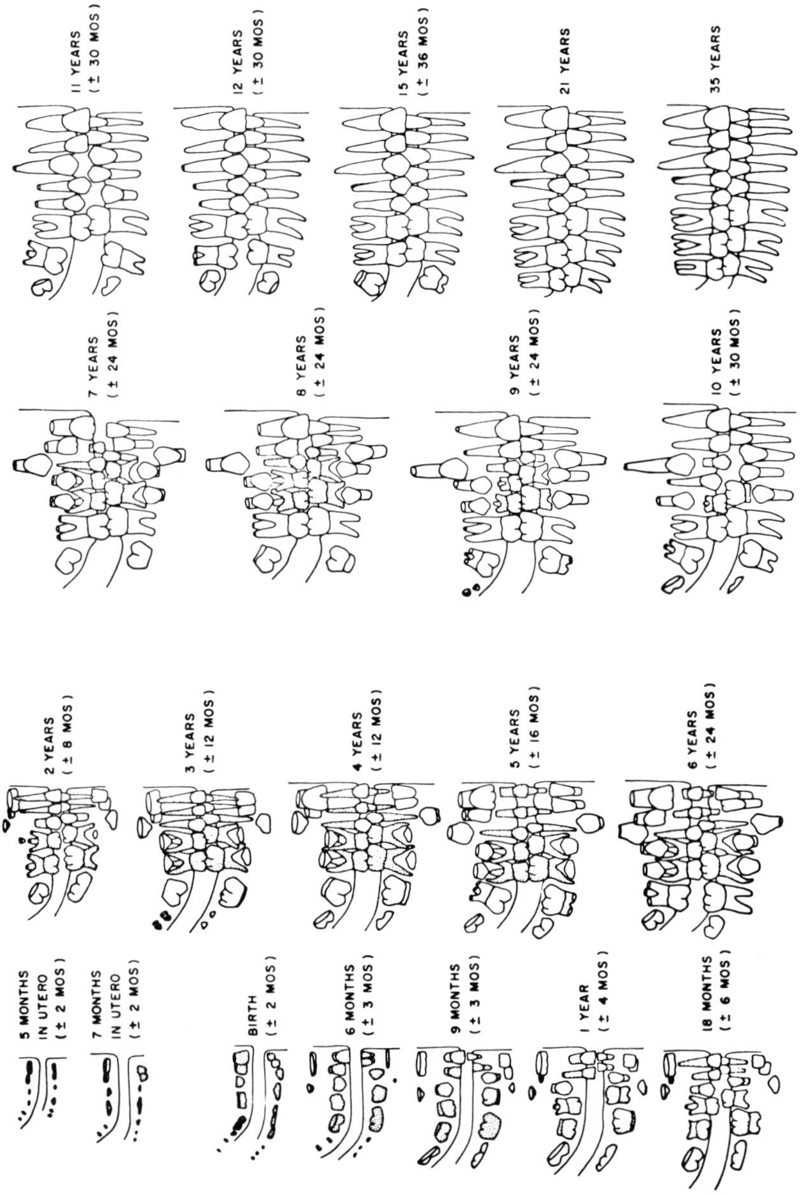

Figure 8.2: Chart of Dental Development and Eruption From Gustafson and Koch (cited in Johanson, 1971). The apex of each triangle indicates the average age for that feature, and the base represents the earliest and latest occurrences, respectively. A to B = intrauterine life, B to C = first year of life, and C to D = two to 16 years of life

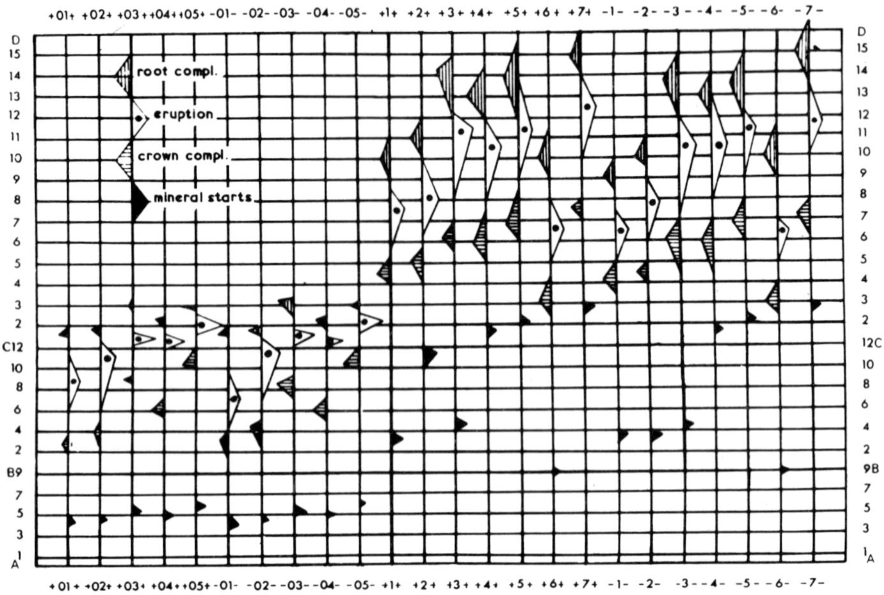

age levels, from five months *in utero* to 35 years of age. Each view shows the developmental stage of the crowns of deciduous and permanent teeth, their degree of root formation and the state of eruption for all the teeth in one half of the mouth. Each view also represents an average age for that stage of development, and gives a range of variation estimated at one standard deviation by Gray and Lamons (1959). Data for males and females are put together. This chart is widely distributed by the American Dental Society and is reprinted in many texts (Morrey and Nelson, 1970; Wheeler, 1974).

A great deal of information on variations in eruption timing has been gained in the last four decades, and Ubelaker (1978) has used this data to update the Massler and Schour chart, mainly by increasing the variations (Figure 8.1). Ubelaker notes that his chart is intended for archaeological use with American Indian remains; however, it was approved by the Workshop of European Anthropologists in 1978 (Ferembach, Schwidetzky and Stloukal, 1980). The Massler and Schour and Ubelaker charts, based as they are on histological sections, have the great advantage that individuals, skulls or radiographs can be compared directly to the charts and assigned an age. Cases that fall between those shown can be approximated. No allowance is made, however, for variations in the sequence of dental eruption. Garn *et al.* (1958) showed that there are three common variations for the eruption of the permanent dentition: (1) the second premolar erupts before the second molar; (2) the second premolar and second molar erupt concurrently; and (3) the second molar precedes the second premolar. Garn *et al.* point out that an individual may calcify the crowns of these teeth in an order different from the actual order of eruption, or can change orders during the process. Barrett *et al.* (1964) noted these variations in Australian Aborigines.

Gustafson and Koch (1971) prepared a schematic chart showing root completion, eruption, crown completion and mineralisation of permanent and deciduous teeth using pooled European and American data (Figure 8.2) (cited in Johanson, 1971). The development of each tooth appears as a series of triangles on a line compared to age. The amount of time variation for each of the four developmental features determines the length of the base of each triangle (Gustafson, 1966). Male and female data are not considered separately. This chart offers the advantage over the Massler and Schour and Ubelaker charts that ranges of eruption timing are shown, though it is not as easily understood due to its diagrammatic nature. This chart was compiled from European data and has been reprinted (Cameron and Sims, 1973; Noble, 1976).

Moorrees *et al.* (1963a and b) produced a schematic chart based on longitudinal radiological studies of Bostonians, but the chart does not include data from other sources (see also Cameron and Sims, 1973). This chart includes data on crown and root development of individual teeth, complete with standard deviations. Like the Gustafson and Koch chart, the diagrammatic nature of this chart does not allow for easy intuitive understanding. It does offer the advantage that males and females are dealt with separately, rather than averaged together, resulting in lower ages and smaller ranges of variation than the Ubelaker update of the Massler and Schour chart.

Secondary Dentin

Secondary dentin begins to form once the crown of a tooth is fully formed, the tooth is in occlusion and the root complete. Bodecker (1925) identified the apposition of secondary dentin, along with several other aspects of dental anatomy, as being related to chronological age. Secondary dentin has also been acknowledged as being a response to dental attrition and is sometimes present on the occlusal surfaces of severely worn teeth. Phillipas and Applebaum (1966; 1967; 1968) performed detailed studies of the pattern and rate of apposition of secondary dentin in upper anterior teeth, however, without the goal of assessing age at death.

Gustafson (1950; 1966) used secondary dentin as one characteristic in his ageing system. Johanson (1971) calculated the amount of secondary dentin present in sectioned teeth and correlated it with age as part of his analysis of Gustafson's system, and notes that quantification is extremely difficult. If caries or extreme wear has damaged a tooth, secondary dentin will be laid down more quickly than in an unaffected tooth. Johanson notes, as do Phillipas and Applebaum, that deposition along the walls rather than the roof of pulp chambers seems to be more a function of age and less a result of pathology. Johanson performed correlation and regression coefficients for secondary dentin and concluded that although apposition did correlate with age, it was of poor value as an age indicator due to a large standard deviation.

Cementum

Cementum covers the entire exposed dentin surface of teeth and increases in thickness throughout the life of an individual. In some cases, hypercementosis will occur as a result of pathology, or possibly due to super-eruption of teeth (Newman and Levers, 1979). Zander and Hurzeler (1958) found that the average thickness of cementum tripled between 11 and 76 years of age. When Johanson (1971) correlated cementum thickness with age he achieved a correlation coefficient of only 0.55 with the considerable standard deviation of 11.47 years, suggesting that cementum is a very poor indicator of age.

In many animals, especially hibernating mammals, cementum is laid down in annual rings. Stott, Sis and Levy (1982) found annulation rings in cross-sections of human teeth stained with alizarin red. These rings could be counted when viewed in photographs taken through a brightfield microscope. Teeth were taken from cadavers of known age. Cross-sectioned teeth were prepared, the rings counted, and the sum added to the estimated eruption time for the particular tooth. For 17 teeth studied, none yielded an age at death more than four years removed from the actual age. For one individual where 10 teeth were prepared, six yielded the correct age. The mean for ten trials was 0.2 years from the actual age, and the standard deviation was 1.8 years. Condon and Charles (1983) examined cementum annulation rings in 130 cadavers, with mixed results due to problems in defining individual rings in specimens. Kay, Rasmussen and Beard (1984) report that cementum annulation rings provide a good means for determining age in *Macaca mulatta*. This method has (so far) not been tested on forensic or archaeological material, but is of considerable promise since it appears to indicate age at death better than established methods.

Discoloration of Teeth

Teeth tend to become darker with age, turning progressively more yellow, or in some cases, brown. This change is in part caused by accumulation of secondary dentin, which is darker in colour than primary dentin, coupled with increased transparency of the primary dentin, thus allowing the darker colour to be seen. With age surface enamel tends to absorb metallic ions, which also discolour teeth (Brudevold, 1957). Dahlberg (1960) considered that although most people's teeth change in colour somewhat as they age, the amount of change is highly variable, depending on diet, dental health care and tobacco consumption. There is also a considerable range of variation in dental colour before any changes occur. Biedow (1963) suggests that tooth coloration be added to Gustafson's ageing method (see below); however, since the factors that control tooth discoloration vary within and between populations and social classes, coloration of teeth is probably a poor indicator of age at best.

Dentin Translucency

Translucency of the dentin is a histologically observable age-related change in the dentin. Dentin is formed in tubules, having a diameter of approximately 3.2μ (Noble, 1976). The opacity of dentin is caused by the difference in refractive index between the crystalline structure of the dentin and the intertubular organic material (about 30 per cent of the dentin) (Kraus, Jordan and Abrams, 1969). Beginning in the third decade of life, the dentin tubules become progressively calcified, and their internal diameter shrinks to 1.2μ or less by the eighth decade. As this shrinkage occurs, the difference in refractive index is lost, thus resulting in translucency. This change takes place first at the root tip and root canal, and migrates coronally and toward the outside of the root.

Attempts to use dentin translucency for ascertaining age at death have met with mixed results. Gustafson (1950; 1966) used it as one of his six criteria. He longitudinally sectioned teeth, ground them, dehydrated them in alcohol, mounted them in Canada balsam and judged the amount of translucency on a scale of 0 to 3. Gustafson points out that only translucency in the root area should be measured, since translucency in the crown can be the sequel of caries or wear. Nalbandian, Gonzales and Sognnaes (1960) noted that translucency was bilateral, was relatively unaffected by pathological changes in the tooth, and was a relatively good indicator of age.

Miles (1963a) tried measuring the length of the root in which translucency occurred, rather than judging it subjectively, but achieved no increase in precision over Gustafson's work. Johnson (1968) tried correlating several measures of translucency against age. The correlation coefficients ranged from 0.36 to 0.71. and he concluded that dentin translucency was not a very good indicator of age. Johanson (1971) used a slightly different histological technique and achieved a correlation coefficient of 0.86 with a comparatively small spread of 7.1 years around the regression line. He concluded that translucency was a good age indicator. Bang and Ramm (1970) measured the amount of translucency in intact, rather than sectioned, teeth, judged from the apex to a point midway between the root canal and the root surface. In this way they disregarded the peripheral area of the root. They claim good accuracy to 75 years of age. Using multiple regression analysis, Maples (1978) chose translucency as one of two variables that most reliably predicted age.

Specific Gravity and Hardness of Teeth

Several dental factors change with the increasing calcification of the dentinal tubules. Kanai (1954, cited in Furuhata and Yamamoto, 1967) determined specific gravity of permanent incisors in a range of people aged from nine to 68 years. Specific gravity was well correlated with age. The total range of variation was from 2.305 to 2.3577. In addition to the precision required for measuring these differences, there was considerable overlap for different specimens within the same age group, making use of this method problematical for assignment of age at death.

Along with an increase in specific gravity, teeth tend to become more brittle with age, which is a combination of increased hardness and decreased elasticity. Kato (1956) investigated age changes in enamel, dentin and cementum and found that the average hardness increased; however, the range of hardness for each 10-year age grouping was much higher than the small increases in average hardness. Furuhata and Yamamoto (1967) give elasticity curves for deciduous and permanent teeth. Permanent teeth in males and females become more elastic until approximately ages 35 and 30, respectively, after which they become progressively less elastic. All teeth become very inelastic after the age of 60. Furuhata and Yamamoto give no sample sizes or statistics. Dental hardness and elasticity are probably insufficient for assigning age at death.

Root Resorption

The roots of deciduous teeth are normally resorbed during the formation and eruption of permanent teeth. Resorption of the roots of permanent teeth is pathological, and the occurrence of this pathology has been related to age by Gustafson (1950; 1966). Johanson (1971) noted that resorption rarely occurs before the age of 50 years, and is often absent altogether. He recorded a correlation coefficient for resorption of only 0.24, with a very large standard deviation, and suggested that root resorption was only of value in age identification for identifying old individuals. Miles (1963b) considered resorption to be the poorest criterion for assigning age at death, and Maples (1978) considered it without merit for distinguishing age.

Recession of the Gingiva

In normal healthy mouths the gingiva attach to teeth in the vicinity of the cemento-enamel junction. Some form of periodontal disease is most likely to be present if the gums have receded from this point. Gustafson used the amount of periodontal recession as part of his ageing system (1950; 1966). Periodontal disease is clearly pathological and is, to some extent, diet-dependent and therefore population-dependent. In modern American and Western European people, periodontal disease is rampant, and responsible for the majority of tooth loss over the age of 40 years (Russell, 1970).

In many primitive or prehistoric skeletal samples, very little periodontal disease is evident, and the aetiology is possibly different (Costa, 1982). Apparent recession of the gingiva may also be a response to super-eruption of teeth (Newman and Levers, 1979). While periodontal disease and recession of the gingival attachment has some correlation with age, it is population- and social class-dependent and therefore a poor indicator of age unless social class and population are known for the individual in question. Periodontal disease can also be manifest as either a generalised resorptive condition where the gingiva recede the same amount throughout the mouth, or as a localised lesion resulting in individualised deep pockets. When only one or a few teeth are used for forensic purposes, it is not always possible to tell if gingival recession is caused by generalised alveolar resorption, super-eruption, or a bony pocket.

Since these changes all have different curves when regressed against age, their predictive value is low. Johanson (1971) found that gingival recession had a correlation coefficient of 0.24, and only used it together with the other criteria used by Gustafson. Maples (1978) regards gingival recession as ineffectual for determining age at death.

Dental Wear

Although the majority of people living in technologically sophisticated cultures exhibit little dental wear regardless of age, wear was considered to be a pathological condition until the early 1950s (Sicher, 1953). Pathological conditions, such as unusually soft teeth or improper occlusion, do cause abnormal dental wear in some people, and some people's teeth do show the effects of paramasticatory wear due to pipe smoking, hair pin chewing or other habits (Molnar, 1972). But for most of human evolution both occlusal and aproximal dental wear was the norm rather than the exception. Most 'primitive' peoples not living in a processed-food society still wear away their teeth as they age. Although Gustafson (1950; 1966) used dental wear as part of his system for ageing forensic specimens, progressive dental wear is most useful in assessing age at death in archaeologically derived skeletal samples.

Broca (1879) devised a four-level system for recording dental wear, which was also used by Leigh (1925; 1928). Murphy described dental wear pictographically (1959a; 1959b). Such systems are useful for recording wear *per se*, but are not sufficiently sensitive for determining age from dental wear. Although Brothwell (1963) suggests using comparative dental wear to assign age at death, Miles (1958; 1962; 1963a; 1963b; 1978) systematised the determination of age at death from dental wear, and his system has been used by others (Lovejoy *et al.* [unpubl.]; Mann, 1975; Nowell, 1975; Ruff, 1981). Miles' method assumes two constants. First, that tooth wear in any given society will be relatively constant over long periods of time, and second, that it will be the same for all members of the society (or at least all the members of each sex within a society) (Campbell, 1939). Additionally, a sufficient number of juvenile specimens must be present.

Miles developed his system using a collection of 190 specimens from an Anglo-Saxon archaeological sample. Before the relative wear of the specimens could be assessed, a 'known age' subsample was developed using specimens with deciduous and mixed dentitions. Age was determined for these specimens by using the Massler and Schour chart (Figure 8.1), although Miles tempered it with information from Gleiser and Hunt (1955), Garn, Lewis and Polacheck (1959) and others (Miles, 1962). By removing loose teeth on some individuals and radiographing others, Miles studied root development, apical closure and eruption status of the immature specimens so that comparisons could be made between his specimens and the Massler and Schour chart. This allowed age determination up to 15 years of age. Relying on unpublished modern data, Miles assumed that third molars generally erupted at 18 years, came into occlusion at 19 years, had closed apices by 20 years, and had observable constriction of the root canals by 22 years of age. Miles assumed that the Anglo-Saxon sample shared its general timing with these modern estimates. Thirty-two individuals were assigned an age using these observations. This became the known age sample (Figure 8.3).

Figure 8.3: Drawings of Longitudinal Sections Through a Generalised Anterior Tooth Showing the Age-Related Parameters used by Gustafson (1950; 1966) at Progressively Older Ages. A = abrasion, S = secondary dentin, P = migration of the periodontal ligament, C = thickness of cementum, R = root resorption, and T = transparency of the dentin (drawing by K. M. Hiiemae)

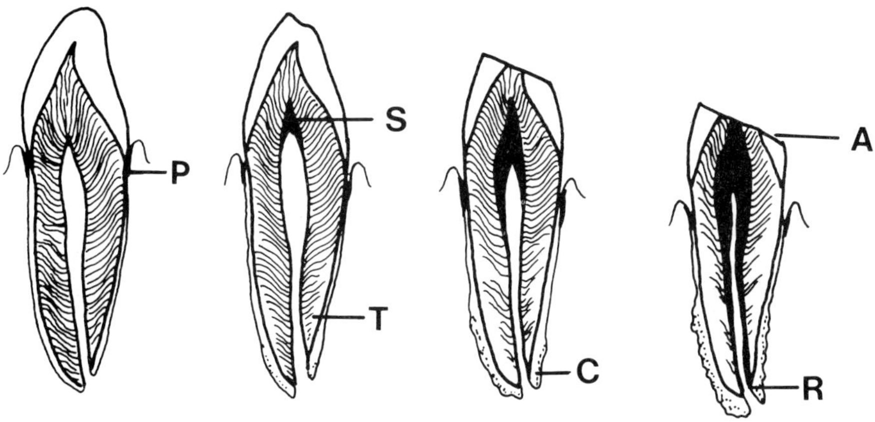

After establishing the known age group, the functional ages of the molars were established. Assuming that first molars appear at six years of age, and second molars at 12 years, when the second molar emerges the amount of wear of the first molar will be six years (Figure 8.4). Therefore, the functional age of the tooth will be six years. Similarly, when the third molar erupts, the second molar will have a functional age of six years, and the first molar 12 years. The functional age of third molars is established by comparing them to similarly worn first or second molars. When a third molar has a functional age of six years, the first molar is aged 18 years, and the second molar 12 years. As each successive functional age is determined, the number of aged individuals in the series increases.

Interpolations are made by comparing worn teeth in the known age sample. A child aged 10 years has first molars whose functional age is four years, and these teeth can be used as comparative standards for other molar teeth with the functional age of four years. Functional ages are added to the time of eruption for each tooth to determine the age of the individual. Eventually, the entire sample can be aged.

This system presents some problems. Eruption timing, especially of the third molars, may not have been the same in archaic populations as it is today, if Begg's (1954) hypothesis concerning interproximal wear is correct. This would alter age assignments, making them one to two years higher. Miles did not sex specimens below the age of 18 years (which is

Figure 8.4: Demonstration of the Miles Ageing System. In the first picture, the difference in wear between the third and second molar is six years and between the second and first molar an additional six years. The functional age of the first molar is therefore 12 years, and the individual would be 18 years of age. The amount of wear on the first molar in the first picture is similar to the second molar in the second picture, which was judged to have a functional age of 13 years, implying that the first molar has a functional age of 19 years. By leap-frogging in this manner, the worn molars in the sample can be assigned functional ages (from Miles, 1963a)

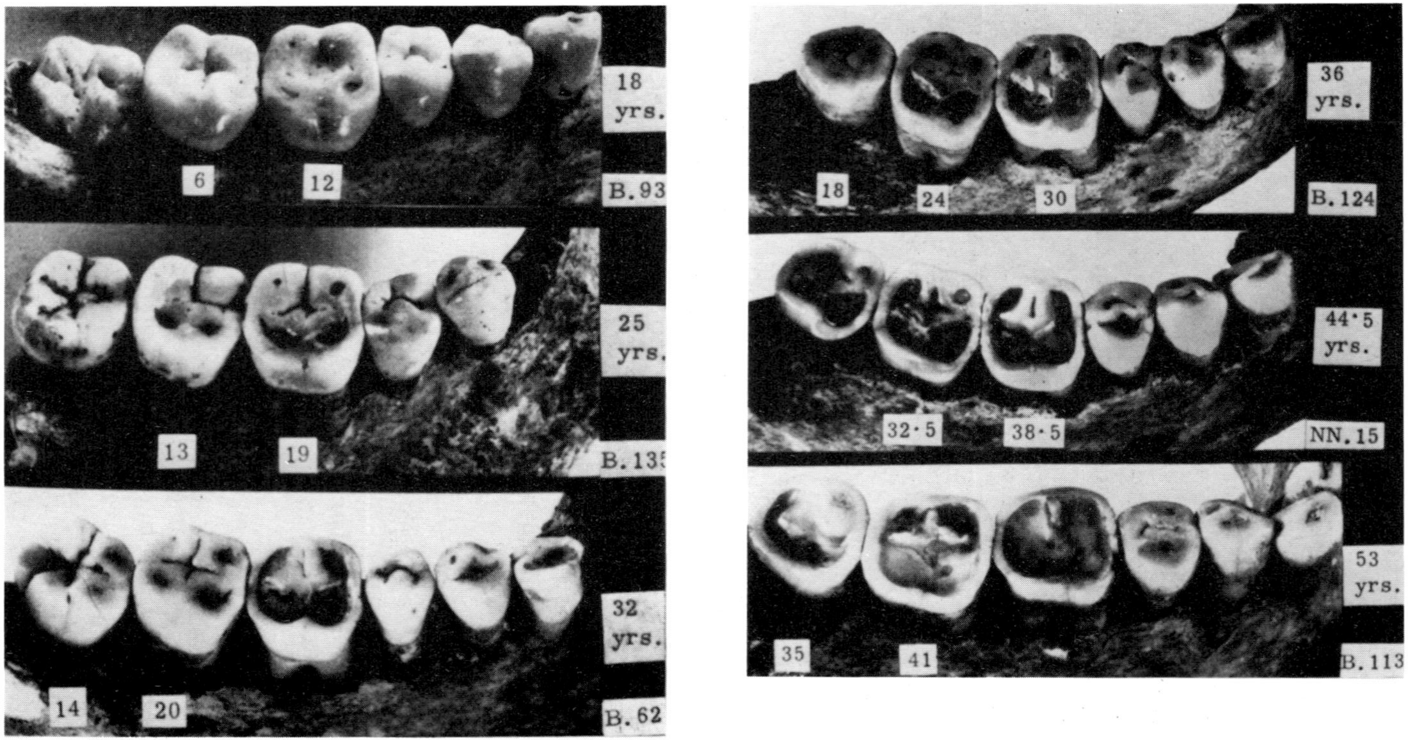

often an impossible task). Since there can be a year's difference in eruption of second molars between males and females, this could alter age assignments. There is a difference in tooth size between the first and second molar, and the second and third molar. While there may be little difference between functional and actual ages for first and second molars, the difference in size between first and third molars may be sufficient for third molars to wear at a different rate, thus confusing the functional age. Indeed, Smith (1984) notes that human third molars wear at one-half the rate of first molars, although she reasons that this is caused by differential muscle load resulting in differing chewing stresses on the first, second and third molars.

Finally, as teeth wear, the rate of wear may accelerate as the enamel is worn through and the softer dentin is exposed. Since this is the common condition in older individuals in high attrition environments, the functional ages of severely worn teeth will appear to be greater than they actually are, causing consistent over-ageing of older individuals. These problems must be considered when ageing archaeological samples, as has been done by Nowell (1975) and Ruff (1981). Mann used a variation of this method on australopithecines (1975), assuming that their eruption timing and sequence were sufficiently similar to our own to assign age at death. Wolpoff (1979) used dental wear to assign ages to the Krapina neanderthal material. Since the number of individuals in the sample is small, and the remains fragmentary, his results should be viewed with caution.

The accuracy of relative wear ageing is dependent on the consistency of the dental wear within the group studied. Lovejoy *et al.* (unpublished) note that wear at the Libben archaeological site was remarkably consistent from one individual to another, giving excellent results. Relative wear ageing can be used to assign age at death to within about two years under the most favourable conditions (Nowell, 1975).

Gustafson's Ageing Method

While all of the characteristics of dental development and anatomy discussed change with age, with the possible exception of annual accretions of cementum, no one character alone provides a reliable method for assigning age after the permanent dentition is fully formed. In 1950 Gustafson described a method for ascertaining age at death in adults that utilised information from six of the criteria already discussed. Gustafson sectioned teeth longitudinally and imbedded them in Canada balsam. He then assigned subjective points on a 0 to 3 scale for abrasion (A), apposition of secondary dentin (S), migration of the periodontal ligament away from the cemento-enamel junction (P), thickness of cementum (C), root resorption (R), and translucency of the dentin (T) (Figure 8.3).

Each of these characters changes with age, although some might be considered pathological (root resorption, periodontal ligament migration), and others influenced by dental pathology (secondary dentin is often laid down as a response to caries). Nevertheless, by adding the points for the six criteria and plotting the sum against age for known age specimens, Gustafson arrived at a correlation coefficient of 0.98 ± 0.01 and a slope of 4.56. He was then able to place unknown specimens on the resulting graph and arrive at an age of death.

While Gustafson's method was a great step forward for forensic odontology, subsequent work has pointed out some flaws in it and has resulted in improvements. Pilz (1959) used Gustafson's method and concluded that more accurate results could be attained when anterior teeth were used, rather than posterior teeth or a random selection from the whole mouth. Pilz was unable to duplicate Gustafson's accuracy. Nalbandian *et al.* (1960) duplicated Pilz' findings, and Miles (1963a, and b), using a more 'intuitive' approach, arrived at the same conclusions.

Gustafson (1966) made minor changes to his system and suggested that the inability of other workers to duplicate his accuracy was due to lack of practice. Gustafson's student, Johanson, revised the system again in 1971. Johanson increased precision by using a range of seven points, rather than Gustafson's four, for each variable. The method of estimating dentin translucency was improved by using a standard thickness for dental sections, and inspecting them before imbedding in Canada balsam. Most importantly, Johanson performed correlations for each of the six variables in the known age sample, and found that two, dentin translucency and secondary dentin, were well correlated with age, while periodontal ligament migration and root resorption were very poorly correlated with age.

A computer program performed multiple correlations for the six variables in the most advantageous way, resulting in a formula where each variable was weighted according to its usefulness in determining age. Surprisingly, Johanson found that the computer could assign age more accurately for his female sample than his male sample, and for his male sample more accurately than for his total sample. This finding suggests that if the sex of the individual from whom a tooth came is known, the result will be more accurate.

Most recently, Burns and Maples (1976), Maples (1978) and Maples and Rice (1979) worked with modifications of the Gustafson method. Burns and Maples added several other factors, such as tooth position, race, sex and periodontal disease to their study. They used a sample of 355 teeth from 267 individuals, as compared to 41 teeth for Gustafson's original (1950) study, and 162 teeth from 46 individuals used by Johanson (1971). Burns and Maples used multiple regression analysis, as well as the simple linear regressions used by Gustafson. The multiple correlations weighted the contributions of the six variables and achieved a correlation coefficient of 0.996, allowing 93 per cent of the anterior teeth to be aged within three years, and 100 per cent within five years, a significant increase over Gustafson's 44 per cent and 58 per cent accuracy for three and five years, respectively.

Maples (1978) refined the multiple regression analysis coupled with the Gustafson method and found that acceptable results could be attained using just two variables, secondary dentin and dentin translucency. Maples and Rice (1979) reanalysed Gustafson's original statistical techniques and found he had reused 19 teeth from his known age sample in his unknown sample, resulting in a slightly higher correlation than other workers achieved without enlisting the aid of computers. Given the availability of small, inexpensive computers, it is now relatively easy to use the Johanson, Maples, or Burns and Maples updates of the Gustafson system to improve the accuracy of assigning age at death in forensic cases.

Teeth remain one of the best methods for ascertaining age at death for forensic and archaeological specimens. Often, the dentition provides the only way of obtaining this information. Sometimes using the dentition is simply the fastest or simplest means to this end. Meindl, Lovejoy and Mensforth (1983) note that functional dental wear is an important palaeodemographic tool, since it shows less statistical bias and inconsistency than other skeletal age indicators. New methods have improved the accuracy of existing techniques within the last three decades, and new techniques have been proposed. Additional research will doubtless improve this accuracy.

References

Ainsworth, N. J. (1925) *Dental Disease in Children*, Med. Research Council, Great Britain Special Series, no. 97

Anderson, D. L., Thompson, G. W. and Popovich, F. (1976) 'Age of Attainment of Mineralization Stages of the Permanent Dentition', *J. Forensic Sci., 21*, pp. 191–200

Bang, G. and Ramm, E. (1970) 'Determination of Age in Humans from Root Dentine Transparency', *Acta Odont. Scand., 28*, p. 3

Banerjee, P. and Mukherjee, S. (1967) 'Eruption of Deciduous Teeth Among Bengalese Children', *Am. J. Phys. Anthropol., 26*, pp. 357–8

Barrett, M. J., Brown, T. and Cellier, K. M. (1964) 'Tooth Eruption Sequence in a Tribe of Central Australian Aborigines', *Am. J. Phys. Anthropol., 22*, pp. 79–89

Begg, P. R. (1954) 'Stone-age Man's Dentition', *Am. J. Orthodont., 40*, p. 298, 373, 462, 517

Biedow, J. (1963) 'A Modified Method of Age Determination of Teeth', paper given at the *Third International Meeting in Forensic Immunology, Medicine, Pathology and Toxicology*, London

Bodecker, C. F. (1925) 'A Consideration of Some of the Changes in the Teeth From Young to Old Age', *Dental Cosmos, 67*, pp. 543–9

Boesen, P., Ericksen, J. H. and Helm, S. (1976) 'Timing of Permanent Tooth Emergence in Two Greenland Eskimo Populations', *Comm. Dent. Oral Epidemiol., 4*, pp. 244–7

Broca, P. (1879) 'Instructions Relatives à l'étude Anthropologique du Système Dentaire', *Bull. Soc. Anthropol. Paris, 2*, pp. 128–65

Brothwell, D. R. (1963) *Digging up Bones*, British Museum, London
Brudevold, F. (1957) 'Aldersforandringer i Tannemaljen (Age Changes in Dental Enamel)', *Norske Tannlaegeforen Tidskrif, 67*, pp. 451–8
Burns, K. R. and Maples, W. R. (1976) 'Estimation of Age from Individual Adult Teeth', *J. Forensic Sci., 21*, pp. 343–56
Cameron, J. M. and Sims, B.G. (1973) *Forensic Dentistry*, Churchill Livingstone, Edinburgh
Campbell, T. D. (1939) 'Food, Food Values and Food Habits of the Australian Aborigines in Relation to Their Dental Condition', *Aust. J. Dent., 43*, pp. 1, 73, 141, 177
Cattell, P. (1928) *Dentition as a Measure of Maturity*, Harvard, Cambridge, Massachusetts
Chagula, W. K. (1960) 'The Age of Eruption of Third Permanent Molars in Male East Africans', *Am. J. Phys. Anthropol., 18*, pp. 77–82
Christensen, G. J. and Kraus, B. S. (1965) 'Initial Calcification of the Human Permanent First Molar', *J. Dent. Res., 44*, pp. 1338–42
Clements, E. M. B., Davies-Thomas, E. and Pickett, K. G. (1953) 'Time of Eruption of Permanent Teeth in British Children in 1947–48', *Brit. Med. J., 1*, p. 1421
Cohen, J. T. (1928) 'The Dates of Eruption of the Permanent Teeth in a Group of Minneapolis Children', *J. Am. Dent. Assoc., 15*, p. 2337
Condon, K. W. and Charles, D. K. (1983) 'Evaluation of the Cemental Annulation Aging Technique', *Am. J. Phys. Anthropol., 60*, p. 183
Costa, R. L. (Jr), (1982) 'Periodontal Disease in the Prehistoric Ipiutak and Tigara Skeletal Remains from Point Hope, Alaska', *Am. J. Phys. Anthropol., 59*, pp. 97–110
Coughlin, J. W. and Christensen, G. J. (1966) 'Growth and Calcification of the Prenatal Human Primary Molars', *J. Dent. Res., 45*, pp. 1541–7
Dahlberg, A. A. (1960) 'Clinical Ageing Patterns in Teeth of Different Population Groups', in *Ageing: Some Social and Biological Aspects. Amer. Assoc. Adv. Sci. Publication No. 65*
—— and Menegez-Bock, R. M. (1958) 'Emergence of the Permanent Teeth in Pima Indian Children', *J. Dent. Res., 37*, pp. 1123–40
Degerbol, M. (1931) 'Om Tandskiftets Forlob Hos Kobenhavnske Kommuneskoleborn', *Meddr. Danm. Antrop., 3*, pp. 237–327 (summary in English)
Demisch, A. and Wartmann, P. (1956) 'Calcification of the Mandibular Third Molar and Its Relation to Skeletal and Chronological Age in Children', *Child Develop., 27*, p. 459
Fanning, E. A. (1962) 'Third Molar Emergence in Bostonians', *Am. J. Phys. Anthropol., 20*, pp. 339–45
Ferembach, D., Schwidetzkey, I. and Stloukal, M. (1980) 'Recommendations for Age and Sex Diagnoses of Skeletons' (English translation by M. G. Praschma and R. I. Sundick), *J. Human Evol., 9*, pp. 517–49
Fulton, J. T. and Price, B. (1954) 'Longitudinal Data on Eruption and Attack of the Permanent Teeth', *J. Dent. Res., 33*, p. 65
Furuhata, T. and Yamamoto, K. (1967) *Forensic Odontology*, Charles C. Thomas, Springfield
Garn, S. M., Lewis, A. B., Koski, K. *et al.* (1958) 'The Sex Difference in Tooth Calcification', *J. Dent. Res., 37*, pp. 561–7
——, Lewis, A. B. and Polacheck, D. L. (1959) 'Variability of Tooth Formation', *J. Dent. Res., 38*, p. 135
Gilster, J. E., Smith, F. H. and Wallace, G. K. (1964) 'Calcification of Mandibular Second Primary Molars in Relation to Age', *J. Dent. Child., 31*, pp. 284–8
Glasstone, S. (1963) 'Regulative Changes in Tooth Germs Grown in Tissue Culture', *J. Dent. Res., 42*, pp. 1364–8
—— (1964) 'Cultivation of Mouse Tooth Germs in a Chemically Defined Protein-Free Medium', *Arch. Oral Biol., 9*, pp. 27–30
Gleiser, I. and Hunt, E. E., (Jr.) (1955) 'The Permanent Mandibular First Molar: Its Calcification, Eruption and Decay', *Am. J. Phys. Anthropol., 13*, pp. 253–83
Gray, S. W. and Lamons, F. P. (1959) 'Skeletal Development and Tooth Eruption in Atlanta Children', *Am. J. Orthodont., 45*, p. 272
Gustafson, G. (1950) 'Age Determination on Teeth', *J. Am. Dent. Assoc., 41*, pp. 45–54
—— (1966) *Forensic Odontology*, Staples Press, London
—— and Koch, G. (1971) 'Age Estimation up to Sixteen Years Based on Development', cited in Johanson, G. (1971) *Odontologisk Revy 22: Suppl. 21*

Hayakawa, K. (1959) 'Estimation of Age in Japanese Females by Roentgenograms of Teeth', *Japanese J. Legal Med. and Criminol., 25* (55), p. 1
Hellman, M. (1943) 'The Phase of Development Concerned With Developing the Permanent Teeth', *Am. J. Orthodont., 29*, p. 507
Helm, S. (1969) 'Secular Trend in Tooth Eruption: A Comparative Study of Danish School Children of 1913 and 1965', *Arch. Oral Biol., 14*, pp. 1177–91
Hunt, E. E. (Jr)., and Glesier, I. (1955) 'The Estimation of Age and Sex of Preadolescent Children from Bones and Teeth', *Am. J. Phys. Anthropol., 13*, pp. 479–87
Hurme, V. O. (1948) 'Standards of the First Six Permanent Teeth', *Child Develop., 19*, pp. 213–31
Johanson, G. (1971) 'Age Determination from Human Teeth', *Odontologisk Revy, 22: Supplement 21*, pp. 1–126
Johnson, C. C. (1968) 'Transparent Dentine in Age Estimation', *Oral Pathol., 25*, pp. 834–8
Kanai, M. (1954) 'On the Increase of Average Specific Gravity of Teeth With Increase in Age', *Ichikawa Gakuho, 2* (1), p. 8. (Japanese. Cited in T. Furuhata and K. Yamamoto (1967) *Forensic Odontology*, Charles C. Thomas, Springfield)
Kato, K. (1956) 'Hardness of the Teeth', *Rinsho Shika, 213*, p. 4 (Japanese. Cited in T. Furuhata and K. Yamamoto (1967) *Forensic Odontology*, Charles C. Thomas, Springfield
Kay, R. F., Rasmussen, D. T. and Beard, K. C. (1984) 'Cementum Annulus Counts Provide a Means of Age Determination in *Macaca mulatta* (Primates, Anthropoidea), *Folia Primatol., 42*, pp. 85–95
Klein, H., Palmer, C. E. and Kramer, M. (1938) 'Studies in Dental Caries. II. The Use of the Normal Curve for Expressing the Age Distribution of Eruption of the Permanent Teeth', *Growth, 1*, p. 386
Kraus, B. S., Jordan, R. E. and Abrams, L. (1969) *Dental Anatomy and Occlusion*, Williams and Wilkins, Baltimore
Kreshover, S. (1940) 'Histopathologic Studies of Abnormal Enamel Formation in Human Teeth', *Am. J. Orthodont. Surg., 26*, pp. 1083–1101
Leigh, R. W. (1925) 'Dental Pathology of the Eskimo', *Dental Cosmos, 67*, pp. 884–98
—— (1928) 'Dental Pathology of Aboriginal California', Univ. of Calif. publ. in *Amer. Archeol. and Ethnol., 23*, pp. 399–440
Logan, W. H. G. and Kronfeld, R. (1933) 'Development of the Human Jaws and Surrounding Structure From Birth to the Age of Fifteen Years', *J. Am. Dent. Assoc., 20*, p. 379
Lombardi, A. V. (1982) 'The Adaptive Value of Dental Crowding', *Am. J. Orthodont., 81*, pp. 38–42
Lovejoy, C. O., Meindl, R. S., Pryzbeck, T. R. *et al.* (no date) 'Chronological Metamorphosis of the Auricular Surface of the Ilium: A New Method For the Determination of Adult Skeletal Age at Death', unpublished manuscript
Mann, A. E. (1975) 'Some Paleodemographic Aspects of the South African Australopithecines', Univ. of Pennsylvania, publ. in *Anthropol. No. 1*, Philadelphia
Maples, W. R. (1978) 'An Improved Technique Using Dental Histology for the Estimation of Adult Age', *J. Forensic Sci., 23*, pp. 764–70
—— and Rice, P. M. (1979) 'Some Difficulties in the Gustafson Dental Age Estimations', *J. Forensic Sci., 24*, pp. 168–72
Massler, M. and Schour, I. (1946) 'The Appositional Life Span of the Enamel and Dentine Forming Cells', *J. Dent. Res., 25*, p. 145
Mayhall, J. T., Belier, P. L. and Mayhall, M. F. (1978) 'Canadian Eskimo Permanent Tooth Emergence Timing', *Am. J. Phys. Anthropol., 49*, pp. 211–16
Meindl, R. S., Lovejoy, C. O. and Mensforth, R. P. (1983) 'Skeletal Age at Death: Accuracy of Determination and Implications for Human Demography', *Human Biol., 55*, pp. 73–87
Miles, A. E. W. (1958) 'The Assessment of Age From the Dentition', *Proc. Roy. Soc. Med., 51*, p. 1057
—— (1962) 'Assessment of the Ages of a Population of Anglo-Saxons From Their Dentitions', *Proc.Roy. Soc. Med., 55*, pp. 881–6
—— (1963a) 'The Dentition in the Assessment of Individual Age in Skeletal Material', in D. R. Brothwell (ed.), *Dental Anthropology*, Macmillan, New York
—— (1963b) 'Dentition in the Assessment of Age', *J. Dent. Res., 42*, p. 255

—— (1978) 'Teeth as an Indicator of Age in Man', in P. M. Butler and K. A. Joysey (eds.), *Development, Function and Evolution of Teeth*, Academic Press, New York
Molnar, S. (1972) 'Tooth Wear and Culture: A Survey of Tooth Functions Among Some Prehistoric Populations', *Curr. Anthropol. 13*, pp. 511–26
Moorrees, C. F. A. (1957) *The Aleut Dentition*, Harvard, Cambridge, Massachusetts
——, Fanning, E. A. and Hunt, Jr., E. E. (1963a) 'Formation and Resorption of Three Deciduous Teeth in Children', *Am. J. Phys. Anthropol., 21*, pp. 205–13
—— (1963b) 'Age Variation of Formation Stages for Ten Permanent Teeth', *J. Dent. Res., 42*, p. 1490
Morrey, L. W. and Nelson, R. J. (1970) *Dental Science Handbook*, US Government Printing Office, Washington
Mucci, R. J. (1982) 'The Role of Attrition in the Etiology of Third Molar Impactation: Confirming the Begg Hypothesis', MA thesis, University of Illinois at Chicago, Chicago
Murphy, T. R. (1959a) 'The Changing Pattern of Dentine Exposure in Human Tooth Attrition', *Am. J. Phys. Anthropol., 17*, pp. 167–78
—— (1959b) 'Compensatory Mechanisms in Facial Height Adjustments of Functional Tooth Attrition', *Aust. Dent. J., 4*, pp. 312–23
Nalbandian, J., Gonzales, F. and Sognnaes, R. F. (1960) 'Sclerotic Age Changes in Root Dentine of Human Teeth as Observed By Optical, Electron and X-Ray Microscopy', *J. Dent. Res., 39*, pp. 598–607
Newman, H. N. and Levers, B. G. H. (1979) 'Tooth Eruption and Function in an Early Anglo-Saxon Population', *J. Roy. Soc. Med., 72*, pp. 341–50
Noble, A. N. (1976) 'The Estimation of Age from the Dentition' in W. Harvey (ed.), *Dental Identification and Forensic Odontology*, Henry Kimpton, London
Nolla, C. M. (1960) 'The Development of the Permanent Teeth', *J. Dent. Child, 27*, pp. 254–66
Nowell, G. W. (1975) 'An Evaluation of the Miles Method of Ageing Using the Tepe Hissar Dental Sample', *Bull. Philadelphia Anthrop. Soc., 27*, pp. 1–25
Philippas, G. G. and Applebaum, E. (1966) 'Age Factors in Secondary Dentin Formation', *J. Dent. Res., 45* , pp. 778–89
—— (1967) 'Age Changes in the Permanent Upper Lateral Incisor', *J. Dent. Res., 46*, pp. 1002–9
—— (1968) 'Age Changes in the Permanent Upper Canine Teeth', *J. Dent. Res., 47*, pp. 411–17
Pilz, W. (1959) 'Das Transparenzphaomen der Zahwurzel als ein Ausdruck der Biomorphose des menschlichen Dentins', *Z. Altersforsch., 13*, pp. 139–52
Richardson, M.E. (1980) 'The Aetiology of Lower Arch Crowding', *J. Irish Dent. Assoc., 26*, pp. 26–31
Robinow, M., Richards, T. W. and Anderson, M. (1957) 'The Eruption of Deciduous Teeth', *Growth, 6*, pp. 127–33
Rose, J. C., Armelagos, G. J. and Lallo, J. W. (1978) 'Histological Enamel Indicator of Childhood Stress in Prehistoric Skeletal Samples', *Am. J. Phys. Anthropol., 49*, pp. 511–16
Ruff, C. B. (1981) 'A Reassessment of Demographic Estimates for Pecos Pueblo', *Am. J. Phys. Anthropol., 54*, pp. 147–56
Russell, A. L. (1970) 'Public Health' in L. W. Morrey and R. J. Nelson (eds.) *Dental Science Handbook*, US Government Printing Office, Washington
Saunders, E. (1837) *The Teeth as a Test of Age, Considered with Reference to the Factory Children: Addressed to the Members of both Houses of Parliament*, H. Renshaw, London
Schour, I. and Massler, M. (1937) 'Rate and Gradient of Growth in Human Deciduous Teeth, With Special Reference to Neonatal Ring', *J. Dent. Res., 16*, p. 349
—— (1941) 'Development of the Human Dentition', *J. Am. Dent. Assoc., 28*, p. 1153
Sicher, H. (1953) 'The Biology of Attrition', *Oral Surg., 6*, pp. 406–12
Smith, B. H. (1984) 'Rates of Molar Wear: Implications for Developmental Timing and Demography in Human Evolution', *Am. J. Phys. Anthropol., 63*, p. 220
Stack, M. V. (1960) 'Forensic Estimation of Age in Infancy by Gravimetric Observations of the Developing Dentition', *J. Forensic Sci. Soc., 1*, p. 49
Steggerda, M. and Hill, T. J. (1942) 'Eruption Time of Teeth Among Whites, Negroes and Indians', *Am. J. Orthodont., 28*, p. 361

Stones, H. H., Lawton, F. E., Bransby, E. R. *et al.* (1951) 'Time of Eruption of Permanent Teeth and Time of Shedding of Deciduous Teeth', *Brit. Dent. J., 90*, p. 1

Stott, G. G., Sis, R. F. and Levy, B. M. (1982) 'Cemental Annulation as an Age Criterion in Forensic Dentistry', *J. Dent. Res., 61*, pp. 814–17

Suk, V. (1919) 'Eruption and Decay of Permanent Teeth in Whites and Negroes', *Am. J. Phys. Anthropol., 2* (o.s), p. 351

Ubelaker, D. H. (1978) *Human Skeletal Remains: Excavation, Analysis, Interpretation*, Taraxacum, Washington

Wheeler, R. C. (1974) *Dental Anatomy, Physiology and Occlusion*, 5th ed, W. B. Saunders, Philadelphia

Wolpoff, M. H. (1979) 'The Krapina Dental Remains', *Am. J. Phys. Anthropol., 50*, pp. 67–114

Yamaji, C. (1958) 'Estimations of Age in Japanese Males by Roentgenograms of Teeth', *Japanese J. Legal Med. and Criminology, 2* (5), p. 34. (Japanese. Cited in T. Furuhata and K. Yamamoto (1967) *Forensic Odontology*, Charles C. Thomas, Springfield)

Zander, A. A. and Hurzeler, B. (1958) 'Continuous Cementum Apposition', *J. Dent. Res., 37*, pp. 1035–44

9 AGE DETERMINATION OF LIVING MAMMALS USING ASPARTIC ACID RACEMISATION IN STRUCTURAL PROTEINS

Patricia M. Masters

The amino acid racemisation reaction converts one optically active form of an amino acid into its opposite enantiomer. The reaction typically measured is in the direction of L- to D-enantiomer because all higher plant and animal organisms on earth synthesise proteins containing exclusively the L-amino acids. The extent of racemisation determined in remnant proteins from fossilised bones, teeth and shell can be used to calculate the ages of the fossils. These applications are discussed in Chapter 2.

Whereas the racemisation reactions typically have half-lives on the order of 10^4 to 10^6 years, the temperature dependence of the reaction rate implies that measurable amounts of racemisation might be occurring on a much shorter time scale. Mammals and birds maintain body temperatures in the range of 37° to 39°C. Many homeotherms, including humans, also live for many decades. In a metabolically stable protein, i.e. one that is synthesised early in development and not broken down and resynthesised during the homeotherm's life span, could one expect to find small amounts of D-enantiomer accumulating if one looked at a relatively rapidly racemising amino acid such as aspartic acid?

In 1974 the Amino Acid Dating Laboratory at the Scripps Institution of Oceanography, University of California, initiated a series of experiments to investigate the extent of aspartic acid racemisation in human structural proteins. Studies have included several calcified dental tissues, the ocular lens, connective tissues and neural tissues. D-aspartic acid does accumulate in these bulk proteins at rates of 0.02 to 0.14 per cent per year. Applications of these results are being made in gerontology, physical anthropology and palaeopathology, forensic sciences, wildlife biology and game management.

Racemisation in Human Enamel and Dentin Proteins

The first demonstration that *in vivo* racemisation does indeed occur was in human tooth enamel proteins (Helfman and Bada, 1975). The extent of aspartic acid racemisation was measured in 23 enamel samples separated from crowns of known-age teeth (Figure 9.1). Several protein standards and haemolysates prepared from people of various ages served as controls. D-aspartic acid was shown to accumulate at a rate of 0.1 per cent per year.

Figure 9.1: The Regression on Age of D/L Aspartic Acid Measurements in Human Tooth Enamel and in Haemolysates from Human Blood. ○, Enantiomeric determination made on the amino acid analyser; △, measurement using a gas chromatographic method; ●, amino acid analyser determination on enamel from an individual whose exact age is not known but is estimated by the dentist at 72 years; ×, haemoglobin; ◊, albumin; □, bovine tendon collagen. Ages of teeth are corrected for date of formation of the crown (Helfman and Bada, 1975)

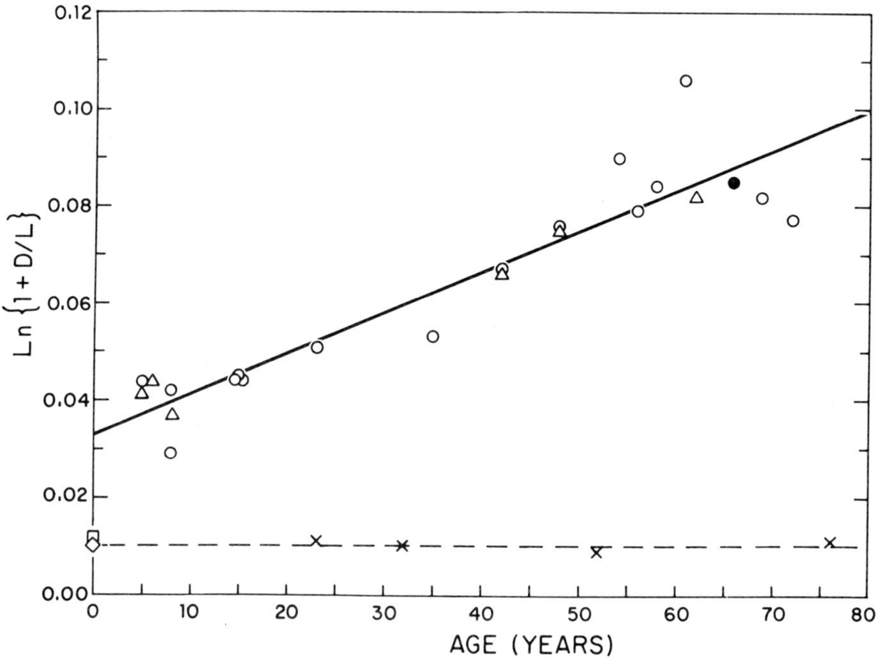

A series of dentin samples were next analysed (Helfman and Bada, 1976). Although more metabolic activity had been expected in dentin and, consequently, lower D/L aspartic acid ratios, the regression on age of aspartic acid racemisation (Figure 9.2) is statistically indistinguishable from the enamel curve. Possibly due to its higher protein content, the analysis of dentin is more straightforward and the error is less than in enamel. The correlation coefficient for dentin is $r = 0.979$, and for enamel $r = 0.921$. In fact, almost all of the error in the dentin D/L ratios could be accounted for by the normal variation in human body temperature of ± 0.7°C. Rates represented by these temperature ranges are indicated by the broken lines in Figure 9.2.

Figure 9.2: Plot of ln [l+(D/L)] for Aspartic Acid Against Age of the Dentin in Years (Solid Line). Actual tooth age is corrected as in Helfman and Bada (1975). The slope of the solid line ($7.87 \times 10^{-4} yr^{-1}$) is equal to k_{asp} for tooth dentin at about 37°C. The broken lines represent the k_{asp} values calculated for a ±0.7°C difference in body temperature. ○, Individual of known age; △, exact ages unknown but estimated by the dentists; ◊, albumin; □, bovine tendon collagen

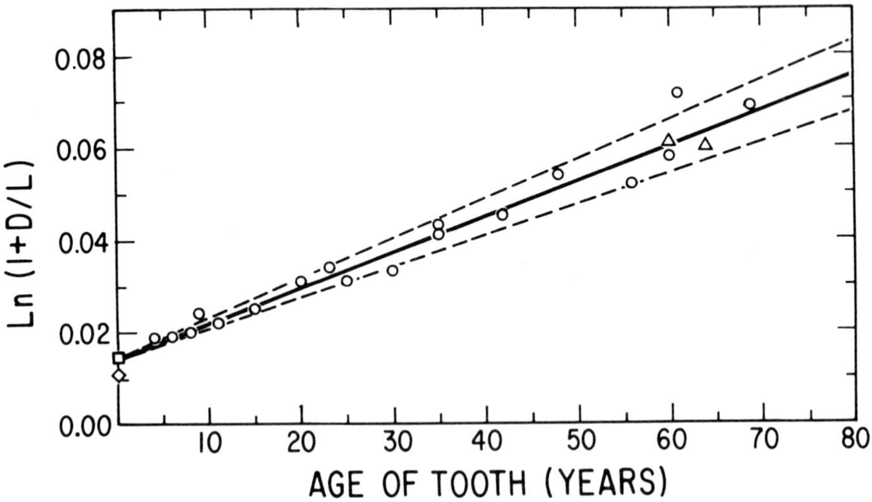

Because the observed D/L ratios in dentin are all low, racemisation is essentially a one-way or irreversible reaction:

$$\text{L-aspartic acid} \rightarrow \text{D-aspartic acid}$$

The low levels of D-aspartic acid present make the reverse reaction negligible. The irreversible rate equation for aspartic acid racemisation in dentin can be written as equation 9.1:

$$\ln(1+D/L) = 7.87\ (\pm 0.39) \times 10^{-4} yr^{-1} t + 0.014$$

The first-order irreversible rate constant, $k = 7.87 \times 10^{-4} yr^{-1}$, then represents the rate of aspartic acid racemisation in bulk dentin protein at an average temperature of 37°C. This k_{asp} is equivalent to an accumulation rate of D-aspartic acid in dentin proteins of approximately 0.1 per cent per year from the time the crown develops to the lifespan of the tooth.

This k_{asp} is also the most accurately measured rate constant yet determined at near-environmental temperatures. For this reason, the dentin k_{asp} and 37°C are the reference values used in Table 2.5, Chapter 2, to calculate the *in situ* temperatures for comparison with meteorological data.

Figure 9.3: D/L Aspartic Acid Contents of Central Nucleus Samples of Normal Human Lenses, ●; and Brunescent Cataracts, ▲

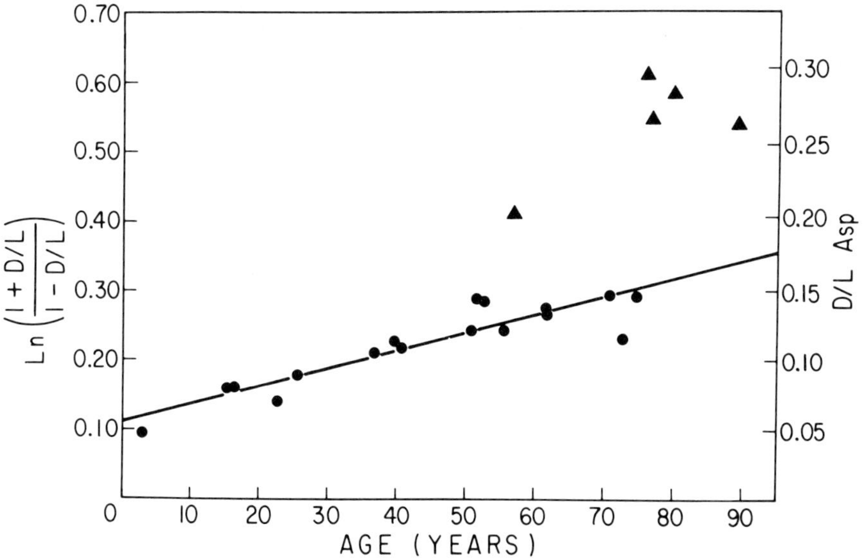

Racemisation in the Ocular Lens

During the investigations of the dental tissues, it was speculated that aspartic acid racemisation may be a ubiquitous reaction throughout the body of any long-lived mammal and could contribute to the ageing process of structural proteins (Helfman, Bada and Shou, 1977). To determine its occurrence in a soft tissue became a primary goal, and the lens appeared to be the ideal system to study. The central nucleus of the lens develops *in utero*, and by the time of birth the cells have differentiated into fibre cells lacking nuclei and protein synthetic activity. While the lens continues to grow throughout life, the new cells are dividing and differentiating only in the outer cortex of the lens. Thus, the proteins of the nuclear region are essentially as old as the individual.

Analyses of 17 normal human lenses (Masters, Bada and Zigler, 1977) ranging in age from 3 to 75 years (Figure 9.3) yielded a regression on age of the following (equation 9.2):

$$\ln \frac{1 + D/L}{1 - D/L} = 2.50\,(\pm 0.29) \times 10^{-3}\text{yr}^{-1}t + 0.112$$

This form of the rate expression is the first-order reversible equation which is used here because D/L asp ratios are high enough for the reverse reaction to become significant. The slope in equation 9.2 corresponds to 2 k_{asp}, and the rate constant for aspartic acid racemisation in the lens nucleus is thus $k_{asp} = 1.25 \times 10^{-3} yr^{-1}$. This rate is nearly 1.5 times faster than the rates observed in calcified dental tissues.

One other finding in this initial study of the lens was that a certain type of cataract almost invariably exhibits extreme amounts of aspartic acid racemisation. While early stages of cataracts (type I or II) are indistinguishable from normal lenses in their D/L ratios, the advance type IV cataract, also called a brunescent cataract due to its brown to black pigmentation, can have D/L aspartic acid ratios twice as high (Figure 9.3) as age-matched normal lenses or early cataracts (Masters, Bada and Zigler, 1977). The mechanism of accelerated racemisation in brunescent lenses has not yet been elucidated.

Application of *in vivo* Racemisation to Biochronological Problems

The strong correlation between D/L aspartic acid ratios and chronological age of tooth dentin prompted the author and her colleagues to consider the uses of *in vivo* racemisation as measurements of age in various situations where an independent estimate of chronological age is needed. The standard deviation of the dentin results ($\pm$ 0.0034) indicates that an age estimate from a suitable tooth will fall with $\pm$ 10 per cent ($\pm$ 2 σ) of the individual's true age 96 per cent of the time.

The first opportunity to test the technique came with the invitation from Dr M. R. Zimmerman to analyse a premolar taken from a frozen Eskimo mummy found by natives on St Lawrence Island, Alaska. The body was discovered in 1972, apparently having eroded from a sea cliff, and was kept in a frozen state until it was transferred to Fairbanks in 1973 for post-mortem examination. Zimmerman's autopsy revealed that the body was that of a woman with aspirated moss fibres in her lungs and a skull fracture. It appeared that she had been buried in a landslide and then suffocated (Zimmerman and Smith, 1975).

Radiocarbon dating of muscle tissue placed the date of death at AD 370 to 390 $\pm$ 90 (Lawn, 1975). This date coincides with the Old Bering Sea cultural phase on St Lawrence Island. Tattoos on the arms, defined by infrared photography, contained artistic motifs resembling those seen in many Old Bering Sea artifacts, thus corroborating the radiocarbon date.

The age at death was estimated initially by morphological and histological techniques. At the time of autopsy, the woman was found to be postmenopausal and to have a high degree of dental attrition and coronary artery disease. An age of 50 to 60 years was suggested. Two years later, osteon density in cortical bone was examined microscopically. The density was found to be within the normal range for modern individuals aged 50 to 59 years.

Four years after the postmortem examination, a preserved premolar was submitted for amino acid racemisation analysis. A D/L asp ratio of 0.053 was determined and from equation 9.1 an age of 48 ± 5 years was calculated for the premolar. Adding five years to correct for the age of the tooth, the woman's age at death was estimated as 53 ± 5 years (Masters and Zimmerman, 1978). All three methods, then, have produced very similar age approximations.

Another series of comparisons between racemisation ages and ages based on skeleton features was carried out with tooth specimens from a medieval cemetery population in Czechoslovakia (Masters and Bada, 1978). The cemetery is dated historically from the seventh or eighth century AD. Stloukal, a physical anthropologist at the National Museum in Prague, is studying the burials and has determined ages on the basis of osteological indicators. The first concern was the amount of postmortem racemisation that had occurred during the approximately 1,000 years of interment. The average annual air temperature in this area is about 9°C so the postmortem component of the D/L ratios would be small relative to the *in vivo* component. Postmortem racemisation was corrected for by analysing a tooth from a young individual (18 to 20 years, an age range where dental and skeletal features can give age estimates of high accuracy). After this 'calibration' was established, other teeth were analysed, and racemisation ages were compared with Stloukal's estimates (Table 9.1). The ages calculated from D/L aspartic acid ratios are in quite close agreement with those based on various osteological criteria.

A second opportunity to work with frozen human remains from the Arctic came with the discovery in 1982 of five Eskimos buried in a mound house outside Barrow, Alaska. Radiocarbon dating placed the time of burial at AD 1510, and the cultural artifacts corroborated a pre-contact date. Age-at-death estimates were made by morphological features at autopsy, osteon density and the pubic symphyses. Aspartic acid racemisation analyses of teeth (Table 9.2) were concordant with the results from the other techniques (Masters, 1984).

In the previous situations, pathologists and anthropologists had available for analysis intact bodies or nearly complete skeletons. Frequently, scientists working with forensic cases must attempt to provide an age at death from very limited skeletal material. Since a racemisation age can be determined on a single tooth, representing the least perishable part of the skeleton, the amino acid technique should be especially useful in forensic applications.

To demonstrate this, the author has analysed tooth specimens provided by local forensic odontologists from county morgue cases. All specimens

Table 9.1: Ages at Death Estimated For Skeletons from Medieval Cemetery in Czechoslovakia

Sample no.	Age estimated by Stloukal (yr)	Racemisation age (yr)
79	18–20	Calibration
80	30–40	33–35
77	40–50	48–50
94	60+	~80

Table 9.2: Aspartic Acid Racemisation Ages on Tooth Dentin from the Barrow Eskimo Remains

Sample	Autopsy age	Osteon density age	Pubic symphyses age	Amino acid age
Northern Body	~25	24	16–24	27±2
Southern Body	50–60?	42±6	36–45	41±4
Skeleton 1	15–20	–	–	20±1
Skeleton 2	~15	–	–	13±1
Skeleton 3	8–9	–	–	–[a]

Note: a. Calibration of the racemisation rate constant, k, using an age of eight to nine years based on stages of tooth development as determined by X-ray of the maxilla and mandible.

were processed blind, and then the racemisation ages were compared with known ages or ages estimated by the odontologist using Gustafson's (1950) tooth wear method. The results are summarised in Table 9.3. With the exception of SD-80-708, racemisation ages agree closely with the known or estimated ages. SD-80-708 was the only skeletonised specimen, found on a hillside in southern San Diego County after an unknown period of exposure. Surface ground temperatures in this region may reach high enough levels to generate considerable amounts of postmortem racemisation.

The amino acid racemisation ageing technique would also be useful in sorting the number of differently aged individuals from a skeletal assembly. One application, of interest to agencies such as the Federal Aviation Administration, would be in identification of victims of aircraft disasters. However, airliner collisions or crashes invariably coincide with intense fires. The temperature dependence of the racemisation reaction would make it an unsuitable method to apply in these circumstances.

Wildlife biology and game management are other fields which may benefit from the development of amino acid racemisation as an *in vivo* age indicator. This laboratory has been working with the National Marine

Table 9.3: Tooth Analyses of Coroner's Cases, San Diego and Los Angeles Counties

Sample	Age	D/L Asp	Calculated age
SD–78–209	16	0.025	17±2
SD–78–705	25	0.034	28±3
SD–80–708	26	0.043	39±4
SD–77–2836	60+	0.068	69±7
SD–78–338	18±2[a]	0.025	17±2
SD–78–456	30±5[a]	0.036	30±3
LA–77–15810	36	0.038	35±4

Note: a. Dental age estimated by R. Siegel, DDS.

Fisheries Service to supply age information on porpoises accidently killed during tuna seining operations. Before the use of the chemical ageing method, biologists had been trying to age porpoise and other toothed cetaceans by the number of growth layers in their teeth. Layer counts on a single tooth can differ significantly depending on the observer and the power of the microscope used. Further more, the number of layers laid down per year was not known and might be expected to vary among the taxa. Thus, growth layer counts, at best, yielded relative ages for the animals.

Preliminary findings (Bada, Brown and Masters, 1980) give a dentinal growth rate of approximately one layer/year in the spotted dolphin, *Stenella attenuata*. In the sperm whale, *Physeter macrocephalus*, however, the rate is about 0.2 layer/year. It would appear that either the layer count is being underestimated or the racemisation rate is considerably faster in sperm whale teeth than in other mammals. The latter possibility is unlikely since temperature is the principal determinant of racemisation rates, and mammals have a very narrow range of body temperatures. Furthermore, aspartic acid racemisation has been investigated in a number of different animal taxa. Tooth samples of known age animals were obtained from a number of zoos. Table 9.4 shows close agreement between the age comparisons, implying that racemisation rates are essentially the same among various primates, ungulates and small cetaceans.

Lenses are also being used to obtain ages for mammals. Figure 9.4 summarises results for some pinnipeds, equids and pongids. The slopes of the regressions are all the same as that for humans but the t=0 value (y-intercept) varies. This could be due to variability in the crystalline composition among different animal taxa (Zigler and Sidbury, 1976). Racemisation ages based on lens analyses would be the only means of ageing mammals without teeth, such as baleen whales.

Table 9.4: Age Estimates of Various Mammalian Species of Known Ages as Determined from the Extent of Aspartic Acid Racemisation in Dentin

Species	D/L aspartic acid	Aspartic acid age (yrs)[a]	Actual age (yrs)[b]
Callorhinus ursinus (fur seal)	0.0361	27	14+
Macaca nemestrina (monkey)	0.0349	26	20+
Macaca nemestrina (monkey)	0.0231	11	4
Macaca nemestrina (monkey)	~0.020	~7	10
Camelus dromedarius (camel)	0.0361	27	20
Equus burchelli bohmi (zebra)	0.0335	24	27
Gorilla gorilla (gorilla)	0.0341	25	30
Zalophus californianus (sea lion)	0.0375	29	25+
Stenella attenuata (porpoise)	0.014	0	0

Notes:
a. Calculated from equation 9.1;
b. Obtained from zoo records.

There has been only one attempt to age a human by the extent of aspartic racemisation in the lens: Charlie Smith, who died on 5 October, 1975, claimed to have been born in Liberia in 1842 and brought to this country on a slave ship. He was sold at auction in Texas in 1854 and freed in 1863 with the Emancipation Proclamation. He worked for many years as a cowboy in the southwest before moving to Florida during the Spanish- American War. He came to public attention in 1955 when he was discovered to be the oldest person on Social Security rolls, aged 113 years. While documentation of his age is lacking, the Social Security Administration's acceptance of 1842 as Charlie Smith's birth year has lent credibility to his claims.

Four days after Charlie Smith's death, the author was contacted by Dr W. Ted Brown of the Division of Gerontology and Geriatrics, Cornell University Medical College. He was en route to Florida to obtain tissue specimens from Charlie Smith's body for various studies, and he inquired if the author would be willing to analyse a lens. Since Charlie Smith was adontic, the lens was the only tissue available for an amino acid racemisation age determination.

The results of the analyses are shown in Table 9.5. An aspartic acid

Figure 9.4: A Plot, in the Form of Equation 9.2, of Aspartic Acid Racemisation in the Lens Nucleus of Several Mammal Species of Known Age. The slopes of the lines are all equal to the k_{asp} value (i.e. $1.25 \times 10^{-3} yr^{-1}$) determined in the human lens nucleus. The abbreviations used are as follows: NSF, northern fur seal; SSL, Steller sea lion; CSL, California sea lion; ZEB, zebra; and GOR, gorilla. The numbers are the Amino Acid Dating Laboratory identification numbers

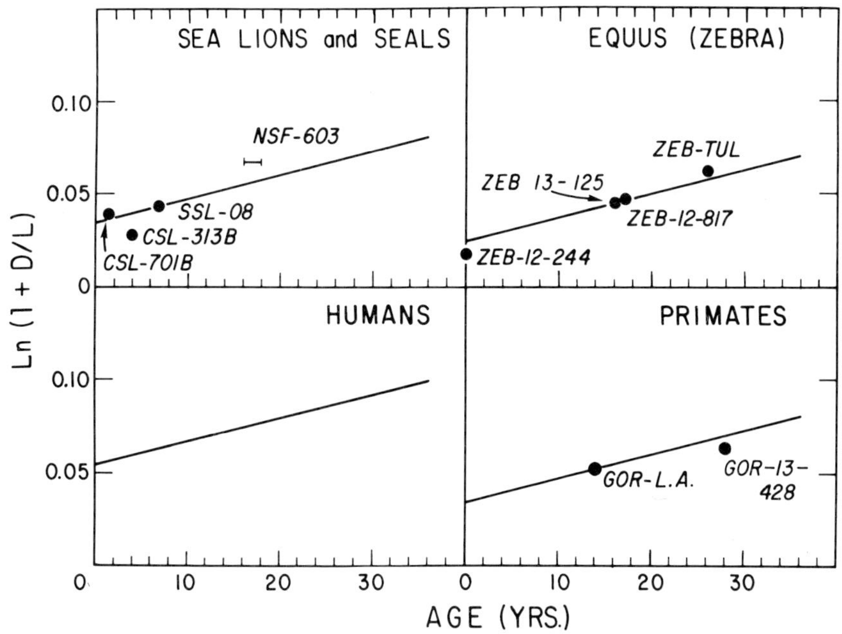

Table 9.5: Lens Nucleus Analyses for Charlie Smith and a Brunescent Cataract[a]

	D/L Ratio	
Amino acid	Charlie Smith	Brunescent
Aspartic acid	0.302	0.323
Phenylalanine	0.310	0.10
Glutamic acid	0.093	0.13
Leucine	0.111	0.067

Note: a.The nucleus of a brunescent cataract from a 67-year-old individual was heated at 100°C in aqueous solution for six hours to enhance the extent of racemisation so that enantiomeric ratios would attain measurable levels in amino acids other than aspartic acid.

ratio of 0.302 falls outside the range of values for normal lenses and is comparable to ratios observed in brunescent cataracts. The lens from Charlie Smith, however, did not display any marked degree of pigmentation. Gas chromatographic analyses on three additional amino acids revealed that these ratios were elevated as well. The normal, uncontaminated order of racemisation rates for these residues is Asp >> Phe ⩾ Glu >> Leu. In the case of the Charlie Smith lens, the relative order is Phe ≅ Asp >> Leu > Glu. When disturbances in the normal order of racemisation rates occur, some form of contamination of the sample is indicated. Postmortem deterioration and bacterial activity may be responsible for these results since several days had elapsed between death and tissue sampling and preservation.

Racemisation in Other Soft Tissues

As part of the investigations of the role of aspartic acid racemisation in the biological ageing process, the Amino Acid Dating Laboratory has screened a variety of connective and neural tissues. Some degree of racemisation is detectable in all of the older samples (Tables 9.6 and 9.7). Table 9.6 lists the aspartic acid rate constants estimated from various tissues or protein fractions of them. The knee cartilage cyanogen bromide residue and the water-insoluble protein from the dura mater have k_{asp} values comparable to that measured in dentin and enamel. Whole tissue tends to have slower rate constants, probably because of the presence of free amino acids and admixtures of younger, more recently synthesised proteins from vascular or other origins.

Fractionation into water-soluble and water-insoluble portions (Masters, Bada and Zigler, 1978) has been shown to provide a means of selecting for more highly racemised protein in lenses. The technique appears to work as well with connective tissue (the dura mater in Table 9.6). Colleagues at the University of Miami (Man *et al.*, 1983) also have been studying aspartic acid racemisation in human cerebral white matter. They have found that trichloroacetic acid precipitation yields a protein fraction that exhibits a rate ($k_{asp} = 1.57 \times 10^{-3}yr^{-1}$) comparable to the water- insoluble fraction from the human lens ($k_{asp} = 1.62 \times 10^{-3}yr^{-1}$). The higher the D/L ratios, the more accurately they can be measured and age differences assessed.

Soft tissues have not been investigated in sufficient number to generate rate constants having the statistical accuracy of the lens and tooth proteins. While the results summarised in Tables 9.6 and 9.7 are preliminary, they do indicate the potential for age determinations on fractions of connective or neural tissue.

A major difficulty in studying animal tissues is the complexity of

Table 9.6: Aspartic Acid Racemisation and Derived Rate Constants in Various Human Collagenous Tissues and in Haemoglobin

	Treatment	Age	D/L Asp	k_{asp}[a] (yr^{-1})
Knee cartilage[b]	—	23	0.023	
	CNBr	23	0.030	6.8×10^{-4}
	CNBr	45	0.046	
Annulus fibrosus[b]	—	14	0.030	4.0×10^{-4}
	—	59	0.049	
Dura mater[c]	—	0	0.010	1.5×10^{-4}
	—	87	0.023	
	WI	0	0.022	9.4×10^{-4}
	WI	87	0.110	
	WS	0	0.024	8.0×10^{-5}
	WS	87	0.031	
Haemoglobin[d]	—	23	0.011	—
	—	76	0.011	

Notes:
a. Estimated from $\ln(1+D/L)old - 1n(1+D/L)young = k\Delta t$);
b. Samples provided by Dr David R. Eyre, The Children's Hospital Medical Center, Boston;
c. Samples from Dr Robert L. Trelstad, Shriners Burn Institute, Boston;
d. Data from Helfman and Bada, 1975.

Table 9.7: Extent of Aspartic Acid Racemisation in Tissues Obtained at Autopsy[a]

	Age	D/L Asp
Neural tissues		
Sciatic nerve	73	0.039
Centrum semiovale	73	0.048
Connective tissues		
Spinal ligament	73	0.040
Central diaphragm	73	0.039

Note:
a. Samples by Dr Judy Myers Suchey, California State University, Fullerton.

cellular and non-cellular protein components. Purification of a particular protein moiety is essential because heterogeneous or unfractionated protein samples cannot give a reliable enantiomeric ratio upon hydrolysis. Any small degree of racemisation is swamped by the predominance of L-enantiomer.

Recently, an enzyme, protein carboxyl methyltransferase, has been discovered to recognise D-aspartyl residues in erythrocyte membrane proteins (McFadden and Clarke, 1982). With a 120-day lifespan, the extent of *in vivo* racemisation in erythrocytes is quite low. However, this substrate specificity of the protein carboxyl methyltransferase demonstrates that racemisation can take place in short-lived cellular proteins (in addition to the structural proteins discussed above). The ubiquity of protein carboxyl methyltransferase activity (Paik and Kim, 1980) implies that aspartic acid racemisation also may be widespread in organisms and that the function of the methylation reaction may be to initiate repair or degradation of the racemised proteins (McFadden and Clarke, 1982).

Summary

Aspartic acid racemisation in structural proteins of homeotherms provides a promising new technique for measuring biological age or age at death. At present, either tooth dentin or the ocular lens nucleus are the samples of choice for enantiomeric analyses and age calculations. An age determination based on dentin should fall within ± 10 per cent ($\pm 2\sigma$) of the organism's true age. Dentin measurements have been applied to a variety of marine and terrestrial mammals and to humans in order to resolve problems in wildlife biology, fisheries management, palaeoanthropology and forensic science. Aspartic acid racemisation may also play some role in the functional deterioration seen at the organismal level during the ageing process.

Acknowledgements

I thank D. Darling and D. Buccigrossi for technical assistance. The research summarised here has been supported by the National Institute on Aging and the National Marine Fisheries Service.

References

Bada, J. L., Brown, S. and Masters, P. M. (1980) 'Age Determination of Marine Mammals Based on Aspartic Acid Racemization in the Teeth and Lens Nucleus', *Report of the International Whaling Commission* (Special Issue 3), pp. 113–18

Gustafson, G. (1950) 'Age Determination on Teeth', *J. Am. Dent. Assoc.*, *41*, pp. 45–54

Helfman, P. M. and Bada, J. L. (1975) 'Aspartic Acid Racemization in Tooth Enamel from Living Humans', *Proc. Natl. Acad. Sci. USA*, *72*, pp. 2891–4

—— (1976) 'Aspartic Acid Racemisation in Dentine as a Measure of Ageing', *Nature*, *262*, pp. 279–81

Helfman, P.M., Bada, J. L. and Shou, M. -Y. (1977) 'Considerations on the Role of Aspartic Acid Racemization in the Aging Process', *Gerontology*, *23*, pp. 419–25

Lawn, B. (1975) 'University of Pennsylvania Radiocarbon Dates XVIII', *Radiocarbon*, *17*, pp. 196–215

Man, E. H., Sandhouse, M. E., Burg, J. *et al.* (1983) 'Accumulation of D-aspartic Acid with Age in the Human Brain', *Science*, *220*, pp. 1407–8

Masters, P. M. (1984) 'Stereochemical Age Determinations From the Barrow Eskimo Remains', *Arctic Anthropology*, *21*, pp. 77–82

—— and Bada, J. L. (1978) 'Amino Acid Racemization Dating of Bone and Shell', in G. F. Carter (ed.), *Archaeological Chemistry II*, Advances in Chemistry Series 171, American Chemical Society, Washington DC, pp. 117–38

——, Bada, J. L. and Zigler, Jr., J. S. (1977) 'Aspartic Acid Racemisation in the Human Lens During Ageing and in Cataract Formation', *Nature, 268*, pp. 71–3
—— (1978) 'Aspartic Acid Racemization in Heavy Molecular Weight Crystallins and Water-insoluble Protein from Normal Human Lenses and Cataracts', *Proc. Natl. Acad. Sci. USA, 75*, pp. 1204–8
—— and Zimmerman, M. R. (1978) 'Age Determination of an Alaskan Mummy: Morphological and Biochemical Correlation', *Science, 201*, pp. 811–12
McFadden, P. N. and Clarke, S. (1982) 'Methylation at D-aspartyl Residues in Erythrocytes: Possible Step in the Repair of Aged Membrane Proteins', *Proc. Natl. Acad. Sci. USA, 79*, pp. 2460–4
Paik, W. K. and Kim, S. (1980) *Protein Methylation*, Wiley, New York, pp. 202–31
Zigler, J. S., Jr. and Sibury, J. B. Jr. (1976) 'A Comparative Study of β-Crystallin from Six Mammals', *Comp. Biochem. Physiol., 53B*, pp. 349–55
Zimmerman, M. R. and Smith, G. S. (1975) 'A Probable Case of Accidental Inhumation of 1,600 Years Ago', *Bull. N.Y. Acad. Med., 51*, pp. 828–37

INDEX